Tomasz W. Dłotko, Yejuan Wang
Critical Parabolic-Type Problems

De Gruyter Series in Nonlinear Analysis and Applications

Volume 34

Tomasz W. Dłotko, Yejuan Wang

Critical Parabolic-Type Problems

DE GRUYTER

Mathematics Subject Classification 2010
35-02, 65-02, 35A25, 35S010, 35Q30, 76D05, 37L05

Authors
Prof. Dr. Tomasz W. Dłotko
University of Silesia in Katowice
Institute of Mathematics
Bankowa 14
40-007 Katowice
Poland
tomasz.dlotko@us.edu.pl

Prof. Dr. Yejuan Wang
Lanzhou University
School of Mathematics and Statistics
222 South Tianshui Road
730000 Lanzhou
People's Republic of China
wangyj@lzu.edu.cn

ISBN 978-3-11-059755-4
e-ISBN (PDF) 978-3-11-059983-1
e-ISBN (EPUB) 978-3-11-059868-1
ISSN 0941-813X

Library of Congress Control Number: 2020934459

Bibliographic information published by the Deutsche Nationalbibliothek
The Deutsche Nationalbibliothek lists this publication in the Deutsche Nationalbibliografie; detailed bibliographic data are available on the Internet at http://dnb.dnb.de.

Typesetting: VTeX UAB, Lithuania
Printing and Binding: LSC Communications, United States

www.degruyter.com

开卷有益。

You cannot open a book without learning something.

三人行，必有我师焉，择其善者而从之，其不善者而改之。

If I am walking with two other men, each of them will serve as my teacher. I will pick out the good points of the one and imitate them, and the bad points of the other and correct them in myself.

（孔子，公元前 551 年--479 年）

(Confucius, 551–479 B.C.)

Acknowledgment

The authors are very grateful to all their friends for the great help in the preparation of this book. First, we mention professors Jan Cholewa, Chunyou Sun, Maria B. Kania, and Shan Ma, the authors of joint publications parts of which were used in the preparation of the present monograph. Without all of them, this publication will not arise.

The authors are especially grateful to the doctor students of Yejuan Wang; Meiyu Sui, Lin Yang, Tongtong Liang, Xiangming Zhu, and Yarong Liu for their extraordinary work on the preparation of the monograph.

Yejuan Wang was supported by the National Science Foundation of China (Grants 41875084, 11571153); she would like to acknowledge the Foundation for the generous support of her research. Tomasz Dlotko would like to acknowledge the Institute of Mathematics of the Polish Academy of Science for hospitality and supporting his scientific activity through the academic year 2016/2017; he held a research position at the Institute. Both authors are grateful to the School of Mathematics and Statistics, Lanzhou University for supporting their international cooperation for many years.

Above all, for their patience and indulgence, the authors thank their families: Maria Dłotko, Weihua Deng, and Yafeng Deng. To them, the authors lovingly dedicate this book.

https://doi.org/10.1515/9783110599831-201

Contents

1 Introduction

A natural way of extending our knowledge is to first solve simpler problems using them next to consider a more complex and involved one. Such idea is used in particular when the Galerkin approximations, being systems of ordinary differential equations (that are "simple objects" to study), are used to solve a partial differential equation in the limit. Here, we propose a similar idea; to approximate a "critical" partial differential problem with a sequence of subcritical approximations that are easy to solve and study inside the semigroup approach. The solutions of such approximations are very easy to obtain and they have excellent regularity properties, just like solutions of the heat equation. Next, we will pass to the limit with the parameter involved in the approximation to obtain a kind of weak solution to the limiting, critical, hard problem.

In a Banach space X, we will consider an abstract evolutionary semilinear Cauchy's problem with sectorial positive operator A in the main part

$$\begin{aligned} &u_t + Au = F(u), \quad t > 0, \\ &u(0) = u_0, \end{aligned} \tag{1.1}$$

with the nonlinear term $F : D(A^b) \to D(A^a)$, $0 \le b - a \le 1$, being Lipschitz continuous on bounded subsets of $D(A^b)$. Here, A^a denotes fractional power of the operator A (see Definition 1.22) and $X^a := D(A^a)$, its domain; $D(A^a) \subset X$ when $a \ge 0$.

When $|b - a| < 1$, such theory was developed in the famous monograph of Dan Henry [86], where problem (1.1) was studied using classical techniques of ordinary differential equations that were modified however to cover equations with an unbounded operator in a Banach space. In that monograph, considered was the case when such approach gives local in time solution $u(t)$ under suitable choice of the fractional power space $D(A^b)$, $\|\phi\|_{X^b} = \|A^b\phi\|_X$ that we set the problem in. Assume further that an a priori estimate for $u(t)$ is available in a Banach space $Y \supset D(A^b)$, $0 < b - a < 1$, and we have an estimate for nonlinearity taken on such arbitrary local solution:

$$\exists_{\text{nondecreasing } g:[0,\infty)\to[0,\infty)} \exists_{\theta\in(0,1)} \|F(u(t))\|_{X^a} \le g(\|u(t)\|_Y)(1 + \|u(t)\|_{X^b}^{\theta}), \tag{1.2}$$

valid for all $t \in (0, \tau_{u_0})$, where τ_{u_0} is the "lifetime" of that solution. In that case (see, e. g., Chapter 4 below) the X^b norm of the local solution is bounded on $[0, \tau_{u_0})$, which allows to extend $u(t)$ globally in time.

If instead of the above estimate only a weaker a priori estimate is available:

$$\exists_{\text{nondecreasing } g':[0,\infty)\to[0,\infty)} \|F(u(t))\|_{X^a} \le g'(\|u(t)\|_Y)(1 + \|u(t)\|_{X^{a+1}}^{1}), \tag{1.3}$$

where θ is replaced with 1 and the distance $(a + 1) - a = 1$, the nonlinearity F is called *critical relative to that a priori estimate*. Note that the nonlinear term in that case is *of the same order, or value*, as the main part operator (up to the present a priori estimate). We will treat also, using a variant of the method known as *vanishing viscosity*

https://doi.org/10.1515/9783110599831-001

technique, originated by E. Hopf, O. A. Oleinik, P. D. Lax in 1950s, the case when one needs to replace the right-hand side of the estimate above with $g(\|u(t)\|_Y)(1+\|u(t)\|^{\eta}_{X^{\alpha+1}})$ where $\eta > 1$ is required, known as the *supercritical case*.

In the monograph [32], joint with Jan Cholewa, the ideas of [86] were analyzed and extended. Also, a number of examples of problems originated in the applied sciences were studied to illustrate such approach. That monograph was dedicated to the long time behavior of solutions, expressed in the language of the *global attractors*. For such studies, the key was also the precise description of the possibility of global in time continuation of the local solutions of equation (1.1) with sectorial positive main part operator A and subordinated to it nonlinear term F. In a series of later publications [48–55], more remaining examples were reported, however, they were basically inside the approach of [32]. It was the paper [47] where the semigroup solutions obtained in the formalism of [32, 86] were first used to approximate solutions of the more complicated, "critical" problems like the 3-D Navier–Stokes equation or the critical quasi-geostrophic equation with exponent $\frac{1}{2}$. The present monograph is mostly devoted to such technique of treating critical problems.

Eberhard Hopf in 1950 was the first who used an idea to *regularize* the equation adding to it viscosity term with small parameter, solving such regularization, and next letting the parameter to zero to obtain a weak solution of the original equation in the limit. E. Hopf was considering Cauchy's problem for the Burgers' equation

$$\begin{aligned} &u_t + \frac{\partial}{\partial x}\left(\frac{u^2}{2}\right) = 0, \quad x \in \mathbb{R},\ t > 0, \\ &u(0,x) = u_0(x), \quad x \in \mathbb{R}, \end{aligned} \tag{1.4}$$

and he proposed the regularization

$$u_{\mu t} + \frac{\partial}{\partial x}\left(\frac{u_\mu^2}{2}\right) = \mu \frac{\partial^2 u_\mu}{\partial x^2}, \quad x \in \mathbb{R},\ t > 0, \tag{1.5}$$

with positive parameter μ. The method of Hopf was almost immediately extended (see, e. g., [147]) to more general equations including

$$u_t + \frac{\partial \phi(u,t,x)}{\partial x} = 0,$$

by Olga Oleinik (through 1954–1957), Avron Douglis (in 1959), Jacques-Louis Lions (in 1969), Peter D. Lax (in 1971) and many others, initiating the method known nowadays as the *vanishing viscosity technique*.

Our strategy, to treat the critical and supercritical problems, is as follows. We will consider the regularization

$$\begin{aligned} &u_t + Au + \epsilon A^{\beta} u = F(u), \quad t > 0, \\ &u(0) = u_0, \end{aligned} \tag{1.6}$$

of the critical/supercritical problem (1.1) *improving the viscosity* to the form A^β, $\beta > 1$, $\epsilon > 0$, where the value of β is taken as large as needed for the nonlinearity to become subcritical with respect to the previous a priori bound (in fact, we need to assume below the condition $\eta\nu = \frac{\eta}{\beta} < 1$):

$$\begin{aligned}\exists_{\text{nondecreasing } g:[0,\infty)\to[0,\infty)} \|F(u(t))\|_{X^\alpha} &\le g(\|u(t)\|_Y)(1 + \|u(t)\|^\eta_{X^{\alpha+1}}) \\ &\le g(\|u(t)\|_Y)(1 + c\|u(t)\|^{\eta\nu}_{X^{\alpha+\beta}}\|u(t)\|^{\eta(1-\nu)}_{X^\alpha}), \qquad (1.7)\end{aligned}$$

for all $t \in (0, \tau_{u_0})$ and $\eta\nu < 1$; note that $\alpha + \beta > \alpha + 1$ and the *moment inequality* is valid. Next, using the approach of Dan Henry, we will solve easily the subcritical regularized problems obtaining a family (parameter ϵ) of its solutions u^ϵ. Using the assumed uniform in β, ϵ, estimate (1.7) valid for u^ϵ, our final step is to pass to the limit in the weakly formulated equation (1.6) to see that the limit u of u^ϵ is a weak solution to the limiting critical/supercritical problem. Depending on the strength of the available uniform a priori estimate, u inherits suitable properties of u^ϵ.

Recall here for completeness the definition of sectorial operator, due to Dan Henry. For $a \in \mathbb{R}$ and $\phi \in (0, \frac{\pi}{2})$, introduce a sector of the complex plain (cf. [86, Definition 1.3.1]):

$$\mathcal{S}_{a,\phi} := \{\lambda \in \mathbb{C} : \phi \le |\arg(\lambda - a)| \le \pi, \lambda \ne a\}. \qquad (1.8)$$

Definition 1.0.1. A linear, closed and densely defined operator $A : X \supset D(A) \to X$ acting in a Banach space X is called sectorial operator in X, if and only if there exist $a \in \mathbb{R}$, $\phi \in (0, \frac{\pi}{2})$ and $M > 0$ such that *the resolvent set* $\rho(A)$ *contains the sector* $\mathcal{S}_{a,\phi}$, and

$$\|(\lambda I - A)^{-1}\|_{\mathcal{L}(X,X)} \le \frac{M}{|\lambda - a|}, \quad \text{for each } \lambda \in \mathcal{S}_{a,\phi}. \qquad (1.9)$$

One of the main advantages of using the above described technique to study critical problems is connected with the fact that proper fractional powers of sectorial positive operators are again sectorial positive operators. We are thus allowed to use the same approach to local solvability of critical problems and of its regularizations.

The latter statement will be easily seen from the next proposition that uses the notion of an *operator of the type* $(\omega, M(\theta))$ acting in a Banach space X.

Definition 1.0.2. We say that A is of type $(\omega, M(\theta))$, $0 \le \omega < \pi$, if the domain $D(A)$ is dense in X, the resolvent set of $-A$ contains the sector $|\arg\lambda| < \pi - \omega$ and the condition $\|\lambda(\lambda + A)^{-1}\| \leqslant M(\theta)$ holds on each ray $\lambda = re^{i\theta}$, $r \in (0, +\infty)$, $|\theta| < \pi - \omega$.

One may easily see that A is of the type $(\omega, M(\theta))$ with $\omega < \frac{\pi}{2}$, if and only if A is a sectorial operator in the sense of Definition 1.0.1 with $a = 0$. A theorem by T. Kato (see [106]) states the following.

Proposition 1.0.3. *If A is of type $(\omega, M(\theta))$ and if $0 < \alpha < \frac{\pi}{\omega}$, then A^α is of type $(\alpha\omega, M_\alpha(\theta))$ with certain positive constant $M_\alpha(\theta)$. Furthermore, the resolvent of A^α is analytic in α and λ in the domain $0 < \alpha < \frac{\pi}{\omega}$, $|\arg \lambda| < \pi - \alpha\omega$.*

Note that when $\omega < \frac{\pi}{2}$ (which corresponds to sectorial operator A), then $\alpha\omega < \frac{\pi}{2}$ for any proper fractional power operator A^α, $\alpha \in (0, 1)$.

Another evident advantage of using our approach is that the approximations we are using are smooth (since they are solutions of the regular dissipative equations). Moreover, the approximating problems are very similar in nature to their critical or supercritical limits. Note that, for regular functions, the difference between $(-\Delta)\phi$ and $(-\Delta)^{1+\epsilon}\phi$ tends to zero as $\epsilon \to 0^+$ (e. g., [130, Proposition 3.1.1]). Moreover, the solutions to our approximations exist globally in time, while for the critical limits they may be only local, with difficult estimation of the lifetime. This property helps, in particular, when one is looking numerically for solutions, since one can be sure that the approximating solution exists and has good properties for arbitrary positive time.

There is also a difficulty connected with such way of constructing solutions of the critical or supercritical problems. When the solution of such problem is not unique, it is not clear if we can achieve *all* existing solutions through our approximation procedure. Solutions obtained inside our approach are called *viscosity solutions* of the critical or supercritical problem. As a role, uniqueness of solutions is true inside classes of more regular solutions. Only in such classes we can be sure to obtain all possible solutions of the critical problems as limits of our approximations. Let us point out, however, that the same disadvantage is present for other existing procedures of obtaining solutions, including the most popular *Galerkin approximation technique*.

The motivation for our studies was the fractional generalization of the celebrated Navier–Stokes equation in space dimensions $N = 2, 3, 4$. We mean here the Dirichlet problem in bounded smooth domain $\Omega \subset \mathbb{R}^N$, written in an abstract form of a differential equation in a Banach space:

$$\begin{aligned} &u_t = -Au - \epsilon A^{\beta_N} u + F(u) + Pf, \quad t > 0, \\ &u(0) = u_0, \end{aligned} \tag{1.10}$$

where A is the Stokes operator, the exponents β_N fulfill

$$\beta_2 > 1, \quad \beta_3 > \frac{5}{4}, \quad \beta_4 > \frac{3}{2}, \tag{1.11}$$

(generally, $\beta_N > \frac{N+2}{4}$, [124]) and P is the projector onto the space of divergence-free functions. The nonlinear term $F(u) := -P(u \cdot \nabla)u$ is standard for the Navier–Stokes equation. Note that the values of the exponents listed in (1.11) were given first by J.-L. Lions in [124, Chapter 1, Remarque 6.11].

Another example of critical problem discussed in the book is the *dissipative quasi-geostrophic equation* with $\alpha = \frac{1}{2}$ which, for fractional exponent $\alpha \in [\frac{1}{2}, 1]$, has the form:

$$\begin{aligned} &\theta_t + u \cdot \nabla\theta + \kappa(-\Delta)^\alpha\theta = f, \quad x \in \mathbb{R}^2, t > 0, \\ &\theta(0, x) = \theta_0(x), \end{aligned} \tag{1.12}$$

where θ represents the potential temperature, $\kappa > 0$ is a diffusivity coefficient, and $u = (u_1, u_2)$ is the *velocity field* determined by θ through the relation:

$$u = \left(-\frac{\partial\psi}{\partial x_2}, \frac{\partial\psi}{\partial x_1}\right), \quad \text{where } (-\Delta)^{\frac{1}{2}}\psi = -\theta, \tag{1.13}$$

or, in a more explicit way,

$$u = (-R_2\theta, R_1\theta), \tag{1.14}$$

where R_i, $i = 1, 2$ are the *Riesz transforms*. The *critical case* for equation (1.12) corresponds to the parameter $\alpha = \frac{1}{2}$, when both the main part operator of the equation and the nonlinear term contains first-order derivatives of the solutions (or their equivalence).

Note also that the simplest to imagine example of the critical problem has the form

$$\begin{aligned} &u_t - \Delta u = -2\Delta u, \quad t > 0, \ x \in \mathbb{R}^N, \\ &u(0, x) = u_0(x), \end{aligned} \tag{1.15}$$

with the "nonlinearity" $F(u) = -2\Delta u$. This is evidently the *backward heat equation*, the equation which leads to classical improperly posed problem. The required for such problem condition (1.3) reads

$$\|F(u)\|_{X^a} \le 2\|\Delta u\|_{X^a} = 2\|u\|_{X^{1+a}},$$

for arbitrary $a \in \mathbb{R}$.

Many different definitions of *criticality* of the semilinear problem

$$\begin{aligned} &u_t + Au = F(u), \\ &u(0) = u_0, \end{aligned}$$

are now present in the literature. In general, they are connected with the "fast growth" of nonlinear term $F(u)$, sometimes relative to the main part operator A. In many papers, the authors are even not specifying the way they understand that label. The notion of *criticality* used in our monograph has its origin in the studies of the 3-D Navier–Stokes equation where it is possible to set the problem locally in a number of phase spaces, while the local solutions will not be in general extended globally in time in those phase spaces. Such type "criticality" was studied in more detail in the monograph [175] in the case of semilinear parabolic equations.

Let us thus specify, how precisely we understand *criticality of the problem* (1.1) in this monograph.

Definition 1.0.4. We are considering the problem (1.1) and assume the operator $A^{1+\epsilon}$ ($\epsilon > 0$, small) is sectorial positive in the Banach space $X^a = D(A^a)$, $a \in \mathbb{R}$ fixed; moreover, let $F : X^{a+1} \to X^a$, and the local Lipschitz condition in X^{a+1} is valid: *there exists a nondecreasing function* $L : \mathbb{R}^+ \to \mathbb{R}^+$, *such that the estimate*

$$\|F(v) - F(w)\|_{X^a} \leq L(r)\|v - w\|_{X^{a+1}} \tag{1.16}$$

holds for each $v, w \in B_{X^{a+1}}(r)$, *where* $B_{X^{a+1}}(r)$ *denotes an open ball in* X^{a+1} *centered at zero with radius* r.

Under such assumptions, the modification of the problem (1.1) (with A replaced with $A^{1+\epsilon}$) is *locally well posed* in the phase space X^{a+1} (compare Chapter 2 for the proof).

Assume further that an a priori estimate in a Banach space $Y \supset D(A^{a+1})$ for the above local solutions is available; $\|u(t)\|_Y \leq$ const, and we have an estimate for nonlinearity acting on arbitrary local solution $u(t)$ of the form:

$$\exists_{\text{nondecreasing } g':[0,\infty)\to[0,\infty)} \ \|F(u(t))\|_{X^a} \leq g'(\|u(t)\|_Y)(1 + \|u(t)\|^1_{X^{a+1}}), \tag{1.17}$$

valid for all $t \in (0, \tau_{u_0})$, where τ_{u_0} is the "lifetime" of that solution.

If the estimate (1.17) *will not be replaced* with a stronger estimate,

$$\exists_{\text{nondecreasing } g:[0,\infty)\to[0,\infty)} \ \|F(u(t))\|_{X^a} \leq g(\|u(t)\|_Y)(1 + \|u(t)\|^{\eta}_{X^{a+1-\epsilon}}), \tag{1.18}$$

where $\eta < 1$, and $\epsilon > 0$ is small, then the problem (1.1) is called *critical with respect to the a priori estimate in* Y.

In connection with the above definition, note that due to the natural embedding of the spaces X^a, we have that $X^{a+1} \subset X^{a+1-\delta}$ whenever $\delta > 0$. Consequently, $\|\phi\|_{X^{a+1-\delta}} \leq \text{const}\|\phi\|_{X^{a+1}}$, with const depending only on a, δ. Note also that in the case of the Navier–Stokes equation in space dimensions 2 and 3, a stronger local Lipschitz condition is available between $X^{1-\epsilon}$ and X with small $\epsilon > 0$ (note that here $a = 0$), which provides local in time solvability in $X^{1-\epsilon}$ (see Chapters 6 and 7 for details).

Note that the above definition strongly depends on the space Y in which an a priori estimate is available. If we are able to find another a priori estimate of the local solutions in a better norm, condition (1.17) above will be often improved and the considered problem will no longer be critical with respect to the new a priori estimate (in Subsection 6.2.3 an example is given in the case of the 2-D Navier–Stokes equation). Consequently, all of its local solutions will be extended to a global one. It is thus important to use the best available a priori estimate for the problem under consideration. As shown by the Navier–Stokes equation in 3-D, we are often unable to improve the existing a priori estimate to a better one.

Finally, we shall explain the nomenclature used in this book. Very briefly, the name *critical problem* is used here for the (general parabolic) problem

$$u_t + Au = F(u), \quad t > 0, \qquad u(0) = u_0,$$

in which the nonlinear term $F(u)$ is controlled by the main part linear operator A together with the best existing a priori estimates of u (usually through a Nirenberg–Gagliardo-type inequality). In classical monographs (e. g., [115, 116]), another classification of nonlinear parabolic equations was used. According to that classification, critical problem we are considering will belong to one of the two classes; *semilinear equations* or *quasi-linear equations*. If A corresponds to $2m$-th order elliptic operator and the nonlinear term F depends on at most $2m-1$ order spatial derivatives of u (like in the case of the Navier-Stokes equation) the problem is semilinear; if A is as previously but the nonlinear term depends also on the highest order spatial derivative (this corresponds to case of the quasi-geostrophic equation and the Hamilton-Jacobi equation discussed further in the book) the problem is quasi-linear.

1.1 Fractional-order derivatives

It seems that the first time the fractional integral was used in 1823 by N. H. Abel to solve the stated *problem of tautochrone*. Abel was solving an integral equation,

$$\int_a^x \frac{\phi(t)}{(x-t)^\mu}\,dt = f(x), \quad x > a,\ 0 < \mu < 1. \tag{1.19}$$

Later, in the years 1832–1837, J. Liouville gave an extension of Abel's studies defining the fractional derivative by the formula

$$D^{-\rho} f(x) = (-1)^{-\rho}\Gamma(\rho)\int_0^\infty f(x+t)t^{\rho-1}\,dt, \quad -\infty < x < \infty,\ \mathrm{Re}\,\rho > 0. \tag{1.20}$$

Next, that notion was studied by B. Riemann (in 1847), H. Holmgren (1865–1866), J. Hadamard (1892), G. H. Hardy and M. Riesz (1915), and many others.

In the twentieth century, systematic studies of the fractional-order derivatives were originated by B. V. Balakrishnan [11] in 1960s and extended by H. Komatsu in a series of papers (among them [106, 107]). Recently, the notion of the fractional derivative was systematically studied in the monograph [148] and in a recent book [130]. It extends the classical derivatives to the more general operators, often having the form of singular integrals, with real (or even complex) exponent α. Certain results extending the classical singular integral definition from R^N to bounded domain were reported in [20], while the notions discussed there in case of bounded domains are not equivalent with the standard Komatsu definition. We refer to the extended discussion of equivalence of various definitions of fractional derivatives presented recently in [109] (see also Chapter 5 below).

The Komatsu definition of the fractional power, which is mostly often used, requires the operator A in a Banach space X to be closed linear and such that its resolvent set contains $(-\infty, 0)$. Moreover, the resolvent needs to satisfy

$$\|\lambda(\lambda + A)^{-1}\| \le M, \quad \lambda > 0. \tag{1.21}$$

Such A is called a *nonnegative operator*. Fractional powers of such operators are, for $\alpha \in (0,1)$ and $\phi \in D(A)$, defined through the Bochner integral:

$$A^\alpha \phi = \frac{\sin(\pi\alpha)}{\pi} \int_0^\infty \lambda^{\alpha-1} A(\lambda + A)^{-1} \phi d\lambda. \tag{1.22}$$

When we are working with functions v in the Schwartz class $\mathcal{S}$ over $\mathbb{R}^N$ and the negative Laplacian (a particular example of the nonnegative sectorial operator), another classical definition of fractional powers is used, which reads

$$(-\Delta)^\beta v(x) = -C \lim_{\epsilon\to 0} \int_{|z|>\epsilon} \frac{v(x-z) - v(x)}{|z|^{N+2\beta}} dz. \tag{1.23}$$

Many different definitions of fractional powers can be found in the literature in that case; see [109] or Section 5.1.

1.2 Book content

The book consists of eleven chapters.

- Chapter 2 is devoted to preliminary information needed in the main part of the monograph. In particular, we recall there the notion of the *sectorial operator* and its basic properties.
- Chapter 3 is devoted to basic facts concerning the abstract evolutionary Cauchy problem with sectorial operator A;

$$\begin{aligned} u_t + Au &= F(u), \quad t > 0, \\ u(0) &= u_0. \end{aligned}$$

- In Chapter 4, we give a sufficient condition for the global in time extendibility of the local solutions constructed in the previous chapter.
- Chapter 5 is devoted to various technical tools and estimates suitable in the studies of the examples discussed further in the book.
- In Chapter 6, we provide an introduction to the study of the Navier–Stokes equation in space dimensions 2 and 3.
- In Chapter 7, a more complete discussion of the fractional generalization of the Navier–Stokes equation is given.
- Chapter 8 is devoted to the Cauchy problem for quasi-geostrophic equation in $\mathbb{R}^2$.
- In Chapter 9, the Dirichlet boundary value problem for the 2-dimensional quasi-geostrophic equation is reported.

- Chapter 10 is devoted to the subcritical and critical Hamilton–Jacobi equation in a bounded domain.
- In the final Chapter 11, some basic properties of solutions to a semilinear heat equation in bounded domain, or in the whole $\mathbb{R}^N$ are discussed.
- The index and the list of references close the monograph.

2 Preliminary concepts

This introductory chapter is devoted to basic tools and notions necessary to present, in the language of semigroups, the existence and uniqueness theory for semilinear evolutionary equations with a sectorial operator in the main part. We discuss, in particular, the notion of sectoriality and provide a number of examples of such operators. Several technical estimates and inequalities are also collected here for completeness, while that collection will be extended in Chapter 5 with technical tools specific in the studies of particular equations reported further in the book. The final part of the chapter contains basic notions of dynamical systems, in particular, the notion of the global attractor.

2.1 Inequalities. Elliptic operators

2.1.1 Sobolev spaces

The notion of *Sobolev spaces* is of fundamental importance in the contemporary theory of partial differential equations. There is also a huge literature devoted to these spaces, including a few monographs dedicated exclusively to Sobolev spaces; [1, 132, 173] or the recent one [7]. In this monograph, we are following closely the notation and definitions of [173], recalling also occasionally the fundamental monograph [1]. Since the basic facts concerning Sobolev spaces are familiar nowadays, we decided not to repeat them here, referring only to the just mentioned source references for more details; see also the monographs [66, 78, 123, 166, 168] partially devoted to Sobolev spaces, and to the notes [57] for eventual beginners in that field.

2.1.2 Some conventions

We did not reserve a special place here to explain the language or special conventions we are using through the book. Perhaps, three points should be specified here, not to lead to some confusion:

- The first one is connected with notation of the constants throughout calculations inside the text. As a role, we use the letter c for a general constant, and that symbol may vary from line to another line without special notification. Moreover, sometimes, we use the symbol c(*certain quantities*) to mark explicitly the quantities the constant, or rather a function, depends on.
- Working with "critical" problems and Sobolev embeddings we need often to use the strongest version of that estimates. It is also convenient, when making calculations, not to specify explicitly the value of exponent s for the embedding like

$$H^s(\Omega) \subset L^\infty(\Omega), \quad \Omega \subset \mathbb{R}^N,$$

https://doi.org/10.1515/9783110599831-002

which needs to satisfy the restriction $s < \frac{N}{2}$. We simply call it $\frac{N}{2}^-$. In general, for $r \in \mathbb{R}$, r^- denotes a number strictly less than r but eventually close to it. Similarly, $r^+ > r$ and r^+ eventually close to r.

- Finally, we are using some names that originated in physics, or in the classical theory of differential equations. Working with an abstract problem

$$\begin{aligned} &u_t + Au = F(u), \quad t > 0, \\ &u(0) = u_0, \end{aligned} \tag{2.1}$$

or its particular case, like the Dirichlet problem for the semilinear heat equation, we are using two names; *base space* and *phase space*. The first one means the space in which the equation is fulfilled. The second one is the space, in which the solutions of the problem are varying (so that, when we apply to it the prescribed by the equation operations, all the components in the equation belong, at least, to the base space). Since, further in the book, we are working in the scales of fractional power spaces, and the problems under consideration may be settled at various "levels" of such scales, we should always specify which pair base space/phase space we are actually talking about (see, e. g., Remark 6.1.6 or Section 9.2 for examples).

2.1.3 Elementary inequalities

Here, we shall set forth several important inequalities which in later chapters we shall use for performing various estimates.

Lemma 2.1.1 (Cauchy inequality). *Let $a, b \in \mathbb{R}$ and let $\varepsilon > 0$. Then*

$$ab \leq \frac{\varepsilon}{2} a^2 + \frac{1}{2\varepsilon} b^2. \tag{2.2}$$

Proof. For the proof, it suffices to note that

$$0 \leq \left(\sqrt{\varepsilon} a - \frac{1}{\sqrt{\varepsilon}} b \right)^2 = \varepsilon a^2 + \frac{1}{\varepsilon} b^2 - 2ab,$$

from which (2.2) follows. □

Lemma 2.1.2 (Young inequality). *Let $\varepsilon > 0$; $a, b \geq 0$; $p, q > 1$; and $\frac{1}{p} + \frac{1}{q} = 1$. Then*

$$ab \leq \varepsilon \frac{a^p}{p} + \frac{1}{\varepsilon^{\frac{q}{p}}} \frac{b^q}{q}. \tag{2.3}$$

Proof. The case $ab = 0$ is obvious. Hence we can assume that $ab > 0$. Recalling that the exponential function is convex, we obtain

$$ab = e^{\ln ab} = e^{\frac{1}{p} \ln a^p + \frac{1}{q} \ln b^q} \leq \frac{1}{p} e^{\ln a^p} + \frac{1}{q} e^{\ln b^q} = \frac{1}{p} a^p + \frac{1}{q} b^q.$$

From this, there easily follows:

$$ab = (\varepsilon^{\frac{1}{p}} a)\left(\frac{b}{\varepsilon^{\frac{1}{p}}}\right) \le \frac{1}{p}(\varepsilon^{\frac{1}{p}} a)^p + \frac{1}{q}\left(\frac{b}{\varepsilon^{\frac{1}{p}}}\right)^q = \varepsilon\frac{a^p}{p} + \frac{1}{\varepsilon^{\frac{q}{p}}}\frac{b^q}{q},$$

which establishes (2.3). □

In the literature, inequality (2.3) often appears in a slightly different form. Thus, we render here the following.

Corollary 2.1.3. *Let $a, b \ge 0$, $\varepsilon > 0$, $m > 1$. Then*

$$ab \le \frac{1}{m}\varepsilon^m a^m + \frac{m-1}{m}\varepsilon^{-\frac{m}{m-1}} b^{\frac{m}{m-1}}.$$

Lemma 2.1.4 (Hölder inequality). *Let $u, v : \Omega \to \mathbb{R}$ be a Lebesgue measurable functions; $p, q > 1$; and $\frac{1}{p} + \frac{1}{q} = 1$. Then*

$$\|uv\|_{L^1(\Omega)} \le \|u\|_{L^p(\Omega)}\|v\|_{L^q(\Omega)}. \tag{2.4}$$

Proof. The case when the right-hand side of (2.4) is infinite or equal to zero is trivial. Otherwise, using the Young inequality with $\varepsilon := \frac{\|v\|_{L^q(\Omega)}}{\|u\|^{p-1}_{L^p(\Omega)}}$, we get

$$\int_\Omega |uv| dx \le \int_\Omega \left(\varepsilon\frac{|u|^p}{p} + \frac{1}{\varepsilon^{\frac{q}{p}}}\frac{|v|^q}{q}\right) dx = \varepsilon\frac{\|u\|^p_{L^p(\Omega)}}{p} + \frac{1}{\varepsilon^{\frac{q}{p}}}\frac{\|v\|^q_{L^q(\Omega)}}{q}$$

$$= \frac{\|v\|_{L^q(\Omega)}}{\|u\|^{p-1}_{L^p(\Omega)}}\frac{\|u\|^p_{L^p(\Omega)}}{p} + \left(\frac{\|u\|^{p-1}_{L^p(\Omega)}}{\|v\|_{L^q(\Omega)}}\right)^{\frac{q}{p}}\frac{\|v\|^q_{L^q(\Omega)}}{q} = \left(\frac{1}{p} + \frac{1}{q}\right)\|u\|_{L^p(\Omega)}\|v\|_{L^q(\Omega)}.$$

Therefore, equation (2.4) is proved. □

Lemma 2.1.5 (Bernoulli inequality). *Let $a, b > 0$, $\rho > \nu \ge 0$, and consider a continuous function $y : [0, \tau_0) \to [0, +\infty)$ differentiable on $(0, \tau_0)$, where $0 < \tau_0 \le +\infty$, and satisfying*

$$y'(t) \le ay^\nu(t) - by^\rho(t), \quad \textit{for all } t \in (0, \tau_0). \tag{2.5}$$

Then

$$\sup_{t\in[0,\tau_0)} \{y(t)\} \le \max\left\{y(0), \left(\frac{a}{b}\right)^{\frac{1}{\rho-\nu}}\right\}. \tag{2.6}$$

Furthermore, if $\tau_0 = \infty$, then

$$\limsup_{t\to+\infty} y(t) \le \left(\frac{a}{b}\right)^{\frac{1}{\rho-\nu}}. \tag{2.7}$$

Proof. The function $h(y) = ay^{\nu} - by^{\rho}$ appearing on the right-hand side of (2.5) is negative at each y larger than $(\frac{a}{b})^{\frac{1}{\rho-\nu}}$, which number is the unique positive root of the equation $h(y) = 0$. Therefore, in the presence of (2.5), $y(t)$ is fully controlled by the initial value $y(0)$ and the quantity $(\frac{a}{b})^{\frac{1}{\rho-\nu}}$, as is stated in the conditions (2.6) and (2.7). □

Lemma 2.1.6. *Let $y \in C^0([0,\infty)) \cap C^1((0,\infty))$, let f be nonnegative and continuous on $[0,\infty)$, and suppose that $\lim_{t\to+\infty} f(t) = M$. If, for some $a > 0$, y satisfies*

$$y'(t) \le -ay(t) + f(t), \quad t \in (0,+\infty), \tag{2.8}$$

then

$$\limsup_{t\to+\infty} y(t) \le \frac{M}{a}.$$

Proof. With our assumptions, f is bounded on $[0,+\infty)$. Fix $\varepsilon > 0$ and a time $t_0 > 0$, such that $f(t) \le M + \varepsilon$ for $t \ge t_0$. Solving the differential inequality (2.8) for $t \ge t_0$, we find

$$\begin{aligned} y(t) &\le y(0)e^{-at} + \left(\int_0^{t_0} + \int_{t_0}^{t}\right) f(s)e^{as}ds e^{-at} \\ &\le y(0)e^{-at} + \sup_{s\in[0,+\infty)} f(s)\frac{e^{a(t_0-t)} - e^{-at}}{a} + (M+\varepsilon)\frac{1 - e^{-a(t-t_0)}}{a} \to \frac{M+\varepsilon}{a} \end{aligned}$$

when $t \to +\infty$. Since ε is arbitrary, the proof is complete. □

Lemma 2.1.7 (Standard Gronwall inequality). *Let y, v, w be three real functions defined and continuous in the interval $[a,b)$, $b \le \infty$, and such that*

$$\begin{aligned} &w(x) > 0, \quad x \in [a,b), \\ &y(x) \le v(x) + \int_a^x w(s)y(s)ds, \quad x \in [a,b). \end{aligned}$$

Then

$$y(x) \le v(x) + \int_a^x w(s)v(s)\exp\left(\int_s^x w(r)dr\right)ds, \quad x \in [a,b).$$

Proof. For $x \in [a,b)$, we define a function

$$z(x) = \int_a^x w(s)y(s)ds.$$

Due to the assumed inequalities,

$$w(x)y(x) \le w(x)v(x) + w(x)\int_a^x w(s)y(s)ds,$$

which can be written equivalently as

$$z'(x) \leq w(x)z(x) + w(x)v(x), \quad x \in [a,b).$$

Since also $z(a) = 0$, then z fulfills an estimate

$$z(x) \leq \int_a^x w(s)v(s)\exp\left(\int_s^x w(r)dr\right)ds, \tag{2.9}$$

which together with the estimate $y(x) \leq v(x) + z(x)$ concludes the proof. □

Lemma 2.1.8 (Asymptotic Gronwall inequality). *Let differentiable function $y : (0,\infty) \to \mathbb{R}$ be integrable on $(0,\infty)$ and satisfy inequality*

$$y'(\tau) \leq \text{const.}y(\tau), \quad \textit{for } \tau > 0. \tag{2.10}$$

Then

$$\limsup_{t\to+\infty} y(t) \leq 0. \tag{2.11}$$

Proof. Fixing $t > 0$, we obtain from (2.10) that

$$\frac{d}{d\tau}\left(y(\tau)e^{\text{const.}(t-\tau)}\right) \leq 0, \quad \text{for } \tau > 0. \tag{2.12}$$

Fixing next $r > 0$, considering any $s \in [t, t+r]$ and integrating (2.12) over the interval $[s, t+r]$, we get

$$y(t+r)e^{-\text{const.}r} \leq y(s)e^{\text{const.}(t-s)} \leq y(s). \tag{2.13}$$

Since inequality (2.13) is valid for $s \in [t, t+r]$, we can integrate it over $s \in [t, t+r]$, so that

$$y(t+r)re^{-\text{const.}r} = \int_t^{t+r} y(t+r)e^{-\text{const.}r}ds \leq \int_t^{t+r} y(s)ds. \tag{2.14}$$

As a consequence of the finite value of the integral $\int_0^\infty y(s)ds$, the upper limit $\limsup_{t\to+\infty}\int_t^{t+r} y(s)ds$ is nonpositive, so that (2.11) follows. □

Lemma 2.1.9 (Uniform Gronwall inequality). *Let differentiable function $y : (0,\tau_0) \to \mathbb{R}$ satisfy conditions*

$$y'(\tau) \leq \text{const.}y(\tau), \quad \textit{for } \tau \in (0,\tau_0), \tag{2.15}$$

and

$$\exists_{r\in(0,\tau_0)}\ \exists_{\text{const.}_1>0}\ \forall_{t\in(0,\tau_0-r)} \quad \int_t^{t+r} y(s)ds \leq \text{const.}_1. \tag{2.16}$$

Then

$$y(t+r) \leq \frac{\text{const.}_1}{r} e^{\text{const.}r}, \quad \textit{for } t \in (0, \tau_0 - r). \tag{2.17}$$

Proof. In order to show (2.17), we shall repeat the proof of Lemma 2.1.8 until (2.14) is established. Then, applying (2.16) to the right-hand side of (2.14), we obtain condition

$$y(t+r) r e^{-\text{const.}r} \leq \int_t^{t+r} y(s)ds \leq \text{const.}_1, \tag{2.18}$$

which is a counterpart of (2.17). □

The latter lemma possess the following stronger version (see [172, p. 89]), in which the estimate is not growing with time.

Lemma 2.1.10 (Generalized Gronwall inequality). *Let $g, h, y : [\tau_0, +\infty) \to (0, +\infty)$ be continuous functions, continuously differentiable on $(\tau_0, +\infty)$. Assume further that for some $r > 0$ and all $t \geq \tau_0$, the following conditions hold:*

$$\begin{cases} y'(t) \leq g(t)y(t) + h(t), \\ \int_t^{t+r} g(s)ds \leq \text{const.}_1, \quad \int_t^{t+r} h(s)ds \leq \text{const.}_2, \\ \int_t^{t+r} y(s)ds \leq \text{const.}_3. \end{cases} \tag{2.19}$$

Then, y fulfills the uniform estimate

$$y(t+r) \leq \left(\frac{\text{const.}_3}{r} + \text{const.}_2\right) e^{\text{const.}_1}, \quad t \geq \tau_0. \tag{2.20}$$

Lemma 2.1.11 (Volterra-type inequality). *Let $\alpha, \beta \in [0, 1)$, $a \geq 0$, $b > 0$, and let $y : [0, \tau) \to \mathbb{R}^+$ be a continuous function satisfying the inequality*

$$y(t) \leq \frac{a}{t^\alpha} + b \int_0^t \frac{1}{(t-s)^\beta} y(s)ds, \quad \textit{for } t \in (0, \tau). \tag{2.21}$$

Then

$$\sup_{t \in [0,\tau)} t^\alpha y(t) \leq a\,\text{const.}(b, \alpha, \beta, \tau), \tag{2.22}$$

where const.(b, α, β, τ) *is a continuous function increasing with respect to τ.*

Proof. The proof occurs in three steps.

Step 1. Choose $t^* = \min\{\frac{\tau}{2}, T^*\}$, where

$$\frac{b {T^*}^{1-\beta}}{2^{1-\alpha-\beta}} \left(\frac{1}{1-\alpha} + \frac{1}{1-\beta}\right) = \frac{1}{2} \tag{2.23}$$

and introduce the following linear transformation Φ, acting on the continuous functions $f : (0,\tau) \to \mathbb{R}^+$:

$$f \overset{\Phi}{\to} \Phi(f)(t) := b\int_0^t \frac{1}{(t-s)^\beta} f(s)ds.$$

Clearly, we can write (2.21) in the form

$$y(t) \le \frac{a}{t^\alpha} + \Phi(y)(t), \quad t \in (0,t^*). \tag{2.24}$$

Since Φ is linear and since $f \le g$ implies $\Phi f \le \Phi g$ for all functions f, g in the domain of Φ, the inequality (2.24) yields

$$\Phi(y)(t) \le a\Phi\left(\frac{1}{t^\alpha}\right) + \Phi^2(y)(t).$$

Consequently,

$$y(t) \le \frac{a}{t^\alpha} + a\Phi\left(\frac{1}{t^\alpha}\right) + \Phi^2(y)(t), \quad t \in (0,t^*). \tag{2.25}$$

Proceeding by induction, we obtain that for all $n \in \mathbb{N}$,

$$y(t) \le \frac{a}{t^\alpha} + a\sum_{k=1}^{n-1} \Phi^k\left(\frac{1}{t^\alpha}\right) + \Phi^n(y)(t), \quad t \in (0,t^*). \tag{2.26}$$

Furthermore, we observe that

$$\begin{aligned}\Phi(y)(t) &\le \|y\|_{L^\infty(0,t^*)} b\int_0^t \frac{1}{(t-s)^\beta} ds \\ &= \|y\|_{L^\infty(0,t^*)} \frac{bt^{1-\beta}}{1-\beta}.\end{aligned} \tag{2.27}$$

The inequalities (2.23) and (2.27) together yield the estimate

$$\|\Phi(y)\|_{L^\infty(0,t^*)} < \text{const.}\|y\|_{L^\infty(0,t^*)}$$

with

$$\text{const.} := \frac{b{t^*}^{1-\beta}}{1-\beta} < 1.$$

Hence,

$$\|\Phi^n(y)\|_{L^\infty(0,t^*)} \le \text{const.}^n \|y\|_{L^\infty(0,t^*)} \to 0 \quad \text{when } n \to \infty. \tag{2.28}$$

Next, using (2.23) again, we obtain the estimate

$$\Phi\left(\frac{1}{t^\alpha}\right) = \left(\int_0^{\frac{t}{2}} + \int_{\frac{t}{2}}^{t}\right) b\frac{1}{(t-s)^\beta}\frac{1}{s^\alpha}ds \le \frac{b}{(\frac{t}{2})^\beta}\frac{(\frac{t}{2})^{1-\alpha}}{1-\alpha} + \frac{b}{(\frac{t}{2})^\alpha}\frac{(\frac{t}{2})^{1-\beta}}{1-\beta}$$

$$= \frac{1}{t^\alpha}\frac{bt^{1-\beta}}{2^{1-\alpha-\beta}}\left(\frac{1}{1-\alpha} + \frac{1}{1-\beta}\right) \le \frac{1}{2}\frac{1}{t^\alpha}, \quad t \in (0, t^*).$$

By induction,

$$\Phi^{k+1}\left(\frac{1}{t^\alpha}\right) = \left(\int_0^{\frac{t}{2}} + \int_{\frac{t}{2}}^{t}\right) b\frac{1}{(t-s)^\beta}\left[\Phi^k\left(\frac{1}{t^\alpha}\right)\right]ds$$

$$\le \left(\int_0^{\frac{t}{2}} + \int_{\frac{t}{2}}^{t}\right) b\frac{1}{(t-s)^\beta}\left[\frac{1}{2^k}\frac{1}{s^\alpha}\right]ds \le \frac{1}{2^{k+1}}\frac{1}{t^\alpha}, \quad t \in (0, t^*). \tag{2.29}$$

From (2.29), there follows

$$\sum_{k=1}^{n-1}\Phi^k\left(\frac{1}{t^\alpha}\right) \le \sum_{k=1}^{n-1}\frac{1}{2^k}\frac{1}{t^\alpha} \le \frac{1}{t^\alpha}, \quad t \in (0, t^*). \tag{2.30}$$

With the aid of (2.28) and (2.30), we can pass to the limit as $n \to \infty$ in (2.26). The result is

$$y(t) \le \frac{a}{t^\alpha} + a\sum_{k=1}^{\infty}\Phi^k\left(\frac{1}{t^\alpha}\right) \le \frac{2a}{t^\alpha}, \quad t \in (0, t^*). \tag{2.31}$$

This proves that

$$\sup_{t\in[0,t^*]} t^\alpha y(t) \le 2a. \tag{2.32}$$

Step 2. Define

$$h(r) = \left(2\left(\frac{\tau}{t^*}\right)^\alpha + \frac{\tau}{t^*}r\right), \quad r \in \mathbb{R},$$

and fix $\delta \in (0, \tau - t^*)$ so that

$$\frac{b\delta^{1-\beta}}{1-\beta} \le \frac{1}{2}. \tag{2.33}$$

We shall prove that, for each $t_0 \in [t^*, \tau - \delta)$, the following implication holds:

$$\text{If } \sup_{t\in[0,t_0]} t^\alpha y(t) \le ar, \quad \text{then} \quad \sup_{t\in[0,t_0+\delta]} t^\alpha y(t) \le ah(r). \tag{2.34}$$

Indeed, if $t_0 \geq t^*$, if $\sup_{t\in[0,t_0]} t^\alpha y(t) \leq ar$, and if $t \in [t_0, t_0 + \delta]$, then from (2.21) there follows

$$y(t) \leq \frac{a}{t^\alpha} + b\left(\int_0^{t_0} + \int_{t_0}^{t}\right)\frac{1}{(t-s)^\beta}y(s)ds$$

$$\leq \frac{a}{t^\alpha} + abr\int_0^{t_0}\frac{1}{(t-s)^\beta s^\alpha}ds + b\sup_{s\in[t_0,t_0+\delta]} y(s)\int_{t_0}^{t}\frac{1}{(t-s)^\beta}ds, \tag{2.35}$$

where keeping in mind that $t^* \leq t_0 \leq t \leq t_0 + \delta < \tau$ and that (2.23) holds,

$$\int_0^{t_0}\frac{1}{(t-s)^\beta s^\alpha}ds = \left(\int_0^{\frac{t_0}{2}} + \int_{\frac{t_0}{2}}^{t_0}\right)\frac{1}{(t-s)^\beta s^\alpha}ds$$

$$\leq \frac{1}{(t-\frac{t_0}{2})^\beta}\frac{1}{1-\alpha}\left(\frac{t_0}{2}\right)^{1-\alpha} + \frac{1}{(\frac{t_0}{2})^\alpha}\frac{1}{1-\beta}\left(\left(t-\frac{t_0}{2}\right)^{1-\beta} - (t-t_0)^{1-\beta}\right)$$

$$\leq \frac{1}{(\frac{t_0}{2})^\beta}\frac{1}{1-\alpha}\left(\frac{t_0}{2}\right)^{1-\alpha} + \frac{1}{(\frac{t_0}{2})^\alpha}\frac{1}{1-\beta}\left(t-\frac{t_0}{2}-(t-t_0)\right)^{1-\beta}$$

$$= \left(\frac{1}{1-\alpha}+\frac{1}{1-\beta}\right)\left(\frac{t_0}{2}\right)^{1-\alpha-\beta} \leq \frac{1}{2b}\frac{t_0^{1-\alpha-\beta}}{t^{*1-\beta}}$$

$$\leq \frac{1}{2b}\frac{t_0^{1-\alpha-\beta}}{t^{*1-\beta}}\left(\frac{t_0}{t^*}\right)^\beta = \frac{1}{2b}\frac{\tau^{1-\alpha}}{t^*}. \tag{2.36}$$

Combining (2.33), (2.35), and (2.36), we obtain

$$y(t) \leq \frac{a}{t^\alpha} + \frac{1}{2}a\frac{\tau^{1-\alpha}}{t^*}r + \frac{1}{2}\sup_{s\in[t_0,t_0+\delta]} y(s), \quad t \in [t_0, t_0+\delta].$$

Next, recalling that $t^* \leq t_0$, we see that

$$\sup_{t\in[t_0,t_0+\delta]} y(t) \leq \frac{2a}{t_0^\alpha} + a\frac{\tau^{1-\alpha}}{t^*}r \leq a\left(\frac{2}{t^{*\alpha}} + \frac{\tau^{1-\alpha}}{t^*}r\right).$$

Since $t \leq t_0 + \delta < \tau$ and since $\sup fg \leq \sup f \sup g$, we have the estimate

$$\sup_{t\in[t_0,t_0+\delta]} t^\alpha y(t) \leq a\left(2\left(\frac{\tau}{t^*}\right)^\alpha + \frac{\tau}{t^*}r\right) = ah(r),$$

and so finally,

$$\sup_{t\in[0,t_0+\delta]} t^\alpha y(t) \leq a\max\{r, h(r)\} = ah(r).$$

Step 3. Given τ, t^*, and δ as above, there exist numbers $\theta \in [0, \delta)$, $k \in \mathbb{N}$, such that $\tau - t^* = k\delta + \theta$. Hence, from (2.32), with the aid of (2.34), we have

$$\sup_{t\in[0,\tau-\delta]} t^\alpha y(t) \le a(\underbrace{h \circ \cdots \circ h}_{k\text{-times}})(2).$$

Therefore, for any sequence $t_n \to t^{*-}$, we have

$$\sup_{t\in[0,t_n-\delta]} t^\alpha y(t) \le a(\underbrace{h \circ \cdots \circ h}_{k\text{-times}})(2), \qquad n \in \mathbb{N}.$$

Again we apply (2.34) and now we obtain

$$\sup_{t\in[0,t_n]} t^\alpha y(t) \le a(\underbrace{h \circ \cdots \circ h}_{k+1\text{-times}})(2), \qquad n \in \mathbb{N}.$$

This completes the proof of Lemma 2.1.11. □

2.1.4 Embedding theorems

In a modern approach to partial differential equations different function spaces, most often of the Sobolev and Hölder type, are simultaneously used. We need to compare the norms of such spaces and these connections are expressed as *embedding theorems*. We start with formulation of the classical version of such a theorem valid for $W^{m,p}$ spaces of natural order m. For the proof of it, we refer to [66, p. 29], [1, p. 97] but we shall follow the presentation of [87].

For $m \in \mathbb{N}$ and $p \in [1, \infty]$, we consider the Sobolev space $W^{m,p}(\Omega)$. It is convenient (see [87]) to introduce the *net smoothness number* characterizing the regularity properties of the elements of $W^{m,p}(\Omega)$:

$$\text{net smoothness } (f) := m - \frac{n}{p} \quad \text{for all } f \in W^{m,p}(\Omega). \tag{2.37}$$

This quantity is decisive for relations between different Sobolev spaces. Let $\Omega \subset \mathbb{R}^N$ be a domain lying on one side of its boundary $\partial\Omega$ with $\partial\Omega$ being a Lipschitz surface. We shall recall now the classical Sobolev embedding theorem.

Sobolev embeddings for natural order spaces

Let $k, m \in \mathbb{N}$, $1 \le p, q \le +\infty$ and Ω be as above. Then the following continuous inclusions take place:

$$W^{m,p}(\Omega) \subset \begin{cases} W^{k,q}(\Omega) & \text{if } m - \frac{n}{p} \ge k - \frac{n}{q},\ q \ge p \\ & (\text{unless } m - \frac{n}{p} = k,\ q = \infty), \\ C^{k+\mu}(\overline{\Omega}) & \text{if } 0 \le \mu \le m - k - \frac{n}{p} < 1 \\ & (\text{unless } m - \frac{n}{p} = k + \mu \in \mathbb{N}). \end{cases} \tag{2.38}$$

Moreover, for bounded Ω *the above inclusions are compact in the case of strict inequalities;* $m - \frac{n}{p} > k - \frac{n}{q}$ *or* $m - \frac{n}{p} > k + \mu$*, respectively.*

Further, the generalization of the embedding theorem to fractional-order spaces $W^{s,p}$ with real $s \geq 0$ will be described. Let us start with some definitions.

We shall consider below bounded domains Ω having the *extension property* as described in [86, p. 38] and recall the definition of the $W^{s,p}$ spaces with $0 \leq s \in \mathbb{R}$.

For $\Omega = \mathbb{R}^N$, let

$$W^{s,p}(\mathbb{R}^N) = \begin{cases} H_p^s(\mathbb{R}^N), & s = 0, 1, 2, \dots, \\ B_{p,p}^s(\mathbb{R}^N) & \text{for } 0 < s \notin \mathbb{R}, \end{cases} \tag{2.39}$$

where $B_{p,p}^s(\mathbb{R}^N)$ are Besov spaces and $H_p^s(\mathbb{R}^N)$ are the spaces of Bessel potentials (cf. [173, Definition 2.3.1]). It is worth noting that $H_p^s(\mathbb{R}^N) \neq B_{p,p}^s(\mathbb{R}^N)$ unless $p = 2$ (see [173, Remark 2.3.3/4]).

For arbitrary domain $\Omega \subset \mathbb{R}^N$ (bounded or not), $-\infty < s < +\infty$, $1 < p < +\infty$ the space $B_{p,p}^s(\Omega)$ consists of (cf. [173, 4.2.1]) restrictions $\phi_{|\Omega}$ (in a sense of distributions) of elements of $B_{p,p}^s(\mathbb{R}^N)$ and is normed by

$$\|f\|_{B_{p,p}^s(\Omega)} = \inf_{\substack{\phi_{|\Omega} = f \\ \phi \in B_{p,p}^s(\mathbb{R}^N)}} \|\phi\|_{B_{p,p}^s(\mathbb{R}^N)}.$$

Simultaneously, $H_p^s(\Omega)$ consists of restrictions $\phi_{|\Omega}$ of elements $\phi \in H_p^s(\mathbb{R}^N)$ and is normed by

$$\|f\|_{H_p^s(\Omega)} = \inf_{\substack{\phi_{|\Omega} = f \\ \phi \in H_p^s(\mathbb{R}^N)}} \|\phi\|_{H_p^s(\mathbb{R}^N)}.$$

We extend naturally (2.39) to

$$W^{s,p}(\Omega) = \begin{cases} H_p^s(\Omega), & s = 0, 1, 2, \dots, \\ B_{p,p}^s(\Omega) & \text{for } 0 < s \notin \mathbb{R}, \end{cases} \tag{2.40}$$

pointing as before that $H_p^s(\Omega) \neq B_{p,p}^s(\Omega)$ unless $p = 2$ (cf. [173, Theorem 4.6.1(b)]).

Next, we shall recall the following two consequences of the *interpolation theory* (cf. [173, 2.4.2]).

For $-\infty < s_0, s_1 < +\infty$, $1 < p_0, p_1 < +\infty$, $0 < \theta < 1$, and $s_0 \neq s_1$,

$$s = (1 - \theta)s_0 + \theta s_1, \quad \frac{1}{p} = \frac{1 - \theta}{p_0} + \frac{\theta}{p_1}, \tag{2.41}$$

we have from [173]

$$[H_{p_0}^{s_0}(\mathbb{R}^N), H_{p_1}^{s_1}(\mathbb{R}^N)]_{\theta,p} = H_p^s(\mathbb{R}^N). \tag{2.42}$$

Also, by [173, Theorem 2.4.1(c)], for the same range of parameters,

$$(B^{s_0}_{p_0,p_0}(\mathbb{R}^N), B^{s_1}_{p_1,p_1}(\mathbb{R}^N))_{\theta,p} = B^s_{p,p}(\mathbb{R}^N), \quad 0 < \theta < 1, \tag{2.43}$$

where s, p are given by (2.41).

Sobolev embeddings for fractional-order spaces

Basing on [173, 4.6.1], we recall the generalization of the Sobolev embedding theorem to fractional-order spaces.

Proposition 2.1.12. *For $0 \le t \le s < +\infty$, $1 < q \le p < +\infty$ and $\Omega = \mathbb{R}^N$ or Ω bounded having the extension property (see [86, p. 38]), the following continuous embeddings are valid:*

$$W^{s,q}(\Omega) \subset W^{t,p}(\Omega) \quad \text{whenever } s - \frac{n}{q} \ge t - \frac{n}{p}. \tag{2.44}$$

When Ω is bounded and $t < s$, then the condition $q \le p$ can be neglected.

Embeddings in $C^k(\overline{\Omega})$ ($W^{k,\infty}(\Omega)$)

Embeddings of the Sobolev spaces in $C^k(\overline{\Omega})$ or $W^{k,\infty}(\Omega)$ are of special importance for our further studies. Using the result of [173, Theorem 4.6.1(e)], we claim an embedding:

$$W^{s,q}(\Omega) \subset C^{k+\mu}(\overline{\Omega}) \tag{2.45}$$

valid whenever $s - \frac{n}{q} \ge k + \mu$ ($s \in \mathbb{R}$, $q \in (1, +\infty)$, $k \in \mathbb{N}$, $\mu \in (0, 1)$), and we require strict inequality $s - \frac{n}{q} > k$ when $\mu = 0$. Here, $\Omega \subset \mathbb{R}^N$ is assumed to be a Lipschitz domain.

Agmon inequalities

Finally, we recall the interpolation estimate due to S. Agmon (cf. [2], [172, p. 50]) valid for bounded domains $\Omega \subset \mathbb{R}^N$ with C^n-smooth boundary $\partial\Omega$;

$$\begin{aligned} &\|\phi\|_{L^\infty(\Omega)} \le \text{const.} \|\phi\|^{\frac{1}{2}}_{H^{\frac{n}{2}-1}(\Omega)} \|\phi\|^{\frac{1}{2}}_{H^{\frac{n}{2}+1}(\Omega)}, && \phi \in H^{\frac{n}{2}+1}(\Omega),\ n \text{ even}, \\ &\|\phi\|_{L^\infty(\Omega)} \le \text{const.} \|\phi\|^{\frac{1}{2}}_{H^{\frac{n-1}{2}}(\Omega)} \|\phi\|^{\frac{1}{2}}_{H^{\frac{n+1}{2}}(\Omega)}, && \phi \in H^{\frac{n+1}{2}}(\Omega),\ n \text{ odd}. \end{aligned} \tag{2.46}$$

2.1.5 Nirenberg–Gagliardo-type inequalities

Many estimates of the present book are based on the *interpolation inequality* that was introduced by L. Nirenberg and E. Gagliardo in 1959. The proof of it can be found in [66, p. 27], [166, p. 111], also in [86, p. 37] or directly in [137]. Recent results extending the classical Nirenberg–Gagliardo estimates to the Besov or Triebel–Lizorkin spaces are reported in [7, Section 5.7]. We follow here the presentation of [66].

Theorem 2.1.13 (Nirenberg–Gagliardo inequality). *Let $m \in \mathbb{N}$, $\Omega \subset \mathbb{R}^N$ be a domain having the C^m smooth boundary $\partial\Omega$. Let $v \in W^{m,r}(\Omega) \cap L^q(\Omega)$ with $1 \le r, q \le +\infty$. Then, for any integer j, $0 \le j < m$ and any number $\theta \in [\frac{j}{m}, 1)$, define*

$$\frac{1}{p} = \frac{j}{n} + \theta\left(\frac{1}{r} - \frac{m}{n}\right) + (1-\theta)\frac{1}{q}. \tag{2.47}$$

If $m - j - \frac{n}{r}$ is not a nonnegative integer, then

$$\|D^j u\|_{L^p(\Omega)} \le \text{const.}\|u\|^{\theta}_{W^{m,r}(\Omega)}\|u\|^{1-\theta}_{L^q(\Omega)}, \tag{2.48}$$

where D^j denotes any partial derivative of the order j and const. *depends on Ω, r, q, m, j, θ. If $m - j - \frac{n}{r}$ is a nonnegative integer, then* (2.48) *holds with $\theta = \frac{j}{m}$.*

Extension of Nirenberg–Gagliardo inequality to fractional spaces

Such an extension of Theorem 2.1.13 follows from the deep interpolation results given in [173] and was formulated in [5].

Proposition 2.1.14. *Let $1 < p, p_0, p_1 < +\infty$, $0 \le s_0 < s_1 < +\infty$, $\theta \in (0,1)$, and s, p be given by* (2.41)*, then*

$$\forall_{v \in W^{s_1,p_1} \cap W^{s_0,p_0}} \; \|v\|_{W^{s^-,p}} \le c_\theta \|v\|^{\theta}_{W^{s_1,p_1}} \|v\|^{1-\theta}_{W^{s_0,p_0}}, \tag{2.49}$$

where

$$s^- \begin{cases} = s & \text{if } s_0, s, s_1 \in \mathbb{N} \text{ or } s_0, s, s_1 \in \mathbb{R} \setminus \mathbb{N}, \\ < s & \text{and arbitrarily close to } s \text{ otherwise.} \end{cases} \tag{2.50}$$

This proposition is valid for an arbitrary domain Ω in $\mathbb{R}^N$ having the extension property (see [86, p. 38]*); in particular, for $\Omega = \mathbb{R}^N$.*

Proof. First, the case $\Omega = \mathbb{R}^N$ will be considered. We shall recall the following inclusions (cf. [173, 2.3.3]):

$$H^{s+\varepsilon}_p(\mathbb{R}^N), W^{s+\varepsilon,p}(\mathbb{R}^N) \subset B^s_{p,p}(\mathbb{R}^N) \subset H^{s-\varepsilon}_p(\mathbb{R}^N), W^{s-\varepsilon,p}(\mathbb{R}^N), \tag{2.51}$$

valid for any $-\infty < s < +\infty$, $1 < p < +\infty$ and $\varepsilon > 0$.

When $s_0, s, s_1 \in \mathbb{N}$ and $H^{s_0}_{p_0}(\mathbb{R}^N)$, $H^s_p(\mathbb{R}^N)$, $H^{s_1}_{p_1}(\mathbb{R}^N)$ are Bessel spaces, then the estimate (2.49) follows directly from (2.42) and the *moments inequality* [173, Theorem 1.3.3(g)], also for all $s_0, s, s_1 \in \mathbb{R} \setminus \mathbb{N}$ we will stay inside the scale of Besov spaces and (2.49) follows from (2.43) and [173, Theorem 1.3.3(g)].

For arbitrary $-\infty < s_0 < s_1 < +\infty$, (2.43) implies

$$\forall_{v \in B^{s_1}_{p_1,p_1}(\mathbb{R}^N) \cap B^{s_0}_{p_0,p_0}(\mathbb{R}^N)} \; \|v\|_{B^s_{p,p}(\mathbb{R}^N)} \le c\|v\|^{\theta}_{B^{s_1}_{p_1,p_1}(\mathbb{R}^N)} \|v\|^{1-\theta}_{B^{s_0}_{p_0,p_0}(\mathbb{R}^N)}. \tag{2.52}$$

For $0 < s_0 < s_1 < +\infty$, we can choose $\varepsilon > 0$ arbitrarily small and such that $s_0 - \varepsilon$, $s_1 - \varepsilon$ and $s - \varepsilon = (1-\theta)(s_0 - \varepsilon) + \theta(s_1 - \varepsilon)$ are not natural numbers, so that by (2.52), for p, s given by (2.41),

$$\forall_{v \in W^{s-\varepsilon,p_0}(\mathbb{R}^N) \cap W^{s_1-\varepsilon,p_1}(\mathbb{R}^N)} \|v\|_{W^{s-\varepsilon,p}(\mathbb{R}^N)} \le c \|v\|^{\theta}_{W^{s_1-\varepsilon,p_1}(\mathbb{R}^N)} \|v\|^{1-\theta}_{W^{s_0-\varepsilon,p_0}(\mathbb{R}^N)}.$$

Using (2.51) to estimate the right-hand side of the inequality above, we obtain

$$\forall_{v \in W^{s_0,p_0}(\mathbb{R}^N) \cap W^{s_1,p_1}(\mathbb{R}^N)} \|v\|_{W^{s-\varepsilon,p}(\mathbb{R}^N)} \le c' \|v\|^{\theta}_{W^{s_1,p_1}(\mathbb{R}^N)} \|v\|^{1-\theta}_{W^{s_0,p_0}(\mathbb{R}^N)}.$$

Since ε was arbitrarily close to 0, this means precisely that

$$\forall_{v \in W^{s_0,p_0}(\mathbb{R}^N) \cap W^{s_1,p_1}(\mathbb{R}^N)} \|v\|_{W^{s^-,p}(\mathbb{R}^N)} \le c' \|v\|^{\theta}_{W^{s_1,p_1}(\mathbb{R}^N)} \|v\|^{1-\theta}_{W^{s_0,p_0}(\mathbb{R}^N)},$$

where $s^- < s$ is arbitrarily close to s.

If $s_0 = 0$, choosing again $\varepsilon > 0$ arbitrarily small and such that $s_1 - \varepsilon$ and $s - \varepsilon$ are not natural numbers we find by (2.43) for p, s given by (2.41), that

$$\forall_{v \in B^{-\varepsilon}_{p_0,p_0}(\mathbb{R}^N) \cap W^{s_1-\varepsilon,p_1}(\mathbb{R}^N)} \|v\|_{W^{s-\varepsilon,p}(\mathbb{R}^N)} \le c \|v\|^{\theta}_{W^{s_1-\varepsilon,p_1}(\mathbb{R}^N)} \|v\|^{1-\theta}_{B^{-\varepsilon}_{p_0,p_0}(\mathbb{R}^N)},$$

(the case $s_0 = 0$ needs to be distinguished because the space $W^{-\varepsilon,p_0}(\mathbb{R}^N)$ was not defined) and using (2.51) to the right-hand side, we get

$$\forall_{v \in W^{0,p_0}(\mathbb{R}^N) \cap W^{s_1,p_1}(\mathbb{R}^N)} \|v\|_{W^{s^-,p}(\mathbb{R}^N)} \le c' \|v\|^{\theta}_{W^{s_1,p_1}(\mathbb{R}^N)} \|v\|^{1-\theta}_{W^{0,p_0}(\mathbb{R}^N)} \tag{2.53}$$

as before. Now the estimate (2.49) will be extended from $\mathbb{R}^N$ to an arbitrary domain $\Omega \subset \mathbb{R}^N$ having the *extension property* ([86, p. 38]). The proof is completed. □

Remark 2.1.15. As a consequence of the Sobolev-type inclusion reported in Proposition 2.1.12,

$$\begin{cases} W^{s,q}(\Omega) \subset W^{t,p}(\Omega) \\ s - \frac{n}{q} \ge t - \frac{n}{p}, \quad s, t \in \mathbb{R},\ 1 < q \le p < +\infty, \end{cases} \tag{2.54}$$

the main estimate of Proposition 2.1.14 will be extended to

$$\|v\|_{W^{t,p}(\Omega)} \le c \|v\|^{\theta}_{W^{s_1,p_1}(\Omega)} \|v\|^{1-\theta}_{W^{s_0,p_0}(\Omega)}, \tag{2.55}$$

where we need to assume that

$$t - \frac{n}{p} < (\le)(1-\theta)\left(s_0 - \frac{n}{p_0}\right) + \theta\left(s_1 - \frac{n}{p_1}\right) \tag{2.56}$$

and

$$\frac{1}{p} \le \frac{1-\theta}{p_0} + \frac{\theta}{p_1}, \tag{2.57}$$

(here, $\theta \in [0,1]$, $0 \le s_0 < s_1 < +\infty$, $1 < p, p_0, p_1 < +\infty$). Note that condition $t \le s = (1-\theta)s_0 + \theta s_1$, needed for validity of the Sobolev embedding, follows from (2.56), (2.57).

For $0 < \theta < 1$, the equality sign in (2.56) is permitted under the additional restrictions described in (2.50).

For $\theta = 1$, the estimate (2.55) is a direct consequence of Proposition 2.1.12, provided that $t \le s_1$, $t - \frac{n}{p} \le s_1 - \frac{n}{p_1}$ and $1 < p_1 \le p < +\infty$. Note, that if Ω is bounded weaker assumptions can be taken, as described in Proposition 2.1.12. The case $\theta = 0$ in (2.54) is quite similar and follows from Proposition 2.1.12 with $t \le s_0$, $t - \frac{n}{p} \le s_0 - \frac{n}{p_0}$, $1 < p_0 \le p < +\infty$ (and, as before, with appropriately weakened assumptions for bounded Ω).

2.1.6 Properties of uniformly strongly elliptic operators

The introductory question in the semigroup approach to differential equations is if the operator appearing in the main part is closable and, what is even more important, how the domain looks of its closed linear extension. The answer usually requires delicate studies of elliptic operators leading to the suitable Calderon–Zygmund-type estimates.

Particular considerations of this type are presented in the following lemma.

Lemma 2.1.16. *The differential Laplace operator $\Delta : L^2(\mathbb{R}^N) \to L^2(\mathbb{R}^N)$ considered on $C_0^\infty(\mathbb{R}^N)$ is closable in $L^2(\mathbb{R}^N)$ and the domain of its closed linear extension is equal to $H^2(\mathbb{R}^N)$.*

Proof. A key step in this proof is to show for $\phi \in C_0^\infty(\mathbb{R}^N)$ the estimate of the type:

$$\text{const.}\|\phi\|_{H^2(\mathbb{R}^N)} \le \|\phi\|_{L^2(\mathbb{R}^N)} + \|\Delta\phi\|_{L^2(\mathbb{R}^N)} \le \text{const.}'\|\phi\|_{H^2(\mathbb{R}^N)}. \tag{2.58}$$

The right inequality in (2.58) is a simple consequence of the definition of the $H^2(\mathbb{R}^N)$-norm. Indeed, for $\phi \in C_0^\infty(\mathbb{R}^N)$ we have

$$\|\phi\|_{H^2(\mathbb{R}^N)} + \|\Delta\phi\|_{L^2(\mathbb{R}^N)} \le \|\phi\|_{H^2(\mathbb{R}^N)} + \sum_{i=1}^{n} \left\| \frac{\partial^2 \phi}{\partial x_i^2} \right\|_{L^2(\mathbb{R}^N)} \le \|\phi\|_{H^2(\mathbb{R}^N)},$$

which shows that (2.58) holds with const.$' = 1$.

Consider next the Fourier transform $\mathcal{F}$, which is known to be an isomorphism on the space $\mathcal{S} \subset C^\infty(\mathbb{R}^N)$ of complex valued functions rapidly decreasing at infinity, and denote by $\mathcal{F}^{-1}$ its inverse on $\mathcal{S}$. For $\phi \in C_0^\infty(\mathbb{R}^N)$, $r, s = 1, \dots, n$, we have $\mathcal{F}\frac{\partial^2\phi}{\partial x_r \partial x_s} = -\xi_r\xi_s\mathcal{F}\phi$ and $\mathcal{F}[(I-\Delta)\phi] = (1+|\xi|^2)\mathcal{F}\phi$. Since in addition, $\mathcal{F}$ preserves the $L^2(\mathbb{R}^N)$ norm, then we obtain

$$\begin{aligned} \left\| \frac{\partial^2\phi}{\partial x_r \partial x_s} \right\|_{L^2(\mathbb{R}^N)} &= \left\| \mathcal{F}\frac{\partial^2\phi}{\partial x_r \partial x_s} \right\|_{L^2(\mathbb{R}^N)} = \|(-\xi_r\xi_s)\mathcal{F}\phi\|_{L^2(\mathbb{R}^N)} \\ &\le \|(1+|\xi|^2)\mathcal{F}\phi\|_{L^2(\mathbb{R}^N)} = \|\mathcal{F}(I-\Delta)\phi\|_{L^2(\mathbb{R}^N)} = \|(I-\Delta)\phi\|_{L^2(\mathbb{R}^N)}. \end{aligned} \tag{2.59}$$

For $r = 1, \dots, n$, using elementary inequality $|\xi_r| \le (1 + |\xi|^2)$ and estimating as in (2.59), we get similarly

$$\begin{aligned}\left\|\frac{\partial\phi}{\partial x_r}\right\|_{L^2(\mathbb{R}^N)} &= \left\|\mathcal{F}\frac{\partial\phi}{\partial x_r}\right\|_{L^2(\mathbb{R}^N)} = \|i\xi_r\mathcal{F}\phi\|_{L^2(\mathbb{R}^N)} \\ &\le \|(1+|\xi|^2)\mathcal{F}\phi\|_{L^2(\mathbb{R}^N)} = \|\mathcal{F}(I-\Delta)\phi\|_{L^2(\mathbb{R}^N)} = \|(I-\Delta)\phi\|_{L^2(\mathbb{R}^N)}. \end{aligned} \tag{2.60}$$

Collecting (2.59) and (2.60), we obtain the crucial estimate

$$\|\phi\|_{H^2(\mathbb{R}^N)} \le (1+n+n^2)\|(I-\Delta)\phi\|_{L^2(\mathbb{R}^N)}, \quad \phi \in C_0^\infty(\mathbb{R}^N), \tag{2.61}$$

which proves the left inequality in (2.58) with const. $= \frac{1}{1+n+n^2}$.

Denote by G the graph of $\Delta : C_0^\infty(\mathbb{R}^N) \to L^2(\mathbb{R}^N)$ in $L^2(\mathbb{R}^N) \times L^2(\mathbb{R}^N)$. Based on (2.58), it is now possible to show that $\mathrm{cl}_{L^2(\mathbb{R}^N)\times L^2(\mathbb{R}^N)} G$ is a graph of some closed linear extension of $\Delta : C_0^\infty(\mathbb{R}^N) \to L^2(\mathbb{R}^N)$ and to determine its domain.

To verify that $\Delta : C_0^\infty(\mathbb{R}^N) \to L^2(\mathbb{R}^N)$ is a graph of a linear operator, take $\{\phi_n\} \subset C_0^\infty(\mathbb{R}^N)$ for which there exist $L^2(\mathbb{R}^N)$-limits $\phi_n \to 0$ and $\Delta\phi_n \to \chi$. From the left inequality (2.58), $\{\phi_n\}$ is a Cauchy sequence in $H^2(\mathbb{R}^N)$ so that $\|\phi_n\|_{H^2(\mathbb{R}^N)} \to 0$. But the right estimate in (2.58) ensures then that $\|\Delta\phi_n\|_{L^2(\mathbb{R}^N)} \to 0$. Therefore, $\chi = 0$ which shows the operator $\Delta : C_0^\infty(\mathbb{R}^N) \to L^2(\mathbb{R}^N)$ is closable.

Consider further $(g_1, g_2) \in \mathrm{cl}_{L^2(\mathbb{R}^N)\times L^2(\mathbb{R}^N)} G$ and take $\{\phi_n\} \subset C_0^\infty(\mathbb{R}^N)$ such that $(\phi_n, \Delta\phi_n) \in G$ and $\phi_n \to g_1$, $\Delta\phi_n \to g_2$ in $L^2(\mathbb{R}^N)$. As a consequence of (2.58), $\{\phi_n\}$ is then a Cauchy sequence in $H^2(\mathbb{R}^N)$. Therefore, g_1 belongs to $H^2(\mathbb{R}^N)$ which shows that the domain of the closed extension of $\Delta : C_0^\infty(\mathbb{R}^N) \to L^2(\mathbb{R}^N)$ is contained in $H^2(\mathbb{R}^N)$.

To prove that the converse inclusion holds, take $g \in H^2(\mathbb{R}^N)$ and a sequence $\{\phi_n\} \subset C_0^\infty(\mathbb{R}^N)$ for which $\phi_n \to g$ in $H^2(\mathbb{R}^N)$. From the estimate (2.58), it is then seen that $\{\Delta\phi_n\}$ is a Cauchy sequence in $L^2(\mathbb{R}^N)$, and hence there exists $g_0 \in L^2(\mathbb{R}^N)$ such that

$$(\phi_n, \Delta\phi_n) \overset{L^2(\mathbb{R}^N)\times L^2(\mathbb{R}^N)}{\to} (g, g_0) \in \mathrm{cl}_{L^2(\mathbb{R}^N)\times L^2(\mathbb{R}^N)} G.$$

The domain of a closed linear extension of $\Delta : C_0^\infty(\mathbb{R}^N) \to L^2(\mathbb{R}^N)$ is thus equal to $H^2(\mathbb{R}^N)$. □

Remark 2.1.17. Let Ω be a bounded domain in $\mathbb{R}^N$. Conditions (2.61), (2.58) ensure validity of the estimate

$$\frac{1}{1+n+n^2}\|\phi\|_{H^2(\Omega)} \le \|\phi\|_{L^2(\Omega)} + \|\Delta\phi\|_{L^2(\Omega)} \le \|\phi\|_{H^2(\Omega)}, \quad \phi \in C_0^\infty(\Omega).$$

As seen from the inequality above, operator

$$\Delta : C_0^\infty(\Omega) \to L^2(\Omega)$$

is closable in $L^2(\Omega)$ and, via density arguments, the domain of its closed linear extensions is equal to $H_0^2(\Omega)$.

Regular elliptic boundary value problems

Consider now the linear $2m$-th order differential operator

$$A = \sum_{|\sigma|\leq 2m} a_\sigma(x)D^\sigma \tag{2.62}$$

whose (complex or real) coefficients a_σ are uniformly continuous in $\overline{\Omega}$ and $\Omega \subset \mathbb{R}^N$ is a bounded domain with the boundary $\partial\Omega$ of the class C^{2m}. Assume that the following *uniform strong ellipticity condition* holds:

$$\exists_{c>0}\ \forall_{x\in\Omega}\ \forall_{\xi\in\mathbb{R}^N-\{0\}}\ (-1)^m \operatorname{Re}\Bigg[\sum_{|\sigma|=2m} a_\sigma(x)\xi^\sigma\Bigg] \geq c|\xi|^{2m}. \tag{2.63}$$

Then, as shown in [66, Theorem 18.1], for $u \in H^{2m}(\Omega) \cap H_0^m(\Omega)$ we have

$$\text{const.}\|u\|_{H^{2m}(\Omega)} \leq \|Au\|_{L^2(\Omega)} + \|u\|_{L^2(\Omega)} \leq \text{const.}'\|u\|_{H^{2m}(\Omega)},$$

which guarantees that

$$A : H^{2m}(\Omega) \cap H_0^m(\Omega) \to L^2(\Omega)$$

is a closed operator in $L^2(\Omega)$. Furthermore,

$$\text{the range of } A + \lambda I \text{ is the whole of } L^2(\Omega) \tag{2.64}$$

provided that $\lambda > \lambda_0$ and λ_0 is chosen sufficiently large (cf. [66, Theorem 18.2]).

This may be generalized for A acting in $L^p(\Omega)$ under more general boundary conditions (A and Ω as above). Take the set $\{B_j,\ j = 1,\dots,m\}$ of m_j-th order ($m_j < 2m$) boundary operators $B_j = \sum_{|\sigma|\leq m_j} b_\sigma^j(x)D^\sigma$, where $b_\sigma^j \in C^{2m-m_j}(\partial\Omega)$ for $|\sigma| \leq m_j$, $j = 1,\dots,m$ and consider

$$A : \{\phi \in C^{2m}(\overline{\Omega}) : B_1\phi_{|\partial\Omega} = \cdots = B_m\phi_{|\partial\Omega} = 0\} \to L^p(\Omega).$$

Since A is uniformly strongly elliptic, then the extended *roots condition* holds, that is,

Roots condition. *For each $x \in \partial\Omega$, any nonzero vector $\xi \in \mathbb{R}^N$ from the tangent hyperplane to $\partial\Omega$ at x and all complex λ lying on the ray $\arg\lambda = \theta$ with arbitrary $\theta \in [\frac{\pi}{2}, \frac{3\pi}{2}]$, the polynomial $p(z) = \sum_{|\sigma|=2m} a_\sigma(x)(\xi + N(x))^\sigma - \lambda$ has exactly m roots $z_1^+(x,\xi,\lambda),\dots,z_m^+(x,\xi,\lambda)$ with positive imaginary parts.*

Assume further that the uniform *normality condition* and *strong complementary condition* are satisfied, that is,

Normality condition. *If $i \neq j$, then $m_i \neq m_j$ and*

$$\Bigg|\sum_{|\sigma|=m_j} b_\sigma^j(x)(N(x))^\sigma\Bigg| \geq \text{const.} > 0, \quad x \in \partial\Omega,\ j = 1,\dots,m,$$

where $N(x) \neq 0$ is the normal to $\partial\Omega$ at x.

Strong complementary condition. *For each $x \in \partial\Omega$, any nonzero vector $\xi \in \mathbb{R}^N$ from the tangent hyperplane to $\partial\Omega$ at x and all complex λ lying on the ray $\arg\lambda = \theta$ with arbitrary $\theta \in [\frac{\pi}{2}, \frac{3\pi}{2}]$ the polynomials $P_j(z)$ $(j = 1, \dots, m)$ where $P_j(z) = \sum_{|\sigma|=m_j} b^j_\sigma(x)(\xi + zN(x))^\sigma$ and $N(x)$ denote the outward normal to $\partial\Omega$ are linearly independent modulo the polynomial $Q(z) = (z - z_1^+(x, \xi, \lambda)) \cdots (z - z_m^+(x, \xi, \lambda))$.*

Definition 2.1.18. If the triple $(A, \{B_j\}, \Omega)$ is such that A is a uniformly strongly elliptic operator, $\partial\Omega$ and coefficients of A, B_j $(j = 1, \dots, m)$ are smooth as it was introduced above and, moreover, the roots condition, normality condition and strong complementary condition hold, then $(A, \{B_j\}, \Omega)$ is called a regular elliptic boundary value problem.

As a particular consequence of the above assumptions on the triple, $(A, \{B_j\}, \Omega)$ $u \in \{\phi \in C^{2m}(\overline{\Omega}) : B_1\phi_{|\partial\Omega} = \dots = B_m\phi_{|\partial\Omega} = 0\}$ fulfills the estimate (cf. [66, Theorem 19.1])

$$\text{const.}\|u\|_{W^{2m,p}(\Omega)} \le \|Au\|_{L^p(\Omega)} + \|u\|_{L^p(\Omega)} \le \text{const.}'\|u\|_{W^{2m,p}(\Omega)}. \tag{2.65}$$

This ensures that A equipped with the domain $D(A) = W^{2m,p}_{\{B_j\}}(\Omega)$, where

$$W^{2m,p}_{\{B_j\}}(\Omega) := \mathrm{cl}_{W^{2m,p}(\Omega)}\{\phi \in C^{2m}(\overline{\Omega}) : B_1\phi_{|\partial\Omega} = \dots = B_m\phi_{|\partial\Omega} = 0\}, \tag{2.66}$$

is a closed operator in $L^p(\Omega)$. Furthermore, as follows from [66, Theorem 19.4], there exist const. > 0 and $\Lambda_0 > 0$ such that for each complex λ with $\operatorname{Re}\lambda < -\Lambda_0$,

$$(\lambda I - A) \text{ takes } W^{2m,p}_{\{B_j\}}(\Omega) \text{ onto } L^p(\Omega) \tag{2.67}$$

and

$$\sum_{j=0}^{2m} |\lambda|^{\frac{(2m-j)}{2m}} \|u\|_{W^{j,p}(\Omega)} \le \text{const.}\|(\lambda I - A)u\|_{L^p(\Omega)}, \quad u \in W^{2m,p}(\Omega). \tag{2.68}$$

See [173, 5.3.4] for the generalization of (2.65), also [173, 5.5.1] for the description of properties of A considered on the fractional order spaces. In the latter case, the spaces $H^s_{p,\{B_j\}}(\Omega)$, $B^s_{p,p\{B_j\}}(\Omega)$, $s > 0$, $p \in (1, +\infty)$ may be introduced by

$$H^s_{p,\{B_j\}}(\Omega) = \{\phi \in H^s_p(\Omega); \quad \forall_{i\in\{j;\, m_j<s-\frac{1}{p}\}} B_i\phi_{|\partial\Omega} = 0\}, \tag{2.69}$$

$$B^s_{p,p,\{B_j\}}(\Omega) = \{\phi \in B^s_{p,p}(\Omega); \forall_{i\in\{j;\, m_j<s-\frac{1}{p}\}} B_i\phi_{|\partial\Omega} = 0\}, \tag{2.70}$$

(e.g. [173, Section 4.3.3]) which allows to generalize (2.66) to the formula

$$W^{s,p}_{\{B_j\}}(\Omega) = \{\phi \in W^{s,p}(\Omega); \forall_{i\in\{j;\, m_j<s-\frac{1}{p}\}} B_i\phi_{|\partial\Omega} = 0\}. \tag{2.71}$$

Compactness of the resolvent

Condition (2.68) ensures that $\lambda I-A$ is an invertible operator in $L^p(\Omega)$ for each complex λ with $\operatorname{Re}\lambda < -\Lambda_0$, and $(\lambda I-A)^{-1}$ is bounded. As a consequence of (2.67), the inverse $(\lambda I-A)^{-1}$ is seen to be defined on the whole of $L^p(\Omega)$, so that the half-plane $\{\lambda \in \mathbb{C};\ \operatorname{Re}\lambda < \Lambda_0\}$ is contained in the resolvent set $\rho(A)$.

From the point of view of further applications to the problems in bounded domains, an important property of the elliptic operators is *compactness of resolvent*.

Definition 2.1.19. Let X be a Banach space and $A : D(A) \to X$ a linear operator in X. We say that A has compact resolvent if $(\lambda I - A)^{-1} : X \to X$ is a compact map for each $\lambda \in \rho(A)$.

Let us consider $\lambda_1, \lambda_2 \in \rho(A)$ and the operators $(\lambda_1 I-A)^{-1}$, $(\lambda_2 I-A)^{-1}$. The *resolvent equation* (see [190, p. 211]) then reads

$$(\lambda_1 I - A)^{-1} = [(\lambda_2 - \lambda_1)(\lambda_1 I - A)^{-1} + I](\lambda_2 I - A)^{-1}, \tag{2.72}$$

which shows the following.

Remark 2.1.20. The resolvent of A is compact if and only if $(\lambda I - A)^{-1}$ is compact for some $\lambda \in \rho(A)$.

Considering again the regular elliptic boundary value problem $(A, \{B_j\}, \Omega)$, we have from (2.68) the estimate

$$\|(\lambda I - A)^{-1}v\|_{W^{2m,p}(\Omega)} \le \text{const.}\|v\|_{L^p(\Omega)}, \quad \operatorname{Re}\lambda < \Lambda_0.$$

Therefore, as a consequence of the Rellich–Kondrachov theorem (cf. [1, Theorem 6.2], Section 2.1.4), we get the compactness of $(\lambda I - A)^{-1}$ in $L^p(\Omega)$ for $\operatorname{Re}\lambda < \Lambda_0$. From this and Observation 2.1.20, we thus obtain the following conclusion.

Proposition 2.1.21. *For* $(A, \{B_j\}, \Omega)$ *being a regular elliptic boundary value problem introduced in Definition* 2.1.18, *the resolvent of A is compact.*

2.2 Sectorial operators

The notion of *sectorial operator* is of great importance in the modern studies of differential equations. A large number of phenomena in the applied sciences are described by systems of equations with sectorial operator in the main part. The problems falling into that class may be treated within a unified theoretical approach.

Let us denote by $S_{a,\phi}$, where $a \in \mathbb{R}$ and $\phi \in (0, \frac{\pi}{2})$, a sector of the complex plain given by the relation (cf. [86, Definition 1.3.1]):

$$S_{a,\phi} := \{\lambda \in \mathbb{C} :\ \phi \le |\arg(\lambda - a)| \le \pi,\ \lambda \ne a\}. \tag{2.73}$$

Definition 2.2.1. Consider a linear, closed and densely defined operator $A : X \supset D(A) \to X$ acting in a Banach space X. Then A is a sectorial operator in X if and only if there exist $a \in \mathbb{R}$, $\phi \in (0, \frac{\pi}{2})$ and $M > 0$ such that

$$\text{the resolvent set } \rho(A) \text{ contains the sector } \mathcal{S}_{a,\phi}, \tag{2.74}$$

$$\|(\lambda I - A)^{-1}\|_{\mathcal{L}(X,X)} \leq \frac{M}{|\lambda - a|}, \quad \text{for each } \lambda \in \mathcal{S}_{a,\phi}. \tag{2.75}$$

Remark 2.2.2. Choose arbitrary $\omega \in \mathbb{R}$. It is easy to see that A is sectorial if and only if $A_\omega := A + \omega I$ is sectorial. Indeed, via the translation's argument, $\mathcal{S}_{a,\phi} \subset \rho(A)$ and (2.75) holds if and only if

$$\mathcal{S}_{a+\omega,\phi} = \{\lambda' \in \mathbb{C} : \phi \leq |\arg(\lambda' - (a + \omega))| \leq \pi, \ \lambda' \neq a + \omega\} \subset \rho(A_\omega)$$

and, simultaneously,

$$\|(\lambda' I - A_\omega)^{-1}\|_{\mathcal{L}(X,X)} \leq \frac{M}{|\lambda' - (a + \omega)|}, \quad \text{for each } \lambda' \in \mathcal{S}_{a+\omega,\phi}.$$

In addition, we also have

$$\operatorname{Re} \sigma(A_\omega) \geq a + \omega,$$

so that it is always possible to choose $\omega > 0$ for which $\operatorname{Re} \sigma(A_\omega) > 0$ and consider A_ω as a *positive operator.*

Equivalent sectoriality conditions

We will list now certain equivalent sectoriality conditions.

Proposition 2.2.3. *Let $A : X \supset D(A) \to X$ be a linear closed and densely defined operator in a Banach space X and consider the operators $A_\omega = A + \omega I$ with $\omega \in \mathbb{R}$. Then the following conditions are equivalent:*

(a) *A_ω is sectorial in X for some $\omega \in \mathbb{R}$,*
(b) *A_ω is sectorial in X for each $\omega \in \mathbb{R}$,*
(c) *There exist $k, \omega \in \mathbb{R}$ such that the resolvent set $\rho(A_\omega)$ of A_ω contains a half-plane $\{\lambda \in \mathbb{C}; \operatorname{Re} \lambda \leq k\}$ and*

$$\|\lambda(\lambda I - A_\omega)^{-1}\|_{\mathcal{L}(X,X)} \leq M \quad \text{for } \operatorname{Re} \lambda \leq k. \tag{2.76}$$

Proof. Certainly, (a) and (b) are equivalent as shown in Remark 2.2.2. Next, the implication (a) $\Longrightarrow$ (c) is formal. Indeed if A_{ω_0} fulfills conditions of Definition 2.2.1 then repeating considerations of Remark 2.2.2, we find that $\mathcal{S}_{0,\phi} \subset \rho(A_{\omega_0} - aI)$ and $\|\lambda'(\lambda' - (A_{\omega_0} - aI))^{-1}\|_{\mathcal{L}(X,X)} \leq M$ for all $\lambda' \in \mathcal{S}_{0,\phi}$, so that (c) is satisfied with $\omega = \omega_0 - a$ and arbitrary $k < 0$.

Let us now proceed to the proof of (c) ⟹ (a), which is essential here (cf. [126, Proposition 2.1.11]). First, let us recall that if A is a closed linear operator in a nontrivial Banach space X then (cf. [135, Chapter 3, Theorem 4.1]):

$$\lambda_0 \in \rho(A) \text{ implies that } \mathcal{B}_{\lambda_0} \subset \rho(A), \tag{2.77}$$

where $\mathcal{B}_{\lambda_0}$ is the following open ball on a complex plane;

$$\mathcal{B}_{\lambda_0} := \left\{\lambda \in \mathbb{C};\ |\lambda - \lambda_0| < \frac{1}{\|(\lambda_0 I - A)^{-1}\|_{\mathcal{L}(X,X)}}\right\}. \tag{2.78}$$

Furthermore,

$$(\lambda I - A)^{-1} = \sum_{n=0}^{\infty} (\lambda_0 - \lambda)^n \left[(\lambda_0 I - A)^{-1}\right]^{n+1} \quad \text{for } \lambda \in \mathcal{B}_{\lambda_0}. \tag{2.79}$$

We may assume without loss of generality that $k < 0$ in (2.76), so that

$$\{\lambda \in \mathbb{C};\ \operatorname{Re}\lambda \le k\} \subset \rho(A_\omega) \tag{2.80}$$

and

$$\|(\lambda I - A_\omega)^{-1}\|_{\mathcal{L}(X,X)} \le \frac{M}{|\lambda|} \le \frac{M}{|\lambda - k|} \quad \text{for } \operatorname{Re}\lambda \le k, \lambda \ne k. \tag{2.81}$$

Define

$$\mathcal{S}_{k,\arctan(2M)} = \{\lambda \in \mathbb{C}:\ \arctan(2M) \le |\arg(\lambda - k)| \le \pi,\ \lambda \ne k\}$$

and take $\lambda \in \mathcal{S}^{<}_{k,\arctan(2M)} := \mathcal{S}_{k,\arctan(2M)} - \{\lambda \in \mathbb{C};\ \operatorname{Re}\lambda \le k\}$. Then

$$|\operatorname{Re}(\lambda - k)| = \frac{|\operatorname{Im}\lambda|}{|\tan(\arg(\lambda - k))|} \le \frac{|\operatorname{Im}\lambda|}{2M}, \tag{2.82}$$

so that, from (2.77) and assumption (c),

$$\lambda \in \mathcal{B}\left(k + i\operatorname{Im}\lambda, \frac{|k + i\operatorname{Im}\lambda|}{M}\right) \subset \rho(A_\omega) \quad \text{for all } \lambda \in \mathcal{S}^{<}_{k,\arctan(2M)}. \tag{2.83}$$

Connecting (2.79), (2.81), (2.82), and following from (2.82) simple inequality,

$$|\operatorname{Im}(\lambda - k)|^2\left(1 + \frac{1}{(2M)^2}\right) \ge |\lambda - k|^2,$$

we obtain

$$
\begin{aligned}
&\|(\lambda I - A_\omega)^{-1}\|_{\mathcal{L}(X,X)} \\
&\le \sum_{n=0}^{\infty} |\lambda - (k + i\operatorname{Im}\lambda)|^n \|((k + i\operatorname{Im}\lambda)I - A_\omega)^{-1}\|_{\mathcal{L}(X,X)}^{n+1} \\
&\le \sum_{n=0}^{\infty} |\operatorname{Re}\lambda - k|^n \frac{M^{n+1}}{|k + i\operatorname{Im}\lambda|^{n+1}} = \frac{M}{|k + i\operatorname{Im}\lambda|} \sum_{n=0}^{\infty} \left[\frac{|\operatorname{Re}\lambda - k|}{|k + i\operatorname{Im}\lambda|} M\right]^n \\
&\le \frac{M}{|\operatorname{Im}\lambda|} \sum_{n=0}^{\infty} \left[\frac{|\operatorname{Re}\lambda - k|}{|\operatorname{Im}\lambda|} M\right]^n \le \frac{M}{|\operatorname{Im}\lambda|} \sum_{n=0}^{\infty} \left[\frac{1}{2M} M\right]^n \\
&= \frac{2M}{|\operatorname{Im}\lambda|} \le \frac{2\sqrt{1 + \frac{1}{(2M)^2}}M}{|\lambda - k|} \quad \text{for } \lambda \in \mathcal{S}^{<}_{k,\arctan(2M)}.
\end{aligned}
\tag{2.84}
$$

Based on assumption (2.80) and (2.83), it is thus seen that $\mathcal{S}_{k,\arctan(2M)} \subset \rho(A_\omega)$ whereas (2.81) and (2.84) imply that

$$
\|(\lambda I - A_\omega)^{-1}\|_{\mathcal{L}(X,X)} \le \frac{2\sqrt{1 + \frac{1}{(2M)^2}}M}{|\lambda - k|} \quad \text{for } \lambda \in \mathcal{S}_{k,\arctan(2M)}. \tag{2.85}
$$

The proof is completed. □

Definition 2.2.4. Let X be a Banach space and $\{T(t)\}$ a C^0-semigroup in X. We say that $\{T(t)\}$ is an analytic semigroup, if for each $v \in X$ the function

$$
(0, +\infty) \ni t \to T(t)v \in X
$$

is analytic.

The well-known result from the semigroup theory is that sectorial operators are precisely the negative generators of *analytic semigroups*. The following theorem holds (see, e. g., [86, Theorem 1.3.4], also [66], [173]).

Theorem 2.2.5. *Let X be a Banach space. Then A is a negative generator of an analytic semigroup $\{e^{-At}\}$ of bounded linear operators $e^{-At} : X \to X$, $t \ge 0$ if and only if A is a sectorial operator in X.*

2.2.1 Two basic estimates characterizing sectorial positive operators

Important in further applications will be the case $\operatorname{Re}\sigma(A) > 0$, that is, when elements of the spectrum have positive real parts. Then the estimate of $\|\cdot\|_{\mathcal{L}(X,X)}$-norm of the semigroup operators shows that the process decays at infinity. In consequence, the powers A^α of A may be defined. We recall the following result (cf. [86, Theorem 1.3.4]).

Theorem 2.2.6. *Let A be a sectorial operator in a Banach space X such that* $\operatorname{Re}\sigma(A) > a > 0$. *Then*

$$\|e^{-At}\|_{\mathcal{L}(X,X)} \le c_0 e^{-at}, \quad t \ge 0, \tag{2.86}$$

$$\|Ae^{-At}\|_{\mathcal{L}(X,X)} \le \frac{c_1}{t} e^{-at}, \quad t > 0. \tag{2.87}$$

In the considerations concerning fractional powers A^α, we use further the results coming from the original papers by H. Komatsu [106], [107], the monograph by H. Triebel [173] and, in Chapter 5, the monograph [130]. The remark below justifies that condition (2.118) formulated further in Section 2.2.3 allows us to use the mentioned results.

Remark 2.2.7. As a consequence of the Hille–Yosida theorem (e. g., [138, Theorem 5.3, p. 20], [173, Section 1.13.1]), condition (2.86) implies in particular that the interval $(-a, +\infty)$ is contained in $\rho(-A)$ and

$$\|(-\lambda I - A)^{-1}\|_{\mathcal{L}(X,X)} \le \frac{M}{\lambda + a}, \quad \lambda > -a.$$

Since

$$\frac{M}{\lambda + a} \le \frac{M}{\lambda + 1}, \quad a \ge 1,\ \lambda \ge 0,$$
$$\frac{M}{\lambda + a} < \frac{M}{a(\lambda + 1)}, \quad a \in (0,1),\ \lambda \ge 0,$$

we obtain the estimate

$$\|(-\lambda I - A)^{-1}\|_{\mathcal{L}(X,X)} \le \max\left\{M, \frac{M}{a}\right\}\frac{1}{\lambda + 1}, \quad \lambda \ge 0. \tag{2.88}$$

Consequently, setting $t = -\lambda$, we then have

$$\begin{aligned} &(-\infty, 0] \subset (-\infty, a) \subset \rho(A) \quad \text{and} \\ &\|(tI - A)^{-1}\|_{\mathcal{L}(X,X)} \le \frac{\text{const.}}{1 + |t|}, \quad t \in (-\infty, 0]; \end{aligned} \tag{2.89}$$

that is, A is a *positive operator* in the sense of [173, Section 1.14.1]. Furthermore, (2.86) and (2.87) ensure together that A fulfills requirements of Definition 2.2.1 with $a = 0$, so that

$$\begin{aligned} &\textit{the resolvent set } \rho(A) \textit{ contains the sector } \mathcal{S}_{0,\phi}, \\ &\textit{and } \|(\lambda I - A)^{-1}\|_{\mathcal{L}(X,X)} \le \frac{M}{|\lambda|}, \quad \textit{for } \lambda \in \mathcal{S}_{0,\phi}, \end{aligned} \tag{2.90}$$

where $\phi \in (0, \frac{\pi}{2})$ and $M > 0$ are fixed (cf. [106, Theorem 12.2], also [66, Problem (5), p. 108]).

The above remark yields the following conclusion.

Corollary 2.2.8. *Consider a linear, closed, and densely defined operator $A : X \supset D(A) \to X$ acting in a Banach space X. Then A is a sectorial operator with* $\mathrm{Re}(\sigma(A)) > 0$ *if and only if both* (2.89) *and* (2.90) *are satisfied.*

2.2.2 Examples of sectorial operators

We start with simple example.

Example 2.2.9. Each bounded linear operator defined on a Banach space X is sectorial.

Proof. If A is a bounded linear operator in X, then $\{\lambda \in \mathbb{C};\ |\lambda| > \|A\|_{\mathcal{L}(X,X)}\} \subset \rho(A)$ and $(\lambda I - A)^{-1} = \sum_{n=0}^{\infty} \frac{A^n}{\lambda^{n+1}}$ (cf. [135, Lemma 4.2]). In particular, the half-plane $\{\lambda \in \mathbb{C};\ \mathrm{Re}\,\lambda \leq -2\|A\|_{\mathcal{L}(X,X)}\}$ is contained in $\rho(A)$ and

$$\|\lambda(\lambda I - A)^{-1}\|_{\mathcal{L}(X,X)} \leq \sum_{n=0}^{\infty} \left(\frac{\|A\|_{\mathcal{L}(X,X)}}{|\lambda|}\right)^n \leq \sum_{n=0}^{\infty} \left(\frac{1}{2}\right)^n = 2.$$

Therefore, A is sectorial as a result of Proposition 2.2.3. □

Example 2.2.10. If X, Y are Banach spaces and A is sectorial in X, B is sectorial in Y, then the product operator $(A, B) : D(A) \times D(B) \to X \times Y$, where $(A, B)(x, y) = (Ax, By)$, is sectorial in $X \times Y$.

Proof. Let S_{a,ϕ_A}, S_{b,ϕ_B} be the sectors chosen respectively for A and B based on Definition 2.2.1. Define $a' = \min\{a; b\}$, $\mathcal{J} = (A, B)$ and $\mathcal{J}_{-a'} = (A - a'I, B - a'I)$. Then $\{\lambda \in \mathbb{C};\ \mathrm{Re}\,\lambda < 0\}$ is a subset of $\rho(\mathcal{J}_{-a'})$ and

$$\|(\lambda I - \mathcal{J}_{-a'})^{-1}\|_{\mathcal{L}(X\times Y, X\times Y)} \leq \frac{\text{const.}}{|\lambda|} \quad \text{for } \mathrm{Re}\,\lambda < 0.$$

Hence (A, B) is sectorial in $X \times Y$ as a result of Proposition 2.2.3. □

In many problems, one needs to deal with the perturbations of a generator A of analytic semigroup. We shall prove below that whenever a perturbation B fulfills an appropriate condition (2.91), the *perturbed operator* $A + B$ will still generate an analytic semigroup (cf. [138, Section 3.2], [86, Theorem 1.3.2]).

Proposition 2.2.11 (Perturbation result). *Let $A : D(A) \to X$ be a sectorial operator in a Banach space X and consider a closed, linear operator $B : D(B) \to X$ such that $D(A) \subset D(B) \subset X$ and B is subordinated to A according to the condition*

$$\|Bv\|_X \leq c\|Av\|_X + c'\|v\|_X, \quad v \in X. \tag{2.91}$$

If the condition (2.91) *holds with any $c \leq M_0$ (where $M_0 > 0$ is defined in* (2.97)*), then the perturbed operator $A + B$ is sectorial.*

Proof. Based on Proposition 2.2.3, take $k, \omega \in \mathbb{R}$ such that, for $A_\omega = A + \omega I$,

$$\{\lambda \in \mathbb{C};\ \operatorname{Re}\lambda \le k\} \subset \rho(A_\omega) \tag{2.92}$$

and

$$\|\lambda(\lambda I - A_\omega)^{-1}\|_{\mathcal{L}(X,X)} \le M \quad \text{for } \operatorname{Re}\lambda \le k. \tag{2.93}$$

Condition (2.91) then reads

$$\|Bv\|_X \le c\|A_\omega v\|_X + c''\|v\|_X, \quad v \in X, \tag{2.94}$$

where $c'' = c' + c\omega$.

The crucial step of the proof is to show that

$$\exists_{k_0 \le k}\ \forall_{\operatorname{Re}\lambda \le k_0}\ 1 \in \rho(B(\lambda I - A_\omega)^{-1}). \tag{2.95}$$

For this, we shall use (2.94) and (2.93) to estimate as follows:

$$\begin{aligned} \|B(\lambda I - A_\omega)^{-1}v\|_X &\le c\|A_\omega(\lambda I - A_\omega)^{-1}v\|_X + c''\|(\lambda I - A_\omega)^{-1}v\|_X \\ &\le c\|(\lambda I - A_\omega)(\lambda I - A_\omega)^{-1}v\|_X + c\|\lambda(\lambda I - A_\omega)^{-1}v\|_X + c''\|(\lambda I - A_\omega)^{-1}v\|_X \\ &\le \left(c(1+M) + \frac{c''M}{|\lambda|}\right)\|v\|_X, \quad v \in X. \end{aligned} \tag{2.96}$$

Under the restrictions,

$$c \le \frac{1}{2(1+M)} =: M_0, \quad 4c''M \le |\lambda| \tag{2.97}$$

we have further from (2.96) the inequality

$$\|B(\lambda I - A_\omega)^{-1}\|_{\mathcal{L}(X,X)} < \frac{3}{4}. \tag{2.98}$$

Since the spectral radius of a bounded linear operator on X does not exceed its $\mathcal{L}(X,X)$-norm (see [190, Chapter VIII, Section 2]), condition (2.98) shows that the number 1 is in the resolvent set of $B(\lambda I - A_\omega)^{-1}$, that is, (2.95) is proved with

$$k_0 := \min\{k, -4c''M\}. \tag{2.99}$$

For $\operatorname{Re}\lambda \le k_0$, we then have

$$\begin{aligned} [\lambda I - (A_\omega + B)]^{-1} &= [(I - B(\lambda I - A_\omega)^{-1})(\lambda I - A_\omega)]^{-1} \\ &= (\lambda I - A_\omega)^{-1}(I - B(\lambda I - A_\omega)^{-1})^{-1}, \end{aligned} \tag{2.100}$$

which, in the presence of (2.93), (2.98) leads to the estimate

$$\begin{aligned} &\|[\lambda I - (A_\omega + B)]^{-1}\|_{\mathcal{L}(X,X)} \\ &\leq \|(\lambda I - A_\omega)^{-1}\|_{\mathcal{L}(X,X)} \|(I - B(\lambda I - A_\omega)^{-1})^{-1}\|_{\mathcal{L}(X,X)} \leq \frac{4M}{|\lambda|}. \end{aligned} \tag{2.101}$$

Since $A_\omega + B = (A+B)_\omega$, the operator $A+B$ fulfills requirements of Proposition 2.2.3(c) which completes the proof of Proposition 2.2.11. □

Example 2.2.12. If A is sectorial in a Banach space X and B is bounded and linear on X, then the perturbed operator $A + B$ is sectorial.

Proof. If B is bounded and linear in X, then (2.91) holds with $c = 0$ and $c' = 1$. Operator $A + B$ is thus sectorial as a result of Proposition 2.2.11. □

The following proposition provides examples of sectorial operators when X is a Hilbert space. An extended discussion of that particular case will be given further in Subsection 5.2.2.

Proposition 2.2.13. *Let $A : H \supset D(A) \to H$ be densely defined, linear, self-adjoint operator in a Hilbert space H. If, in addition, A is bounded below in H, that is,*

$$\exists_{m\in\mathbb{R}} \forall_{x\in H} \langle Ax, x\rangle_H \geq m\|x\|_H^2, \tag{2.102}$$

then A is a sectorial operator in H.

Proof. We will prove directly validity of Definition 2.2.1. Let us recall that the spectrum of a self-adjoint operator is contained in the real axis. Moreover, since A is bounded below (2.102), then $\sigma(A)$ must be contained in the interval $[m, \infty)$. This implies, in particular, that the sector

$$\mathcal{S}_{m,\frac{\pi}{4}} := \left\{\lambda \in \mathbb{C} : \frac{\pi}{4} \leq |\arg(\lambda - m)| \leq \pi,\ \lambda \neq m\right\} \tag{2.103}$$

is contained in the complement of $\sigma(A)$. Therefore, $\mathcal{S}_{m,\frac{\pi}{4}}$ is a subset of $\rho(A)$ and (2.74) is satisfied. We shall now prove, for $\lambda \in \mathcal{S}_{m,\frac{\pi}{4}}$, validity of the estimate (2.75).

Let $\lambda \in \mathcal{S}_{m,\frac{\pi}{4}}$ and take $\lambda' := \lambda - m$. Only the following two cases are possible:

Case 1: $\lambda = \lambda' + m$, where $\mathrm{Re}(\lambda') < 0$. In this case, since $A - mI$ must be symmetric and nonnegative whereas $-2\,\mathrm{Re}(\lambda') > 0$, we obtain that

$$\begin{aligned} \|(\lambda I - A)x\|_H^2 &= \|(\lambda' I - (A - mI))\|_H^2 \\ &= |\lambda'|^2\|x\|_H^2 - 2\,\mathrm{Re}(\lambda')\langle(A - mI)x, x\rangle_H + \|(A - mI)x\|_H^2 \\ &\geq |\lambda'|^2\|x\|_H^2. \end{aligned} \tag{2.104}$$

Case 2: $\lambda = \lambda' + m$, where $|\mathrm{Im}(\lambda')| \geq |\mathrm{Re}(\lambda')|$. We then have

$$\begin{aligned}\|(\lambda I - A)x\|_H^2 &= \|(\lambda' I - (A - mI))\|_H^2 \\ &= |\mathrm{Im}(\lambda')|^2 \|x\|_H^2 + \|(\mathrm{Re}(\lambda')I - (A - mI))x\|_H \\ &\geq |\mathrm{Im}(\lambda')|^2 \|x\|_H^2 \geq \frac{|\lambda'|^2}{2} \|x\|_H^2. \end{aligned} \tag{2.105}$$

As a result of (2.104) and (2.105), it is seen that

$$\|(\lambda I - A)x\|_H \geq \frac{|\lambda - m|}{\sqrt{2}} \|x\|_H, \quad \text{for each } \lambda \in \mathcal{S}_{m,\frac{\pi}{4}}, \quad x \in H, \tag{2.106}$$

which is the counterpart of (2.75). The proof is completed. □

Remark 2.2.14. Let us recall that the linear operator A in a Hilbert space H is *self-adjoint* whenever it is symmetric and its range $R(A)$ coincides with H. Note also that, although boundedness of A from below suffices to prove that A satisfying assumptions of Proposition 2.2.13 is sectorial, the most important will be the case when A is *positive definite*, that is, the constant m in condition (2.102) is positive. This latter requirement is connected with the need of considerations of fractional powers of A.

Example 2.2.15. An unbounded operator $I - \Delta$ considered in $L^2(\mathbb{R}^N)$ with the domain $D(I - \Delta) = H^2(\mathbb{R}^N)$ is positive definite and sectorial.

Proof. For functions $\phi, \psi \in C_0^\infty(\mathbb{R}^N)$, take an open ball $B_{\mathbb{R}^N}(r)$ containing their supports and use twice the integration by parts formula. We then obtain

$$\begin{aligned}\int_{\mathbb{R}^N} (I - \Delta)\phi\, \psi dx &= \int_{\mathbb{R}^N} \phi\psi dx - \int_{B_{\mathbb{R}^N}(r)} \Delta\phi\, \psi dx \\ &= \int_{\mathbb{R}^N} \phi\psi dx - \int_{B_{\mathbb{R}^N}(r)} \phi\, \Delta\psi dx = \int_{\mathbb{R}^N} \phi (I - \Delta)\psi dx. \end{aligned} \tag{2.107}$$

Furthermore, boundedness of $(I - \Delta)$ from below is a consequence of the following estimate:

$$\begin{aligned}\int_{\mathbb{R}^N} (I - \Delta)\phi\, \phi dx &= \int_{\mathbb{R}^N} \phi^2 dx - \int_{B_{\mathbb{R}^N}(r_{\mathrm{supp}\,\phi})} \Delta\phi\, \phi dx \\ &= \int_{\mathbb{R}^N} \phi^2 dx + \int_{B_{\mathbb{R}^N}(r_{\mathrm{supp}\,\phi})} |\nabla\phi|^2 dx \\ &\geq \int_{\mathbb{R}^N} \phi^2 dx, \quad \phi \in C_0^\infty(\mathbb{R}^N). \end{aligned} \tag{2.108}$$

Consider next the Fourier transform $\mathcal{F}$ and recall that $\mathcal{F}$ is an isomorphism on the set $\mathcal{S} \subset C^\infty(\mathbb{R}^N)$ of rapidly decreasing functions. Denote by $\mathcal{F}^{-1}$ its inverse on $\mathcal{S}$, take

$\phi \in C_0^\infty(\mathbb{R}^N)$ and define the element

$$h := \left(\mathcal{F}^{-1}\frac{1}{1+|\xi|^2}\mathcal{F}\right)\phi. \tag{2.109}$$

It is seen that $h \in \mathcal{S}$ and

$$\begin{aligned}\mathcal{F}[(I-\Delta)h] &= \mathcal{F}h - \mathcal{F}[\Delta h]\\ &= \left(\frac{1}{1+|\xi|^2}\right)\mathcal{F}\phi + |\xi|^2\left(\frac{1}{1+|\xi|^2}\right)\mathcal{F}\phi = \mathcal{F}\phi.\end{aligned} \tag{2.110}$$

Also, by application of $\mathcal{F}^{-1}$ to both sides of (2.110),

$$(I-\Delta)h = \phi. \tag{2.111}$$

Since $C_0^\infty(\mathbb{R}^N)$ is dense in $H^2(\mathbb{R}^N)$ conditions (2.107), (2.108) justify that $I-\Delta$ is a symmetric and positive definite operator on $L^2(\mathbb{R}^N)$. Formulas (2.110), (2.111) show further that the range $R(I-\Delta)$ contains $C_0^\infty(\mathbb{R}^N)$ and, therefore, is dense in $L^2(\mathbb{R}^N)$. This proves directly that $1 \in \rho(\Delta)$. Since Laplacian considered in $L^2(\mathbb{R}^N)$ on the domain $H^2(\mathbb{R}^N)$ is closed, then the resolvent operator $(I-\Delta)^{-1}$ is bounded and closed, that is, its domain $D((I-\Delta)^{-1})$ is closed in $L^2(\mathbb{R}^N)$. This shows that

$$R(I-\Delta) = D((I-\Delta)^{-1}) = L^2(\mathbb{R}^N) \tag{2.112}$$

and, as a consequence of Proposition 2.2.13 (cf. Remark 2.2.14), $I-\Delta : H^2(\mathbb{R}^N) \to L^2(\mathbb{R}^N)$ is sectorial in $L^2(\mathbb{R}^N)$. □

Example 2.2.16. As shown directly in [86, Section 1.6], the negative Laplacian considered in $L^p(\mathbb{R}^N)$ with the domain $D(-\Delta) = W^{2,p}(\mathbb{R}^N)$ is sectorial in $L^p(\mathbb{R}^N)$ and elements of its spectrum have positive real parts.

Example 2.2.17. $-\Delta : H^2(\Omega) \cap H_0^1(\Omega) \to L^2(\Omega)$, where $\partial\Omega \in C^2$, is sectorial and positive definite.

Proof. Using elliptic theory (2.64) choose $\lambda_0 > 0$ so that $(-\Delta) + \lambda_0$ is onto $L^2(\Omega)$. Integration by parts shows that $(-\Delta) + \lambda_0$ is symmetric and positive definite, and hence sectorial as a result of Proposition 2.2.13. Therefore, itself $(-\Delta)$ is sectorial as an immediate consequence of Definition 2.2.1. From the Poincaré inequality, it is seen that $-\Delta$ is positive definite in $L^2(\Omega)$. □

Example 2.2.18. $\Delta^2 : H^4(\Omega) \cap H_0^2(\Omega) \to L^2(\Omega)$, where $\partial\Omega \in C^4$, is sectorial and positive definite.

Proof. The proof is the same as in Example 2.2.17, although positive definiteness of Δ^2 follows from the Smoler inequality (cf. [154, Theorem 11.11]):

$$\|\Delta\phi\|_{L^2(\Omega)} \geq \mu_1\|\phi\|_{L^2(\Omega)}, \quad \phi \in H_0^2(\Omega), \tag{2.113}$$

where μ_1 denotes the smallest positive eigenvalue of $-\Delta$ on Ω with the Dirichlet boundary conditions. □

Example 2.2.19. If the triple $(A, \{B_j\}, \Omega)$ forms a regular elliptic boundary value problem as described in Section 2.1.6 then, for some $\lambda_0 > 0$, operator $A + \lambda_0 I$ acting in $L^p(\Omega)$ with the domain $D(A + \lambda_0 I) = W^{2m,p}_{\{B_j\}}(\Omega)$ is sectorial and $\operatorname{Re}\sigma(A + \lambda_0 I) > 0$.

Proof. From (2.67) and (2.68), it is seen that the resolvent $(\lambda I - A)^{-1}$ exists for all $\lambda \in \mathbb{C}$ with $\operatorname{Re}\lambda \leq -\Lambda_0$ and

$$\|(\lambda I - A)^{-1}\|_{\mathcal{L}(L^p(\Omega),L^p(\Omega))} \leq \frac{\text{const.}}{|\lambda|}$$

which, based on Proposition 2.2.3, is sufficient for validity of Definition 2.2.1. As a further result of Proposition 2.2.3, $A + \omega I$ is sectorial in $L^p(\Omega)$ for each $\omega \in \mathbb{R}$. Therefore, $\lambda_0 > 0$ may be chosen, for which $A + \lambda_0 I$ becomes positive in $L^p(\Omega)$, that is, the condition $\operatorname{Re}\sigma(A + \lambda_0 I) > 0$ is satisfied (cf. Remark 2.2.2). □

Example 2.2.20. Since $(\Delta^2, \{\frac{\partial}{\partial N}, \frac{\partial \Delta}{\partial N}\}, \Omega)$ forms a regular elliptic boundary value problem, then Δ^2 considered with the domain $D(\Delta^2) = H^4_{\frac{\partial}{\partial N},\frac{\partial\Delta}{\partial N}}(\Omega)$ ($\partial\Omega$ of the class C^4) is sectorial in $L^2(\Omega)$. More precisely, integration by parts shows that $\Delta^2 + \delta I$ ($\delta > 0$) is symmetric and positive definite so that $\Delta^2 + \delta I$ is a self-adjoint operator in $L^2(\Omega)$.

Example 2.2.21. Consider in $[L^p(\Omega)]^d$ a second-order vector value operator of the form (cf. [41], [4, Section 3]):

$$Au = \sum_{i,j=1}^{n} (A_{ij}(x)u_{x_i})_{x_j} + \sum_{i=1}^{n} B_i(x)u_{x_i} \tag{2.114}$$

acting on $u = (u_1, \dots, u_d)$, where $\Omega \subset \mathbb{R}^N$ is a bounded smooth domain, and $A_{ij} = [a^{\mu\nu}_{ij}]$, $B_i = [b^{\mu\nu}_i]$ are symmetric $N \times N$ matrices. Assume that:

- For some $c_0 > 0$ and any real nN-tuple $q = (q^\mu_i)$ with $\mu = 1, \dots, N$, $i = 1, \dots, n$:

$$\sum_{i,j=1}^{n} \sum_{\mu,\nu=1}^{N} a^{\mu\nu}_{ij}(x) q^\mu_i q^\nu_j \geq c_0 |q|^2, \quad x \in \overline{\Omega},$$

- $A_{ij} = A_{ji}$, for all $i, j = 1, \dots, n$,
- components $a^{\mu\nu}_{ij}$ of A_{ij}-matrices belong to $C^1(\overline{\Omega})$ and B_i-components $b^{\mu\nu}_i$ are the elements of $C^0(\overline{\Omega})$.

Then $-A$ with the domain $W^{2,p}(\Omega) \cap W^{1,p}_0(\Omega)$ is sectorial in $L^p(\Omega)$ (cf. [41, p. 608]).

Remark 2.2.22. An important class of examples of sectorial positive operators will be introduced in the next subsection. It will be shown in Proposition 2.2.31 that the proper fractional powers of sectorial positive operators are again sectorial positive operators. Such operators act as the standard "linear main part" in the examples discussed further in the book.

2.2.3 Fractional powers of operators

When introducing fractional powers of linear operators, two classes of operators suitable for defining such notion are usually considered. The first one, class of *positive operators*, is smaller but contains most of the relevant examples of operators appearing in applications. The second, larger, class of *nonnegative operators* will be also introduced, mostly because it contains an additionally important example of the negative Laplace operator densely defined in $L^p(\mathbb{R}^N)$, $1 < p < \infty$. Comparison of those two classes will be given in Subsection 5.2.1. Here, we concentrate mainly on the smaller class of positive operators.

We recall the definition of *positive operator* in a Banach space X (e. g., [173, 130]).

Definition 2.2.23. A closed linear operator $A : X \supset D(A) \to X$ with dense domain is called positive if $(-\infty, 0] \subset \rho(A)$ and the condition

$$\exists_{M>0}\forall_{\lambda>0}\ \|(\lambda + A)^{-1}\| \leq \frac{M}{1+\lambda}, \quad \lambda > 0, \tag{2.115}$$

is satisfied.

In particular, a positive definite operator in a Hilbert space is positive in the above sense.

Another widely used class of *nonnegative operators* was introduced by B. V. Balakrishnan in 1960 and studied in a series of papers by Hikosaburo Komatsu (e. g., [106, 107]) allowing to define its fractional powers. We recall next the Balakrishnan definition of fractional power of nonnegative operator.

Definition 2.2.24. Let A be a nonnegative operator that means a closed linear densely defined operator in a Banach space X, such that its resolvent set $\rho(A)$ contains $(-\infty, 0)$ and the resolvent satisfies

$$\|\lambda(\lambda + A)^{-1}\| \leq M, \quad \lambda > 0. \tag{2.116}$$

Then, for $\eta \in (0, 1)$, $\phi \in D(A)$,

$$A^\eta \phi = \frac{\sin(\pi\eta)}{\pi} \int_0^\infty s^{\eta-1} A(s + A)^{-1} \phi ds. \tag{2.117}$$

There are extensions of the above definition valid for the powers $\eta \geq 1$ (e. g., [130]), and for negative powers η (e. g., [86]); see Subsection 5.2.3.

Simple comparison of the definitions of *sectorial operator* and *positive operator* shows that under the additional requirement $\operatorname{Re}\sigma(A) > 0$ sectorial operator is in particular positive operator. Not to overburden further presentation, and since sectorial operators are our main interest here, we will assume further that in a Banach space X

$$A \text{ is a sectorial operator in } X \text{ with } \operatorname{Re}\sigma(A) > 0, \tag{2.118}$$

and formulate the results for *sectorial positive* operators. Note finally that the arbitrary sectorial operator will be "improved" to sectorial positive operator by adding to it a multiply of identity cI with sufficiently large real c.

Theorem 2.2.6 allows to define, for $\alpha \in (0, +\infty)$, the operators $A^{-\alpha} : X \to X$ by the integral formula (cf. [86], [66]):

$$A^{-\alpha}v = \frac{1}{\Gamma(\alpha)} \int_0^{+\infty} t^{\alpha-1} e^{-At} v dt. \tag{2.119}$$

Proposition 2.2.25. *Suppose that* (2.118) *holds.* $A^{-\alpha}$, $\alpha \in (0, +\infty)$, *are well-defined linear bounded operators on* X *giving a one-to-one correspondence between* X *and the range* $R(A^{-\alpha})$. *Also,* $A^{-1} : D(A) \to X$ *coincides with the inverse of* A *and*

$$A^{-\alpha}A^{-\beta} = A^{-(\alpha+\beta)}, \quad \textit{for } \alpha, \beta > 0.$$

As seen above, each $A^{-\alpha}$, $\alpha > 0$, is invertible. We denote further these inverse operators by A^{α} and use symbol X^{α} for the domain of definition of A^{α}, that is, $X^{\alpha} := R(A^{-\alpha})$. We also extend the notion of the power operator for the case when $\alpha = 0$, taking $A^0 := I$ (identity map) on $X^0 := X$.

The following properties of fractional powers are well known (cf. [66, Part 2, Section 14], [86, Section 1.4]).

Proposition 2.2.26. *Let* (2.118) *be satisfied.* X^{α} $(\alpha \in [0, +\infty))$ *with the norm* $\|v\|_{X^{\alpha}} := \|A^{\alpha}v\|_X$ *are Banach spaces whereas* $A^{\alpha} : X^{\alpha} \to X$ *are linear closed and densely defined operators in* X *satisfying*

$$A^{\alpha}A^{\beta} = A^{\beta}A^{\alpha} = A^{\alpha+\beta}, \quad \alpha, \beta \geq 0. \tag{2.120}$$

Furthermore, X^{α} *is a dense subset of* X^{β} *for* $\alpha \geq \beta \geq 0$, *the inclusions*

$$X^{\alpha} \subset X^{\beta}, \quad \alpha > \beta \geq 0, \tag{2.121}$$

are dense, continuous, and additionally, they are compact provided that A *has compact resolvent.*

The following property is also useful (cf. [118, p. 294]):

Remark 2.2.27. Let A be a sectorial positive operator in X and $\alpha, \beta \in \mathbb{R}$. Then, for $v \in D(A^{\alpha})$, $D(A^{\alpha})$ with natural norm,

$$A^{\beta}v \in D(A^{\alpha-\beta}), \quad \text{or} \quad A^{\beta} : D(A^{\alpha}) \to D(A^{\alpha-\beta}) \quad \text{is an isometry.} \tag{2.122}$$

In fact, for $v \in D(A^{\alpha})$, by the definition of the domain of linear operator, we have $A^{\alpha}v \in X$. For $\beta \leq \alpha$ or $\beta > \alpha$ and $\alpha \geq 0$, A^{β} is well-defined on $D(A^{\alpha})$. Consequently,

$$A^{\beta}v = A^{\beta-\alpha}A^{\alpha}v \in D(A^{\alpha-\beta}).$$

For $\beta > \alpha$ and $\alpha < 0$, we extend an operator A^β to be well-defined on $D(A^\alpha)$ setting

$$A^\beta v := A^{\beta-\alpha} A^\alpha v, \quad v \in D(A^\alpha).$$

Since also $A^\gamma : X \to D(A^{-\gamma})$, the first part of the statement (2.122) follows.

To justify the second part, observe that

$$\|A^\beta v\|_{D(A^{\alpha-\beta})} = \|A^{\alpha-\beta} A^\beta v\|_X = \|A^\alpha v\|_X = \|v\|_{D(A^\alpha)}.$$

Proposition 2.2.28. *Assume that* (2.118) *holds. For $\alpha \geq 0$, $t \geq 0$, $A^\alpha e^{-At}$ is a bounded linear operator on X, such that*

$$\begin{aligned} \|A^\alpha e^{-At}\|_{\mathcal{L}(X,X)} &\leq c_\alpha \frac{e^{-at}}{t^\alpha}, \quad t > 0, \\ A^\alpha e^{-At} &= e^{-At} A^\alpha \quad \text{on } X^\alpha; \end{aligned} \tag{2.123}$$

here, $a > 0$ is such that $\operatorname{Re} \sigma(A) > a$.

For sectorial operator A with $\operatorname{Re}\sigma(A) > 0$, the following rule of rising power A^β to a power α holds (cf. [106, Theorem 10.6], [181]).

Proposition 2.2.29. *Fix $a \in \mathbb{R}$ and $\phi \in (0, \frac{\pi}{2})$. Let A satisfy conditions of the Definition 2.2.1 and* $\operatorname{Re}(\sigma(A)) > 0$. *Then we have*

$$(A^\beta)^\alpha = A^{\alpha\beta}, \tag{2.124}$$

for arbitrary $\beta \in (0, \frac{\pi}{\phi})$ and $\alpha > 0$.

We remark that (2.124) may not be true in general. For example, $(A^2)^{\frac{1}{2}}$ may not be equal to A for A^2 and A generating C^0-semigroups (cf. [189] for details). However, if A satisfies (2.118), then half of the angle opening ϕ of its sector $\mathcal{S}_{a,\phi}$ fulfills $0 < \phi < \frac{\pi}{2}$. Therefore, Proposition 2.124 yields in particular the formula:

$$(A^2)^\alpha = A^{2\alpha}, \quad \text{for each} \alpha > 0. \tag{2.125}$$

Below we show that *sectoriality* of the operator A remains valid for A restricted to any of the fractional-order spaces X^β, $\beta \geq 0$, provided it holds for A considered on the basic phase space X. This property will be useful in the studies of the smooth solutions to the abstract Cauchy problem (3.1).

Proposition 2.2.30. *Let X^1 be a domain of a sectorial operator $A_{|_X} : X^1 \subset X \to X$ acting in a Banach space X and satisfying* $\operatorname{Re}\sigma(A_{|_X}) > 0$. *Let $\beta > 0$ and $A_{|_{X^\beta}} : X^{1+\beta} \subset X^\beta \to X^\beta$, equipped with the domain $X^{1+\beta} \subset X^1$, denote restriction of $A_{|_X}$ to the fractional power space $X^\beta \subset X$ so that*

$$A_{|_X} x = A_{|_{X^\beta}} x \quad \text{for } x \in X^{1+\beta}. \tag{2.126}$$

Then $A_{|_{X^\beta}}$ is a sectorial operator in X^β and $\operatorname{Re}\sigma(A_{|_{X^\beta}}) > 0$.

Proof. From Proposition 2.2.26, the inclusion $X^{1+\beta} \subset X^\beta$ is dense for each $\beta > 0$. $A_{|_{X^\beta}}$ is thus linear and densely defined operator in X^β, $\beta > 0$. Furthermore, $A_{|_{X^\beta}}$ is closed in X^β, which is the direct consequence of closedness of $A_{|_X}$ in X.

Let us now estimate the resolvent of $A_{|_{X^\beta}}$ in X^β. Since $A_{|_X}$ is sectorial in X, there are $a \in \mathbb{R}$ and $\phi \in (0, \frac{\pi}{2})$ (a, ϕ fixed from now on) such that the sector $\mathcal{S}_{a,\phi}$ is contained in the resolvent set $\rho(A_{|_X})$ of $A_{|_X}$ and

$$\|(\lambda I - A_{|_X})^{-1}x\|_X \le \frac{M}{|\lambda - a|}\|x\|_X, \quad \lambda \in \mathcal{S}_{a,\phi},\ x \in X. \tag{2.127}$$

We shall show for the restriction $A_{|_{X^\beta}}$ of $A_{|_X}$ that

$$\mathcal{S}_{a,\phi} \subset \rho(A_{|_{X^\beta}}),$$
$$\|(\lambda I - A_{|_{X^\beta}})^{-1}y\|_{X^\beta} \le \frac{M}{|\lambda - a|}\|y\|_{X^\beta}, \quad \lambda \in \mathcal{S}_{a,\phi},\ y \in X^\beta.$$

If $\lambda \in \mathcal{S}_{a,\phi}$, then to each $y \in X^\beta$ corresponds a unique $x \in X^1$ satisfying $(\lambda I - A_{|_X})x = A^\beta_{|_X}y$. Applying $A^{-\beta}_{|_X}$ to both sides of the latter equality and noting that

$$A^{-\beta}_{|_X}(\lambda I - A_{|_X})x = (\lambda I - A_{|_X})A^{-\beta}_{|_X}x, \quad x \in X^1, \tag{2.128}$$

we conclude, in the presence of (2.126), that the equation $(\lambda I - A_{|_{X^\beta}})\tilde{x} = y$ has a unique solution $\tilde{x} \in X^{1+\beta}$ for each $y \in X^\beta$. Consequently, the inverse $(\lambda I - A_{|_{X^\beta}})^{-1}$ is defined on X^β and, by (2.126),

$$(\lambda I - A_{|_{X^\beta}})^{-1}y = (\lambda I - A_{|_X})^{-1}y, \quad y \in X^\beta. \tag{2.129}$$

Based on (2.129), (2.128), and the estimate (2.127), we verify finally that

$$\begin{aligned}\|(\lambda I - A_{|_{X^\beta}})^{-1}y\|_{X^\beta} &= \|(\lambda I - A_{|_X})^{-1}y\|_{X^\beta}\\ &= \|A^\beta_{|_X}(\lambda I - A_{|_X})^{-1}y\|_X = \|(\lambda I - A_{|_X})^{-1}A^\beta_{|_X}y\|_X\\ &\le \frac{M}{|\lambda - a|}\|A^\beta_{|_X}y\|_X = \frac{M}{|\lambda - a|}\|y\|_{X^\beta}, \quad \lambda \in \mathcal{S}_{a,\phi},\ y \in X^\beta.\end{aligned}$$

Similar calculations show that $\operatorname{Re}\sigma(A_{|_{X^\beta}}) > 0$ if $\operatorname{Re}\sigma(A_{|_X}) > 0$. The proof is complete. □

It is interesting to note that the operation of taking proper fractional power acts *inside* the class of sectorial positive operators.

Proposition 2.2.31. *Let A be a sectorial positive operator in a Banach space X. Any proper fractional power A^α, $\alpha \in (0,1)$, is sectorial positive in X.*

For the proof, it is convenient to use a notion, due to H. Komatsu [106, p. 288], of an *operator of the type $(\omega, M(\theta))$ in a Banach space X.*

Definition 2.2.32. We say that A is of type $(\omega, M(\theta))$, $0 \le \omega < \pi$, if the domain $D(A)$ is dense in X; the resolvent set of $-A$ contains the sector $|\arg \lambda| < \pi - \omega$ and the condition $\|\lambda(\lambda + A)^{-1}\| \le M(\theta)$ holds on each ray $\lambda = re^{i\theta}$, $r \in (0, +\infty)$, $|\theta| < \pi - \omega$.

One may easily verify that A is of the type $(\omega, M(\theta))$ with $\omega < \frac{\pi}{2}$ *if and only if* A is a sectorial operator in the sense of Definition 2.2.1 with $a = 0$. An interesting observation of T. Kato (see [106, p. 320] also [187, p. 97]) states the following.

Proposition 2.2.33. *If A is of type $(\omega, M(\theta))$ and if $0 < \alpha < \frac{\pi}{\omega}$, then A^α is of type $(\alpha\omega, M_\alpha(\theta))$ with certain positive constant $M_\alpha(\theta)$. Furthermore, the resolvent $(\lambda + A^\alpha)^{-1}$ is analytic in α and λ in the domain $0 < \alpha < \frac{\pi}{\omega}$, $|\arg \lambda| < \pi - \alpha\omega$.*

Characterization of X^α as interpolation space

Fix $\alpha \in (0, 1)$ and $v \in D(A)$. Recalling (2.119), we find that

$$\begin{aligned} A^\alpha v = A^{-(1-\alpha)} A^{1-\alpha} A^\alpha v &= \frac{1}{\Gamma(1-\alpha)} \int_0^{+\infty} t^{-\alpha} e^{-At} A^{1-\alpha} A^\alpha v dt \\ &= \frac{1}{\Gamma(1-\alpha)} \int_0^{+\infty} t^{-\alpha} e^{-At} A v dt. \end{aligned} \tag{2.130}$$

Since A is a negative generator of an analytic semigroup such that $\operatorname{Re} \sigma(A) > 0$, the resolvent of $-A$ may be written in the integral form:

$$(sI + A)^{-1} w = \int_0^{+\infty} e^{-st} e^{-At} w dt, \quad w \in X,\ s \ge 0. \tag{2.131}$$

Using (2.131) and the equality

$$t^{-\alpha} = \frac{1}{\Gamma(\alpha)} \int_0^{+\infty} s^{\alpha-1} e^{-ts} ds$$

we obtain further from (2.130)

$$\begin{aligned} A^\alpha v &= \frac{1}{\Gamma(1-\alpha)\Gamma(\alpha)} \int_0^{+\infty} \left(\int_0^{+\infty} s^{\alpha-1} e^{-ts} ds \right) e^{-At} A v dt \\ &= \frac{1}{\Gamma(1-\alpha)\Gamma(\alpha)} \int_0^{+\infty} s^{\alpha-1} \left(\int_0^{+\infty} e^{-ts} e^{-At} A v dt \right) ds \\ &= \frac{1}{\Gamma(1-\alpha)\Gamma(\alpha)} \int_0^{+\infty} s^{\alpha-1} (sI + A)^{-1} A v ds \\ &= \frac{1}{\Gamma(1-\alpha)\Gamma(\alpha)} \int_0^{+\infty} s^{\alpha-1} A (sI + A)^{-1} v ds. \end{aligned} \tag{2.132}$$

Formula (2.132) coincides for real α satisfying inequality $0 < \alpha < 1$ with the definition of fractional powers A^α of positive operators stated in [173, Remark 1.15.1/2]. Following further the approach [173, 1.15.3], that is, extending (2.132) to the formula (cf. [173, 1.15.1(1)]) valid for α from the whole complex plane, we obtain the description of X^α as the complex *interpolation spaces* under the additional assumption that the purely imaginary powers A^{it} of A (with real t) are known to be uniformly bounded. The result [173, Theorem 1.15.3] then states the following.

Proposition 2.2.34. *The following interpolation formula holds:*

$$[X^\alpha, X^\beta]_\theta = X^{(1-\theta)\alpha+\theta\beta}, \quad \text{for } \alpha, \beta \geq 0,\ \theta \in (0,1) \tag{2.133}$$

together with corresponding moments inequality

$$\|v\|_{X^{(1-\theta)\alpha+\theta\beta}} \leq c_\theta \|v\|^\theta_{X^\beta} \|v\|^{1-\theta}_{X^\alpha}, \quad v \in X^\alpha \cap X^\beta,\ \alpha, \beta \geq 0,\ \theta \in (0,1), \tag{2.134}$$

provided A is sectorial operator in a Banach space X such that $\mathrm{Re}\,\sigma(A) > 0$ *and* $\|A^{it}\|_{\mathcal{L}(X,X)} \leq \mathrm{const.}(\varepsilon)$ *for all $t \in [-\varepsilon, \varepsilon]$. Here, $[\cdot,\cdot]_\theta$ denotes complex interpolation space (cf. [173, 1.9.3]).*

Since $X = X^0$ and $D(A) = X^1$, then choosing $\alpha = 0$, $\beta = 1$ in Proposition 2.2.34, we conclude the following.

Corollary 2.2.35. *Under the assumptions of Proposition 2.2.34, X^θ ($\theta \in (0,1)$) are intermediate spaces between X and $D(A)$, that is,*

$$X^\theta = D(A^\theta) = [X, D(A)]_\theta \tag{2.135}$$

and

$$\|v\|_{X^\theta} \leq c_\theta \|Av\|^\theta_X \|v\|^{1-\theta}_X, \quad v \in D(A). \tag{2.136}$$

Some applications of interpolation theory

The latter result is important in applications. To see the action of Corollary 2.2.35, let us reconsider some examples introduced earlier in Section 2.2.2.

Take in the Hilbert space $L^2(\mathbb{R}^N)$ positive definite, self-adjoint operator $A = I - \Delta$ with the domain $D(A) = H^2(\mathbb{R}^N)$ (cf. Example 2.2.15). In this Hilbert case, A^{-it} are unitary operators and, in particular, Proposition 2.2.34 holds (cf. [173, 1.18.10]). Hence, condition (2.135) implies that

$$X^\alpha = [L^2(\mathbb{R}^N), H^2(\mathbb{R}^N)]_\alpha = H_2^{2\alpha}(\mathbb{R}^N), \quad \alpha \in (0,1),$$

where the second equality is written based on the characterization given in [173, 2.4.2 (11)]. In the general case of $A = I - \Delta$ on $L^p(\mathbb{R}^N)$, we can follow the calculations of [173, 2.5.3] that lead to the formula

$$(I - \Delta)^\alpha \phi = \mathcal{F}^{-1}(1 + |\xi|^2)^\alpha \mathcal{F}\phi, \quad \phi \in S(\mathbb{R}^N),\ \alpha > 0 \tag{2.137}$$

($\mathcal{F}$ being the Fourier transform and $S(\mathbb{R}^N)$ denoting the set of rapidly decreasing functions). Since $S(\mathbb{R}^N)$ is dense in $H_p^{2\alpha}(\mathbb{R}^N) = \{\phi \in L^p(\mathbb{R}^N);\ \|\phi\|_{H_p^{2\alpha}(\mathbb{R}^N)} = \|\mathcal{F}^{-1}(1 + |\xi|^\alpha)\mathcal{F}\phi\|_{L^p(\mathbb{R}^N)} < \infty\}$ and $(I - \Delta)^\alpha$ is closed, we must have the inclusion $H_p^{2\alpha}(\mathbb{R}^N) \subset D((I-\Delta)^\alpha)$. Based on the *lift property* (cf. [173, Theorem 2.3.4]), we observe further that (2.137) holds for $\phi \in H_p^{2\alpha}(\mathbb{R}^N)$ and, moreover,

$$
\begin{aligned}
X^\alpha = (I-\Delta)^{-\alpha}(L^p(\mathbb{R}^N)) &= (I-\Delta)^{-\alpha}(\mathcal{F}^{-1}(1+|\xi|^2)^\alpha \mathcal{F}(H_p^{2\alpha}(\mathbb{R}^N))) \\
&= H_p^{2\alpha}(\mathbb{R}^N), \quad \alpha > 0.
\end{aligned} \tag{2.138}
$$

The above result can be generalized onto matrix elliptic operators on the whole of $\mathbb{R}^N$ whose top-order coefficients are merely assumed to be bounded and uniformly continuous matrix valued functions (cf. [140], [8]).

Characterization of the domains $X^\alpha = D(A^\alpha)$, $\alpha \in (0,1)$, of fractional powers of strongly elliptic operators on bounded domains $\Omega \subset \mathbb{R}^N$, is still in progress; compare [187, Chapter 16]. There are also examples of sectorial operators A which does not satisfy the condition $D(A^{\frac{1}{2}}) = [X, D(A)]_{\frac{1}{2}}$; see, for example, [187, p. 537]. Simultaneously, the knowledge of the exact form of the domain X^α is vital in the studies of particular equations, all the more for the critical problems considered further in this book. Through the last 35 years (see, e. g., [133, 187]), such characterization, under natural regularity assumptions on the coefficients, has been obtained in case of the second order strongly elliptic operators; similar results valid for general classes of sectorial operators are not available yet (see the comment in [187, p. 562]).

We will introduce an extra assumption and collect sufficient conditions under which such characterization is available.

Characterization-condition

A regular elliptic boundary value problem $(A, \{B_j\}, \Omega)$ is said to satisfy the *characterization condition*, whenever the equality holds:

$$
D(A^\alpha) = [L^p(\Omega), W^{2m,p}_{\{B_j\}}(\Omega)]_\alpha \subset W^{2m\alpha,p}(\Omega), \quad \alpha \in (0,1), \tag{2.139}
$$

where $[\cdot,\cdot]_\alpha$ denotes *complex interpolation space*.

The characterization-condition holds in particular if the imaginary powers of A are bounded. The first equality in (2.139) is then a consequence of Proposition 2.2.34 (cf. [173, Section 1.15.3]). It is also known (cf. [173, Section 4.3.1]) that for bounded $\Omega \subset \mathbb{R}^N$ satisfying the cone condition,

$$
[L^p(\Omega), W^{2m,p}(\Omega)]_\alpha = W^{2m\alpha,p}(\Omega), \quad \alpha \in (0,1).
$$

Directly from the definition of the complex interpolation space (cf. [173, Section 1.9.1]), we thus have the embedding

$$
[L^p(\Omega), W^{2m,p}_{\{B_j\}}(\Omega)]_\alpha \subset [L^p(\Omega), W^{2m,p}(\Omega)]_\alpha, \quad \alpha \in (0,1),
$$

leading to the inclusion in (2.139).

In case of the higher 2m-order elliptic operators ($m > 1$), classical results guaranteeing validity of the characterization-condition require that both the coefficients of the operators A, B_j; $j = 1, 2, \dots, m$, and the boundary $\partial\Omega$ are of the class C^∞ (cf. [149], [173]). One may compare [8], [81], [133], [140], [157], and in particular [187, Chapter 16] for the recent developments within this field. That problem still attracts attention of many mathematicians.

Below we describe four known sufficient conditions for validity of the characterization condition. These cases cover all the examples reported later in Chapter 5.

Case I

We start with the results of [149] (cf. also [173]) concerning the regular elliptic boundary value problems $(A, \{B_j\}, \Omega)$ of arbitrary order 2m under the additional assumption, which will be valid throughout the whole of Case I that the coefficients a_σ of A are in $C^\infty(\overline{\Omega})$, the coefficients b^j_σ of B_j belong to $C^\infty(\partial\Omega)$, and the boundary $\partial\Omega$ is of the class C^∞.

Certainly, A is sectorial in $L^p(\Omega)$ and we may assume here without loss of generality that A is positive. For such problems, from [173, Theorem 4.9.1], it follows that

$$\|A^{it}\|_{\mathcal{L}(L^p(\Omega),L^p(\Omega))} \le \text{const.} e^{\text{const.}_1|t|}, \quad t \in \mathbb{R}.$$

Therefore, using (2.135) we obtain in this case that

$$X^\alpha = [L^p(\Omega), W^{2m,p}_{\{B_j\}}(\Omega)]_\alpha, \quad \alpha \in (0,1); \tag{2.140}$$

(cf. [173, Section 4.9.2], also the source paper [149]). However, expected equality

$$[L^p(\Omega), W^{2m,p}_{\{B_j\}}(\Omega)]_\alpha = H^{2m\alpha}_{p,\{B_j\}}(\Omega), \quad \alpha \in (0,1) \tag{2.141}$$

is not always true (cf. [173, 4.3.3 (7)]) but holds under the additional requirement that

$$2m\alpha - \frac{1}{p} \ne m_j \quad \text{for all } j = 1, \dots m. \tag{2.142}$$

Here, m_j denotes the order of B_j, whereas $H^{2m\alpha}_{p,\{B_j\}}(\Omega)$ is defined as in (2.69) by the formula

$$H^{2m\alpha}_{p,\{B_j\}}(\Omega) := \{\phi \in H^{2m\alpha}_p(\Omega);\ \forall_{i\in\{j;\ m_j<2m\alpha-\frac{1}{p}\}}\ B_i\phi_{|\partial\Omega} = 0\}. \tag{2.143}$$

If (2.142) is violated, we have merely the proper inclusion

$$[L^p(\Omega), W^{2m,p}_{\{B_j\}}(\Omega)]_\alpha \subset H^{2m\alpha}_{p,\{B_j\}}(\Omega), \quad \alpha \in (0,1); \tag{2.144}$$

and the structure of $[L^p(\Omega), W^{2m,p}_{\{B_j\}}(\Omega)]_\alpha$ in this case is more complicated (cf. [149], [173, 4.3.3 (9)]).

Let A be a sectorial, positive operator on $L^q(\Omega)$ given by a regular $2m$-th order elliptic boundary value problem $(A, \{B_j\}, \Omega)$ with C^∞-smooth data. The domain of A satisfies then inclusions

$$W_0^{2m,q}(\Omega) \subset D(A) = W_{\{B_j\}}^{2m,q}(\Omega) \subset W^{2m,q}(\Omega). \tag{2.145}$$

Using the characterization of the space X^α given in (2.140), (2.144), we find for all $\alpha \in [0,1]$ that

$$X^\alpha \subset W^{2m\alpha,q}(\Omega), \quad \text{if } \alpha \in [0,1] \text{ and either } q \geq 2 \text{ or } 2m\alpha \in \mathbb{N} \tag{2.146}$$

(for restriction on parameters in (2.146) recall that $H_q^s(\Omega) \subset B_{q,q}^s(\Omega)$ if and only if $q \geq 2$; see [173, Remark 2.3.3/4, Theorem 4.6.1(b)]).

We can extend the embedding (2.146) to all values of $\alpha \geq 0$ using regularity theory of elliptic operators. Based on the isomorphism property of A (cf. [173, Theorem 5.5.1(b)] with $\lambda = 0$), we have the following condition:

$$\begin{aligned} &\text{if } s \geq 0,\ u \in D(A) \text{ and } Au \in W^{s,q}(\Omega), \text{ then } u \in W^{2m+s,q}(\Omega) \\ &\text{and } \|u\|_{W^{2m+s,q}(\Omega)} \leq \text{const.} \|Au\|_{W^{s,q}(\Omega)}. \end{aligned} \tag{2.147}$$

Condition (2.147) allows to justify inductively for $k = 1, 2, \ldots$ that

$$\begin{aligned} &\text{if } \alpha \in [k-1, k) \text{ and either } q \geq 2 \text{ or } 2m\alpha \in \mathbb{N}, \\ &\text{then } X^\alpha \subset W^{2m\alpha,q}(\Omega), \end{aligned} \tag{2.148}$$

which ensures validity of (2.146) for all $\alpha > 0$. Indeed, the case $k = 1$ is obvious from (2.146). For the induction, assume that (2.148) holds for k and take $u \in X^\alpha$ where $\alpha \in [k, k+1)$. Then $\alpha = k + \theta_\alpha$ with $\theta_\alpha \in [0,1)$, $u \in X^\alpha \subset D(A)$ and $Au \in X^{k-1+\theta_\alpha} \subset W^{2m(k-1+\theta_\alpha),q}(\Omega)$. Therefore, from (2.147), $u \in W^{2m(k+\theta_\alpha),q}(\Omega) = W^{2m\alpha,q}(\Omega)$, which completes the induction.

As a consequence of the above considerations, we obtain the following.

Proposition 2.2.36. *Let A be a sectorial operator on $L^q(\Omega)$, $1 < q < +\infty$, given by a regular elliptic boundary value problem $(A, \{B_j\}, \Omega)$ satisfying the characterization-condition. Then, for $\alpha \in [0,1]$ the following inclusions hold:*

$$X^\alpha \subset \begin{cases} W^{t,p}(\Omega) & \text{if } 2m\alpha - \frac{n}{q} \geq t - \frac{n}{p},\ 2 \leq q \leq p < +\infty, \\ C^{k+\mu}(\overline{\Omega}) & \text{if } 2m\alpha - \frac{n}{q} \geq k + \mu,\ k \in \mathbb{N},\ \mu \in (0,1). \end{cases} \tag{2.149}$$

In the first embedding, the condition $q \leq p$ will be neglected when $2m\alpha > t$. Furthermore, the assumption $q \geq 2$ may be changed to $q > 1$ whenever $2m\alpha \in \mathbb{N}$. The second embedding holds also with $\mu = 0$, provided that strict inequality $2m\alpha - \frac{n}{q} > k$ is satisfied. Also, (2.149) holds with all $\alpha > 0$ provided that Ω is a bounded C^∞-domain and coefficients of A and B_j ($j = 1, \ldots, m$) are infinitely many times continuously differentiable in $\overline{\Omega}$ or $\partial\Omega$, respectively.

Proof. The proof is a direct consequence of (2.146), Proposition 2.1.12, and the inclusion (2.45). For the requirement $q \geq 2$, note that for fractional $2m\alpha$ and $q \neq 2$ we have by (2.40), $W^{2m\alpha,q}(\Omega) = B^{2m\alpha}_{q,q}(\Omega)$, and hence (2.146) fails for $q < 2$ (cf. [173, Remark 2.3.3/4, Theorem 4.6.1(b)]). At last, to include all $\alpha \geq 0$ more regularity of the data, needed for validity of (2.147), is required. □

Remark 2.2.37. Note that, for $\alpha = k + \frac{r}{2m}$ ($k \in \mathbb{N}$, $r = 0, 1, \ldots, 2m-1$), in the presence of (2.140), (2.141), we have

$$\|u\|_{X^\alpha} = \|A^{\frac{r}{2m}} A^k u\|_{L^q(\Omega)} \leq c\|A^k u\|_{H^r_q(\Omega)} = c\|A^k u\|_{W^{r,q}(\Omega)}. \tag{2.150}$$

Furthermore,

$$\|A^k u\|_{W^{r,q}(\Omega)} \leq c'\|u\|_{W^{2mk+r,q}(\Omega)}, \tag{2.151}$$

where (2.151) follows inductively from [173, Theorem 5.3.4]. Connecting (2.150), (2.151), and (2.149), we thus observe that $\|\cdot\|_{W^{2m\alpha,q}(\Omega)}$ is for such α an equivalent norm in the space X^α.

Remark 2.2.38. It should be also emphasized that the structure of X^α ($\alpha \geq 0$) strongly involves boundary conditions. In the above considerations, related to a regular $2m$-th order elliptic boundary value problem $(A, \{B_j\}, \Omega)$ with C^∞ data, we have $D(A) = W^{2m,q}_{\{B_j\}}(\Omega)$, $X^\theta \subset W^{2m\theta,q}_{\{B_j\}}(\Omega)$ ($\theta \in [0,1]$) so that we obtain the equivalence:

$$\begin{aligned} &\phi \in X^{k+\theta} \quad (k \in \mathbb{N},\ \theta \in [0,1)) \quad \text{if and only if} \\ &\phi, A\phi, \ldots, A^{k-1}\phi \in W^{2m,q}_{\{B_j\}}(\Omega) \quad \text{and} \quad A^k\phi \in H^{2m\theta}_{q,\{B_j\}}(\Omega). \end{aligned}$$

An illustration may serve here as the second-order Dirichlet problem $(-\Delta, I, \Omega)$ with $\partial\Omega \in C^\infty$. In this simple example, the fractional power spaces $X^{k+\theta}$ ($k \in \mathbb{N}$, $\theta \in (\frac{1}{2mq}, 1)$) may be characterized as follows:

$$X^{k+\theta} = \{\phi \in W^{2m(k+\theta),q}(\Omega);\ \phi = \Delta\phi = \cdots = \Delta^k\phi = 0 \text{ on } \partial\Omega\}.$$

Remark 2.2.39. Often it is inconvenient to assume the C^∞-smoothness of the data. Under more natural smoothness assumption that $(A, \{B_j\}, \Omega)$ is a regular elliptic boundary value problem (Definition 2.1.18) as a consequence of the subordination inequality of [66, Lemma 17.1, p. 177] (cf. also [86, Section 6.4, Exercise 11]), a version of Proposition 2.2.36 will be proved (cf. [86, Theorem 1.6.1]). However, under these assumptions, the inequalities describing the range of parameters in (2.149) should be sharp; that is, $2m\alpha - \frac{n}{q} > t - \frac{n}{p}$ or $2m\alpha - \frac{n}{q} > k + \mu$, respectively. We lose also the nice characterization of X^α as the complex interpolation space intermediate between X and $D(A)$.

Case II

Also when X is a Hilbert space description of fractional-order spaces X^α is rather complete. If A is a self-adjoint positive operator in a Hilbert space X, then for $\alpha, \beta \geq 0$ (cf. [173, 1.18.10]):

$$[D(A^\alpha), D(A^\beta)]_\theta = D(A^{\alpha(1-\theta)+\beta\theta}), \quad \theta \in (0,1),$$

that is, Proposition 2.2.34 holds. We remark that in this case A^{it} are *unitary operators*.

Following the results of [81], if $\Omega \subset \mathbb{R}^N$ is a bounded domain with $C^{2m+\varepsilon}$ boundary ($\varepsilon > 0$), the coefficients a_σ of A are ε-Hölder continuous in $\overline{\Omega}$, and coefficients b^j_σ of the boundary operators B_j belong to $C^{2m-m_j+\varepsilon}(\overline{\Omega})$. Then for any $\theta \in (0,1)$, such that $2\theta - \frac{1}{2}$ is not equal m_j for any $j = 1, \dots, m$, we have the equality (cf. [81, Sections 4, 3], also [126, p. 114])

$$(L^2(\Omega), D(A))_{\theta,2} = H^{2m\theta}_{2,\{B_j\}}(\Omega).$$

Moreover, since (cf. [173, Remark 1.19.3])

$$(L^2(\Omega), D(A))_{\theta,2} = [L^2(\Omega), D(A)]_\theta,$$

we have finally the characterization

$$D(A^\theta) = [L^2(\Omega), D(A)]_\theta = H^{2m\theta}_{2,\{B_j\}}(\Omega),$$

where $\theta \in (0,1)$ and $2m\theta - \frac{1}{2}$ is not equal to any m_j (m_j, $j = 1, \dots, m$, denoting, as before, the order of B_j). Again, if $2m\theta - \frac{1}{2} = m_j$ for some $j = 1, \dots, m$, then we have only the proper inclusion $D(A^\theta) \subset H^{2m\theta}_{2,\{B_j\}}(\Omega)$ (see [81, p. 449] for the structure of $(L^2(\Omega), B^{2m}_{2,2,\{B_j\}})_{\theta,2}$ in that case).

Setting in particular $A = \Delta^2 + \rho I$ on $L^2(\Omega)$ ($\rho > 0$, $\partial\Omega \in C^{4+\varepsilon}$) and choosing $D(A) = \{\phi \in H^4(\Omega) : \frac{\partial\phi}{\partial N} = \frac{\partial(\Delta\phi)}{\partial N} = 0 \text{ on } \partial\Omega\}$, we obtain

$$D(A^\theta) = (L^2(\Omega), D(A))_{\theta,2} = H^{4\theta}_{2,\{\frac{\partial}{\partial N}, \frac{\partial(\Delta)}{\partial N}\}}(\Omega), \tag{2.152}$$

except the values $\theta \in \{\frac{1}{4}, \frac{3}{4}\}$.

A variant of the characterization of interpolation space $[L^2(\Omega), D(A)]_\theta$ belongs to [80, Theorem 8.1]. It is assumed there that the boundary $\partial\Omega$ belongs to C^{2m} while the coefficients of B_j are in $C^{2m}(\overline{\Omega})$ (cf. [80] for the precise assumptions).

An extended discussion of the domains of fractional powers of sectorial operators corresponding to second order uniformly strongly elliptic operators in bounded smooth domains $\Omega \subset \mathbb{R}^N$ with Dirichlet or Neumann boundary conditions was also reported in [118] in case of the Hilbert scales.

Case III

We now discuss the result of [140] concerning the second-order elliptic boundary value problem. Consider a sectorial operator on $L^p(\Omega)$, $1 < p < +\infty$, corresponding to

$$Au = -\sum_{i,j=1}^{n} a_{ij}(x)\frac{\partial^2 u}{\partial x_i \partial x_j} + \sum_{i=1}^{n} b_i(x)\frac{\partial u}{\partial x_i} + c(x)u, \tag{2.153}$$

where $\Omega \subset \mathbb{R}^N$ is either a bounded domain with $\partial\Omega \in C^2$ or $\Omega = \mathbb{R}^N$. We set $D(A) = W^{2,p}(\Omega) \cap W_0^{1,p}(\Omega)$ and assume that:

(i) The matrix $[a_{ij}(x)]_{n\times n}$ is real, symmetric, and uniformly elliptic:

$$\exists_{a_0>0}\ \forall_{x\in\overline{\Omega}}\ \forall_{\xi\in\mathbb{R}^N}\ a_0|\xi|^2 \le \sum_{i=1}^{n} a_{ij}(x)\xi_i\xi_j \le a_0^{-1}|\xi|^2.$$

(ii) For some $\alpha \in (0,1)$, $a_{ij} \in C^\alpha(\overline{\Omega})$ and, if $\Omega = \mathbb{R}^N$ the limit $a_{ij}^\infty := \lim_{|x|\to+\infty} a_{ij}(x)$ exists, fulfilling

$$|a_{ij}(x) - a_{ij}^\infty| \le C|x|^{-\alpha} \quad \text{for } x \in \mathbb{R}^N,\ |x| \ge 1.$$

(iii) Coefficients b_i, c admit the representations:

$$b_i = \sum_{j=1}^{N} b_i^j(x), \quad c(x) = \sum_{j=1}^{N} c^j(x), \quad i = 1, \dots, n,$$

with $b_i^j \in L^{r_j}(\Omega)$, $c^j \in L^{s_j}(\Omega)$, $p \le r_j \le +\infty$, $p \le s_j \le +\infty$, $r_j > n$, $s_j > \frac{n}{2}$, and

$$b_i \in L^{r'}(\Omega), \quad c \in L^{s'}(\Omega), \quad i = 1, \dots, n, \quad 1 \le r' < n, \quad 1 \le s' < \frac{n}{2}.$$

Note that for bounded Ω the condition (iii) holds if, in particular, b_i and c are in $L^\infty(\Omega)$.

Under the assumptions (i)–(iii), for sufficiently large $\delta > 0$, the operator $A + \delta I$ considered on $W^{2,p}(\Omega) \cap W_0^{1,p}(\Omega)$, $1 < p < +\infty$, fulfills the estimate [140]:

$$\|A^{it}\|_{\mathcal{L}(L^p(\Omega),L^p(\Omega))} \le Ce^{C' t}, \quad t \in \mathbb{R}, \tag{2.154}$$

so that the characterization condition holds. The extension of the above result can be found in [8]. Also, the recent paper [157] should be mentioned where the case of Neumann boundary condition is considered restricted, however, to the space of functions with zero average.

Finally, imaginary powers of the *Stokes operator* A_q were studied in [76, 77]. In each of the four cases, $\Omega = \mathbb{R}^N$ or Ω a half-space or $\Omega \subset \mathbb{R}^N$ a bounded domain with $\partial\Omega \in C^{2+\mu}$, $0 < \mu < 1$ or Ω an exterior domain with $\partial\Omega \in C^{2+\mu}$ and for any $1 < q < +\infty$, we have an estimate

$$\|A_q^{is}\|_{\mathcal{L}(L^q(\Omega),L^q(\Omega))} \le Ce^{\varepsilon|s|},$$

with $\varepsilon > 0$ and $C = C(\Omega, q, \varepsilon)$ independent of s. Therefore, for any of the above choices of Ω the characterization condition is satisfied for the Stokes operator.

Case IV

At least we recall the notion of an *operator with bounded H^∞ calculus*, due to Alain McIntosh [42]. Let $0 < \omega \le \pi$; then an operator T in a Banach space X is *of type* ω if it is closed and densely defined with

$$\sigma(T) \subset S_\omega \cup \{\infty\},$$

and fulfills an estimate $\|(T - zI)^{-1}\| \le C_\theta \frac{1}{|z|}$ whenever $z \in \mathbb{C} \setminus S_\theta^0$ where $S_\theta^0 = \{z \in \mathbb{C}; z \ne 0, |\arg z| < \theta\}$.

Many analytic functions of such operators (in particular fractional powers) will be introduced through the following construction. Denote by $H(S_\mu^0)$ the space of all holomorphic functions on S_0^μ and let $\psi(\xi) = \frac{\xi}{(1+|\xi|)^2}$. For $0 < \omega < \pi$, consider the following subspace of $H(S_\mu^0)$:

$$H^\infty(S_\mu^0) = \{f : (S_\mu^0) \to \mathbb{C}; f \text{ is analytic and } \|f\|_\infty < \infty\}$$

where $\|f\|_\infty = \sup\{|f(z)|; z \in S_\mu^0\}$. Further, let

$$\Psi(S_\mu^0) = \{f \in H(S_\mu^0); \exists_{s \in \mathbb{R}^+} f\psi^{-s} \in H^\infty(S_\mu^0)\}.$$

For operator T of type ω and analytic function $\psi \in \Psi(S_\mu^0)$, the value $\psi(T) \in \mathcal{L}(X)$ is defined through the contour integral

$$\psi(T) = \frac{1}{2\pi i} \int_\Gamma (T - \zeta I)^{-1} \psi(\zeta) d\zeta, \tag{2.155}$$

where the contour Γ is defined by the function

$$g(t) = \begin{cases} -te^{-i\theta}, & -\infty < t \le 0, \\ te^{i\theta}, & 0 \le t < \infty. \end{cases}$$

Next, for $f \in H^\infty(S_\mu^0)$ we set

$$f(T) = (\psi(T))^{-1} (f\psi)(T), \quad \text{where } \psi(\zeta) = \frac{\zeta}{(1 + |\zeta|)^2}.$$

It was shown in [133] that the following two properties are equivalent for an operator T defined on a Hilbert space:

1. $\|f(T)\| \le C\|f\|_\infty$ for all $f \in H^\infty(S_\mu^0)$,
2. $\{T^{iy}; y \in \mathbb{R}\}$ is a C^0 group and $\|T^{iy}\| \le C_\mu e^{\mu|y|}$, μ and C_μ are positive constants.

The second property coincides with boundedness of the purely imaginary powers. In the case of operators defined on reflexive Banach spaces, the property 1 is stronger than 2 (cf. [187, Theorem 16.5]) and the characterization-condition is satisfied. In particular, elliptic operators satisfy property 1 in $L^p(\Omega)$, $1 < p < \infty$, spaces. We refer to [187, Chapter 16] for extended characterization of the domains of fractional powers of elliptic operators in $L^p(\Omega)$, in case of bounded domains with Lipschitz or C^2 boundary.

Domains of fractional powers of the perturbed operator

Let us finally come back to Proposition 2.2.11 in which perturbation result for sectorial operators was stated. Having shown that the perturbed operator is sectorial, it is important to observe how the perturbation influences the domains of its fractional powers. For perturbations described in Proposition 2.2.11, it is easy to conclude that the corresponding fractional power spaces remain unchanged.

Corollary 2.2.40. *Under the assumptions of Proposition* 2.2.11 *and additional requirements that both A and $A + B$ are positive, the following equality holds:*

$$D((A + B)^\alpha) = D(A^\alpha), \quad \alpha \in (0, 1). \tag{2.156}$$

Proof. Since $D(A) \subset D(B)$, we have $D(A + B) = D(A)$. Using then Corollary 2.2.35, we obtain

$$D(A^\alpha) = [X, D(A)]_\alpha = [X, D(A + B)]_\alpha = D((A + B)^\alpha), \quad \alpha \in (0, 1).$$

The proof is complete. □

2.3 Elements of stability theory

We include here also, following the monograph [32], elements of the studies of asymptotic behavior of solutions for nonlinear partial differential equations. Recent developments in the theory of dynamical systems show the unquestionable advantage of treating such solutions as an abstract flow on an appropriately selected phase space (e. g., [13], [32], [83], [86], [172], [187]).

In this section, we will recall some contemporary ideas of the theory of dynamical systems. Our purpose here is to lay a foundation for analyzing and describing the long time dynamics of infinite dimensional differential equations.

2.3.1 Strongly continuous semigroups and stability of sets

We begin with the notion, basic for all our investigations, of a C^0-*semigroup* (*strongly continuous semigroup*).

Definition 2.3.1. Let V be a metric space. A one parameter family $\{T(t)\}$ of maps $T(t) : V \to V, t \geq 0$, is called a C^0-semigroup if:
- T(0) is the identity map on V;
- $T(t + s) = T(t)T(s)$ for all $t, s \geq 0$;
- the function

$$[0, \infty) \times V \ni (t, x) \to T(t)x \in V$$

is continuous at each point $(t, x) \in [0, \infty) \times V$.

It is known that for the semigroups of bounded linear operators in a Banach space X the condition (iii) holds if and only if, at any element $x \in X$,

$$T(t)x \to x \quad \text{when } t \to 0^+,$$

which is basically a consequence of the Banach *uniform boundedness property.*

For purposes of convenience, we shall forthwith introduce several concepts closely related to Definition 2.3.1 above. These concepts will frequently appear in our further considerations.

- The semigroup $\{T(t)\}$ is said to be *compact* if $T(t) : V \to V$ is a compact map for each $t > 0$, that is, each $T(t)$ takes bounded sets into precompact sets.
- The semigroup $\{T(t)\}$ is called *completely continuous* if it is compact and if for each bounded set $B \subset V$ and each number $t > 0$ the union $\bigcup_{s\in[0,t]} T(t)B$ is bounded in V.
- Let W_1, W_2 be two subsets of V. We say that W_2 is $\{T(t)\}$*-attracted* by W_1 if

  $$d(T(t)W_2, W_1) \to 0 \quad \text{as } t \to +\infty,$$

  where, for each $t \geq 0$,

  $$d(T(t)W_2, W_1) := \sup_{w_2\in T(t)W_2} \inf_{w_1\in W_1} \operatorname{dist}_V(w_2, w_1).$$

Of course, it is true that the quantity $d(T(t)W_2, W_1)$ enters into the construction of the Hausdorff distance between the two sets $T(t)W_2$ and W_1. This is a matter we will not pursue here. However, we note that, roughly speaking, the number $d(T(t)W_2, W_1)$ measures how much the set $T(t)W_2$ lies outside the set W_1.

- Given any two subsets $W_1, W_2 \subset V$, we say that W_1 *absorbs* W_2 under $\{T(t)\}$ if there exists a number $t_0 \geq 0$ such that $T(t)W_2 \subset W_1$ for all $t \geq t_0$. Notice that W_1 attracts W_2 if and only if each open neighborhood $\mathcal{N}_{W_1}$ of W_1 in V absorbs W_2.
- An element $v \in V$ is called an *equilibrium point* for $\{T(t)\}$ if $T(t)v = v$ for all $t \geq 0$. Extending this notion, we say that a set $\mathcal{A} \subset V$ is $\{T(t)\}$*-invariant* if $T(t)\mathcal{A} = \mathcal{A}$ for all $t \geq 0$. Also, we will call $\mathcal{A} \subset V$ *positively* $\{T(t)\}$*-invariant* if $T(t)\mathcal{A} \subset \mathcal{A}$ for all $t \geq 0$.
- For any set $B \subset V$, the two sets $y^+(B)$ and $\omega(B)$ defined by

  $$y^+(B) := \bigcup_{t\geq 0} T(t)B$$
  $$\omega(B) := \bigcap_{s\geq 0} \operatorname{cl}_V \bigcup_{t\geq s} T(t)B$$

  are called, respectively, the *positive orbit* and the *ω-limit set* of B. Thus, $\omega(B)$ consists of all points $v \in V$ for which there exist positive numbers $t_n \nearrow +\infty$ and points $v_n \in B$ with $T(t_n)v_n \to v$.

- For any point $v \in V$, we let S_v^- denote the set of all functions $\phi : (-\infty, 0] \to V$ such that $\phi(0) = v$ and such that $T(t)\phi(s) = \phi(t+s)$ whenever $-\infty < s \le -t \le 0$. Here, we allow the possibility that S_v^- is empty, that S_v^- consists of exactly one element ϕ, or that S_v^- consists of more than one element ϕ. By a *negative orbit* through a given point $v \in V$, we mean any set

$$\gamma_\phi^-(v) = \bigcup_{t \ge 0} \{\phi(-t)\},$$

where $\phi \in S_v^-$. Of course, there may or may not exist a nonempty negative orbit.
- For each point $v \in V$, a *complete orbit* through v is any set

$$\gamma_\phi(v) := \gamma^+(v) \cup \gamma_\phi^-(v),$$

where $\phi \in S_v^-$. Since we allow a negative orbit to be empty, we remark that a complete orbit $\gamma_\phi(v)$ is invariant if and only if its component $\gamma_\phi^-(v)$ is nonempty.

Definitions describing *stability properties* of invariant sets will be given next.

Definition 2.3.2. Let $\mathcal{A} \subset V$ be nonempty and $\{T(t)\}$-invariant. We say that:
- $\mathcal{A}$ is stable if and only if for each open neighborhood U of $\mathcal{A}$ there exists an open neighborhood W of $\mathcal{A}$ such that $T(t)W \subset U$ for all $t \ge 0$;
- $\mathcal{A}$ is asymptotically stable if and only if $\mathcal{A}$ is stable and attracts each point lying in some open neighborhood of $\mathcal{A}$;
- $\mathcal{A}$ is uniformly asymptotically stable if and only if $\mathcal{A}$ is stable and attracts some open neighborhood of itself.

When $\mathcal{A}$ is compact, in order to conclude that $\mathcal{A}$ is uniformly asymptotically stable, it suffices to show that $\mathcal{A}$ attracts some one of its open neighborhoods. More precisely, we have the following.

Remark 2.3.3. Let $\{T(t)\}$ be a C^0-semigroup in a metric space V. If $\mathcal{A}$ is compact and $\{T(t)\}$-invariant and if $\mathcal{A}$ attracts at least one of its own open neighborhoods, then $\mathcal{A}$ is stable.

Proof. For the sake of argument, suppose that $\mathcal{A}$ is not stable. Then there exists a neighborhood $\mathcal{N}_\mathcal{A}$ of $\mathcal{A}$ having the property that to each neighborhood W of $\mathcal{A}$ there corresponds a point $w \in W$ and a number $t_w \ge 0$ such that $T(t_w)w \notin \mathcal{N}_\mathcal{A}$. From this and from the compactness of $\mathcal{A}$, there follows the existence of two sequences $t_n \to t_0 \in [0, +\infty]$ and $w_n \to w_0 \in \mathcal{A}$ such that

$$T(t_n)w_n \notin \mathcal{N}_\mathcal{A} \quad \text{for all } n \in \mathbb{N}. \tag{2.157}$$

In the case that $t_0 = +\infty$, (2.157) implies that there is no integer $n_0 > 0$ such that the bounded set $\{w_n, n \ge n_0\}$ is attracted to $\mathcal{A}$. This is absurd.

In the case that $t_0 < +\infty$, we have $T(t_n)w_n \to T(t_0)w_0$ and, since $\mathcal{A}$ is invariant, we must conclude that $T(t_0)w_0 \in \mathcal{A}$. Hence, the neighborhood $\mathcal{N}_\mathcal{A}$ contains all but finitely many elements of the set $\{T(t_n)w_n\}$. In view of (2.157), this also is absurd.

Thus in fact, $\mathcal{A}$ is stable. □

Compact, uniformly asymptotically stable subsets of the *phase space* are usually called *local attractors*. That is, we have the following.

Definition 2.3.4. A set $\mathcal{A} \subset V$ is called a local attractor for the semigroup $\{T(t)\}$ on V if and only if:
- $\mathcal{A}$ is nonempty, compact, and invariant with respect to $\{T(t)\}$;
- $\mathcal{A}$ attracts some open neighborhood $\mathcal{N}_\mathcal{A}$ of $\mathcal{A}$.

For many infinite dimensional problems, the notion of local attractor is, from a certain point of view, inadequate. The notion of global attractor involves stability properties stronger than those formulated in Definition 2.3.4. Specifically, we have the following.

Definition 2.3.5. By a global attractor for $\{T(t)\}$, we mean a nonempty, compact, $\{T(t)\}$-invariant set $\mathcal{A} \subset V$, which attracts every bounded subset of V.

Remark 2.3.6. The global attractor, if it exists, is unique and also is maximal in the class of bounded invariant subsets of V.

Proof. Indeed, let $\mathcal{A}$, $\mathcal{A}_1$ be compact invariant subsets of the phase space V, and suppose that $\mathcal{A}$, $\mathcal{A}_1$ each attract all the bounded subsets of V. Then, keeping in mind that $\mathcal{A}$, $\mathcal{A}_1$ are both bounded, we have

$$d(\mathcal{A}, \mathcal{A}_1) = d(T(t)\mathcal{A}, \mathcal{A}_1) \to 0 \quad \text{as } t \to +\infty,$$
$$d(\mathcal{A}_1, \mathcal{A}) = d(T(t)\mathcal{A}_1, \mathcal{A}) \to 0 \quad \text{as } t \to +\infty.$$

Consequently,

$$d(\mathcal{A}, \mathcal{A}_1) = d(\mathcal{A}_1, \mathcal{A}) = 0.$$

But both $\mathcal{A}$, $\mathcal{A}_1$ are closed. Therefore, $\mathcal{A} = \mathcal{A}_1$. One also can see that each bounded invariant subset of V must be contained in the global attractor. □

Remark 2.3.7. The global attractor $\mathcal{A}$ is minimal in the class of all those closed bounded sets B in V which attract bounded sets.

Proof. Indeed, consider any closed bounded set $B \subset V$ which attracts all bounded subsets of V. Then, arguing in a manner similar to the proof above for Observation 2.3.6, we obtain $d(\mathcal{A}, B) = 0$. From this, it follows that $\mathcal{A} \subset B$. □

Before formulating our next observation, we want to recall the notion of connectedness. Two sets $K, L \subset V$ are said to be separated if and only if

$$L \cap \mathrm{cl}_V K = K \cap \mathrm{cl}_V L = \emptyset. \tag{2.158}$$

A set $S \subset V$ is called *connected* if and only if S cannot be decomposed into two separated sets K, L.

Remark 2.3.8. The global attractor $\mathcal{A}$ is connected if and only if there exists a connected bounded set $B \subset V$ such that $\mathcal{A} \subset \mathcal{B}$.

Proof. If $\mathcal{A}$ is connected, then taking $B = \mathcal{A}$, we trivially have a bounded connected set $B \subset V$ such that $\mathcal{A} \subset B$.

The converse assertion requires more argumentation. Specifically, suppose that B is a bounded connected subset V which contains $\mathcal{A}$. From the continuity of $T(t)$, it follows that the image $T(t)B$, $t \geq 0$, is connected.

By hypothesis, $\mathcal{A}$ is the global attractor for $\{T(t)\}$. Hence, $\mathcal{A}$ attracts every bounded set in V. It follows that, for each open neighborhood $\mathcal{N}_\mathcal{A}$ of $\mathcal{A}$ there exists a number $t_{\mathcal{N}_\mathcal{A}} > 0$ such that

$$\mathcal{A} \subset T(t)B \subset \mathcal{N}_\mathcal{A} \quad \text{for all } t \geq t_{\mathcal{N}_\mathcal{A}}. \tag{2.159}$$

For the sake of argument, suppose that $\mathcal{A}$ is not connected, that is, $\mathcal{A} = K \cup L$ where $K, L \subset V$ are nonempty and separated. We known that $\mathcal{A}$ is closed and we know that K, L satisfy (2.158). Hence, the sets K, L must each be closed. Thus, $\mathcal{A}$ decomposes into a sum of two disjoint nonempty closed sets K, L.

Since K, L are disjoint and closed, there must exist two open sets $\mathcal{U}_K, \mathcal{U}_L \subset V$ such that

$$\mathcal{U}_K \cap \mathcal{U}_L = \emptyset, \quad K \subset \mathcal{U}_K,\ L \subset \mathcal{U}_L. \tag{2.160}$$

Let $\mathcal{N}_\mathcal{A}$ be the particular open neighborhood of $\mathcal{A}$ given by $\mathcal{N}_\mathcal{A} := \mathcal{U}_K, \mathcal{U}_L$. We know that $T(t)B$ is connected and that $\mathcal{U}_K$, $\mathcal{U}_L$ are separated. Therefore, with the aid of (2.159) we have either $\mathcal{A} = K \cup L \subset T(t)B \subset \mathcal{U}_K$ or $\mathcal{A} = K \cup L \subset T(t)B \subset \mathcal{U}_L$. But this last statement contradicts (2.160). The proof is complete. □

We remark that, if a metric space V is connected, then the global attractor $\mathcal{A}$ for a C^0-semigroup $T(t) : V \to V$, $t \geq 0$ is connected. However, this same statement is not necessarily true for discrete semigroups (cf. [76]).

Our goal now is to establish conditions guaranteeing the *existence* of the global attractor $\mathcal{A}$. To that end, we will introduce a class of semigroups having the property that, for each compact invariant set $B \subset V$, asymptotic stability is equivalent to uniform asymptotic stability.

2.3.2 Existence of a global attractor

In that which follows, we will set forth conditions,formulated in [83], which guarantee the existence of a global attractor. These conditions relate to the notions of *dissipativeness* and *asymptotic smoothness* for $\{T(t)\}$.

Definition 2.3.9. The semigroup $\{T(t)\}$ is called point dissipative if and only if there exists a nonempty, bounded set $B \subset V$ which attracts every point in V. The semigroup $\{T(t)\}$ is called bounded dissipative if and only if there exists a nonempty, bounded set $B \subset V$ which attracts every bounded subset of V.

Definition 2.3.10. The semigroup $\{T(t)\}$ is called asymptotically smooth if and only if each nonempty, closed, bounded, positively invariant set $W \subset V$ contains a nonempty, compact subset C which attracts W.

Clearly, if the semigroup $\{T(t)\}$ has a global attractor $\mathcal{A}$ in V, then $\{T(t)\}$ must be dissipative in the sense of Definition 2.3.9. Furthermore, under the same conditions, $\{T(t)\}$ is of necessity asymptotically smooth. Indeed, consider any nonempty, closed, bounded set $W \subset V$ such that

$$T(t)W \subset W \quad \text{for } t \geq 0. \tag{2.161}$$

Certainly, $\mathcal{A} \cap W$ is a closed subset of the compact set $\mathcal{A}$. Hence, $\mathcal{A} \cap W$ is compact. Moreover, since $\mathcal{A}$ attracts bounded sets, condition (2.161) ensures that $\mathcal{A} \cap W$ is nonempty and attracts W. Thus, as asserted, $\{T(t)\}$ is asymptotically smooth. We summarize these remarks as follows.

Remark 2.3.11. If $\{T(t)\}$ is a C^0-semigroup on a metric space V and if $\{T(t)\}$ has a global attractor $\mathcal{A}$, then $\{T(t)\}$ is bounded dissipative and asymptotically smooth.

At this moment, we want to establish several important properties of *ω-limit sets for bounded sets* in the case that the semigroup $\{T(t)\}$ is asymptotically smooth (cf. [83, Lemma 3.2.1]).

Proposition 2.3.12. *Let $\{T(t)\}$ be a C^0-semigroup acting on a metric space V. If $\{T(t)\}$ is asymptotically smooth, if B is a nonempty subset of V, and if, for some number $t_B \geq 0$, the set*

$$\bigcup_{s \geq t_B} T(s)B$$

is bounded, then $\omega(B)$ is nonempty, compact, and invariant. Furthermore, $\omega(B)$ attracts B.

Proof. We start with the proof that $\omega(B)$ is compact. Since $\bigcup_{s\geq t_B} T(s)B$ is positively invariant and the maps $T(t)$ $(t \geq 0)$ are continuous, we have

$$T(t)\mathrm{cl}_V \bigcup_{s\geq t_B} T(s)B \subset \mathrm{cl}_V \bigcup_{s\geq t_B} T(s)B, \quad t \geq 0.$$

Since the semigroup is asymptotically smooth, there exists a nonempty, compact set $C \subset \mathrm{cl}_V \bigcup_{s \geq t_B} T(s)B$ attracting $\mathrm{cl}_V \bigcup_{s \geq t_B} T(s)B$. Hence C attracts B and, therefore,

$$\forall_{\substack{\mathcal{N}_C\text{-bounded, open}\\ \text{neighborhood of } C}} \exists_{\substack{\mathcal{N}_C'\text{-bounded, open}\\ \text{neighborhood of } C}} \exists_{t_{\mathcal{N}_C} \geq 0} \quad \mathrm{cl}_V \bigcup_{t \geq t_{\mathcal{N}_C}} T(t)B \subset \mathrm{cl}_V \mathcal{N}_C' \subset \mathcal{N}_C. \tag{2.162}$$

Condition (2.162) ensures that $\omega(B)$ is contained in each open neighborhood of C which (since C is closed) implies the inclusion $\omega(B) \subset C$. Furthermore, using (2.162) and the compactness of C, one may show the existence of sequences $t_n \nearrow +\infty$ and $\{v_n\} \subset B$ for which $\{T(t_n)v_n\}$ is convergent. The set $\omega(B)$ is thus nonempty and, since $\omega(B)$ is closed and C is compact, $\omega(B)$ must be compact.

Next, we shall prove that

$$d(T(t)B, \omega(B)) \to 0 \quad \text{when } t \to +\infty. \tag{2.163}$$

Suppose that (2.163) is violated, that is, there exist $\varepsilon > 0$, a sequence $\{t_n\}$ increasing to infinity and a sequence $\{v_n\} \subset B$ such that

$$\inf_{y \in \omega(B)} \mathrm{dist}_V(T(t_n)v_n, y) > \varepsilon, \quad n \in \mathbb{N}. \tag{2.164}$$

Then $\{T(t_n)v_n\}$ can not have a convergent subsequence; otherwise, such a limit point would belong to $\omega(B)$ and (2.164) would not be true. However, from (2.162) it follows that for any bounded, open neighborhood $\mathcal{N}_C$ of C almost all elements of $\{T(t_n)y_n\}$ are contained in $\mathcal{N}_C$. Since C is compact, this allows one to choose from $\{T(t_n)y_n\}$ a convergent subsequence, which contradicts (2.164).

To prove that $\omega(B)$ is invariant, note first from the definition of $\omega(B)$ and continuity of $T(t)$ that the set $\omega(B)$ is positively invariant. Consider further any point $v_0 \in \omega(B)$. Then $T(t_n)v_n \to v_0$ in V for some sequences $t_n \nearrow +\infty$ and $\{v_n\} \subset B$. Fix $t \geq 0$ and define

$$w_n := T(t_n - t)v_n, \quad n \geq n_t.$$

Certainly

$$T(t)w_n \to v_0, \tag{2.165}$$

whereas, by (2.163) and the compactness of $\omega(B)$, there exists a subsequence $\{w_{n'}\}$ of $\{w_n\}$ convergent to some $w_0 \in \omega(B)$. This and (2.165) ensure that $T(t)w_0 = v_0$ which proves the required inclusion $\omega(B) \subset T(t)(\omega(B))$. The proof is complete. □

Corollary 2.3.13 (cf. [113, Proposition 2.2]). *Let $\{T(t)\}$ be a C^0-semigroup on a metric space V, and suppose that $\{T(t)\}$ has a global attractor $\mathcal{A}$. Then:*
- *$\mathcal{A}$ is the union of the ω-limit sets of all bounded subsets of V,*
- *$\mathcal{A}$ is the union of the ω-limit sets of all compact subsets of V,*

- *$\mathcal{A}$ is the union of all bounded, invariant complete orbits through $v \in V$,*
- *$\mathcal{A}$ is the union of all precompact, invariant complete orbits through $v \in V$.*

Proof. First, note from Observation 2.3.11 that $\{T(t)\}$ must be asymptotically smooth. Since $\mathcal{A}$ attracts bounded sets, for each bounded set $B \subset V$ there is a $t_B \geq 0$ such that $T(t_B)(\gamma^+(B))$ is bounded. Therefore, $\omega(B)$ is compact, invariant, and consequently, $\omega(B) \subset \mathcal{A}$. Moreover, $\omega(\mathcal{A}) = \mathcal{A}$, and that completes the proof of (i). Similar reasoning shows (ii). To prove (iii) (and (iv)), note first that, for each $v \in \mathcal{A}$, a negative orbit through v is nonempty. Hence, for each $v \in \mathcal{A}$ there exists invariant, bounded, and precompact complete orbit through v. Therefore, $\mathcal{A}$ is contained in the union of such orbits. Obviously, each bounded, invariant complete orbit must lie within $\mathcal{A}$. Thus (iii) (and (iv)) are established. □

In Definition 2.3.10 above, we introduced the notion of asymptotically smooth semigroup, and thus, in effect, singled out such semigroups for special attention. Our motive for doing this is embodied in the next observation.

Remark 2.3.14. Let $\{T(t)\}$ be a C^0-semigroup in a metric space V and let $\mathcal{A} \subset V$ be a nonempty, compact, invariant set. If $\{T(t)\}$ is asymptotically smooth, then the following two conditions are equivalent:
(a) $\mathcal{A}$ is asymptotically stable;
(b) $\mathcal{A}$ is uniformly asymptotically stable.

Proof. From Definition 2.3.2, it trivially follows that $(b) \rightarrow (a)$. Consequently, we need only prove that $(a) \longrightarrow (b)$.

By hypothesis, $\mathcal{A}$ is asymptotically stable. Hence, there exists at least one bounded open neighborhood U of $\mathcal{A}$ such that $\mathcal{A}$ attracts points of U. Consider another open neighborhood U' of $\mathcal{A}$ such that $\mathrm{cl}_V U' \subset \mathcal{U}$. Since $\mathcal{A}$ is stable, there must exist an open neighborhood $\mathcal{U}''$ of $\mathcal{A}$ such that

$$W := \mathrm{cl}_V \bigcup_{t\geq 0} T(t)U'' \subset \mathrm{cl}U' \subset U.$$

Clearly, W is nonempty, closed, bounded, and positively invariant. Therefore, $\omega(W) \subset W$ and, by virtue of Proposition 2.3.12, $\omega(W)$ is compact and invariant. Also, $\omega(W)$ attracts W. Hence, in particular,

$$\omega(W) \text{ attracts } U''. \tag{2.166}$$

Also, since $\mathcal{A}$ is a compact, invariant subset of U'',

$$\mathcal{A} \subset \omega(W). \tag{2.167}$$

It is clear from the preceding considerations that $\mathcal{A}$ attracts points of $\omega(W)$. Thus, the continuity of $\{T(t)\}$ and the stability of $\mathcal{A}$ yield the statement that, for each open neighborhood $\mathcal{N}_{\mathcal{A}}$ of $\mathcal{A}$,

$$\exists_{\substack{\mathcal{N}'_{\mathcal{A}}\text{-an open}\\ \text{neighborhood of } \mathcal{A}}} \forall_{v\in\omega(W)} \exists_{t_v\geq 0} \exists_{\varepsilon_v>0}$$

$$T(t_v)B_V(v,\varepsilon_v) \subset \mathcal{N}'_{\mathcal{A}} \subset \gamma^+(\mathcal{N}'_{\mathcal{A}}) \subset \mathcal{N}_{\mathcal{A}}, \tag{2.168}$$

where $B_V(v,\varepsilon_v)$ denotes the open ball in V centered at v and having radius ε_v. Let $\mathcal{N}_{\mathcal{A}}$ be fixed. From the compactness of $\omega(W)$, there follows

$$\exists_{k\in\mathbb{N}} \exists_{v_1,\dots,v_k\in\omega(W)}\ \omega(W) \subset B_V(v_1,\varepsilon_{v_1}) \cup \cdots \cup B_V(v_k,\varepsilon_{v_k}). \tag{2.169}$$

Introduce

$$t_{\max} := \max\{t_{v_1},\dots,t_{v_k}\}.$$

The assertions (2.169), (2.168), and the invariance of $\omega(W)$ imply

$$\omega(W) = T(t_{\max})(\omega(W)) \subset T(t_{\max})\left(\bigcup_{j=1}^{k} B_V(v_j,\varepsilon_{v_j})\right) = \bigcup_{j=1}^{k} T(t_{\max})B_V(v_j,\varepsilon_{v_j})$$

$$= \bigcup_{j=1}^{k} T(t_{\max}-t_{v_j})T(t_{v_j})B_V(v_j,\varepsilon_{v_j}) \subset \gamma^+(\mathcal{N}'_{\mathcal{A}}) \subset \mathcal{N}_{\mathcal{A}}.$$

All of this implies that $\omega(W)$ is contained in each open neighborhood of $\mathcal{A}$. Since $\mathcal{A}$ is compact, we obtain $\omega(W) \subset \mathcal{A}$ which, by virtue of (2.167), leads us to the equality $\omega(W) = \mathcal{A}$. Taking into account (2.166), we see that $\mathcal{A}$ attracts its open neighborhood U''. Thus, $\mathcal{A}$ is uniformly asymptotically stable and the proof is complete. □

The theorem stated below, due to J. K. Hale [83], gives sufficient conditions for the existence of a global attractor.

Theorem 2.3.15. *Let $\{T(t)\}$ be a C^0-semigroup on a metric space V. If $\{T(t)\}$ is point dissipative, asymptotically smooth, and keeps orbits of bounded sets bounded, then $\{T(t)\}$ has a global attractor in V.*

Proof. The proof occurs in two steps.

Step 1. We shall first show the existence of a bounded set $\mathcal{O} \subset V$ such that each compact set $C \subset V$ has an open neighborhood $\mathcal{N}_C$ which is absorbed by $\mathcal{O}$.

By hypothesis, there exists a nonempty bounded set $W_0 \subset V$ which attracts points of V. Let $\mathcal{N}_{W_0}$ be any bounded open neighborhood of W_0. Using the continuity of $\{T(t)\}$ and the fact that $\mathcal{N}_{W_0}$ absorbs points of V, we conclude that

$$\forall_{v\in V} \exists_{\tau_v\geq 0} \exists_{B_V(v,\varepsilon_v)}\ T(\tau_v)B_V(v,\varepsilon_v) \subset \mathcal{N}_{W_0}, \tag{2.170}$$

where $B_V(v,\varepsilon_v) \subset V$ denotes the open ball in V centered at v and having radius ε_v. Next, choose $t_{\mathcal{N}_{W_0}} \geq 0$ such that

$$\mathcal{O} := \bigcup_{t\geq t_{\mathcal{N}_{W_0}}} T(t)(\mathcal{N}_{W_0})$$

is bounded. By our assumptions, $\mathcal{O}$ is positively invariant and absorbs points of V. Moreover, from (2.170) we observe that

$$\forall_{v\in V}\ \exists_{t_v:=\tau_v+t_{\mathcal{N}_{W_0}}\geq 0}\ \exists_{B_V(v,\varepsilon_v)}\ T(t)B_V(v,\varepsilon_v)\subset\mathcal{O}\quad\text{for } t\geq t_v. \tag{2.171}$$

Consider now any compact set $C\subset V$. Certainly $C\subset\bigcup_{v\in C}B_V(v,\varepsilon_v)$ so that, from the compactness of C,

$$C\subset B_V(v_1,\varepsilon_{v_1})\cup\cdots\cup B_V(v_k,\varepsilon_{v_k})=:\mathcal{N}_C \tag{2.172}$$

for some $k\in\mathbb{N}$ and $v_1,\dots,v_k\in C$. With the aid of (2.171), (2.172), we obtain

$$T(t)C\subset T(t)\mathcal{N}_C=\bigcup_{j=1}^{k}T(t)(B_V(v_j,\varepsilon_{v_j}))\subset\mathcal{O}\quad\text{for } t\geq\max\{t_{v_1},\dots,t_{v_k}\}.$$

Step 2. We shall now construct a compact invariant set $\mathcal{A}$ which attracts each bounded subset of V.

Let $B\subset V$ be a bounded set. By our assumptions, Proposition 2.3.12 ensures that $\omega(B)$ is compact and attracts B, that is,

$$\forall_{\substack{\mathcal{N}_{\omega(B)}\ -\ \text{an open}\\ \text{neighborhood of }\omega(B)}}\ \exists_{t_B\geq 0}\ T(t)B\subset\mathcal{N}_{\omega(B)}\quad\text{for } t\geq t_B. \tag{2.173}$$

However, as shown in Step 1, there exists some open neighborhood $\mathcal{N}_{\omega(B)}$ of $\omega(B)$ absorbed by $\mathcal{O}$. From this result and condition (2.173), we obtain

$$\forall_{\substack{B\subset V\\ B\text{-bounded}}}\ \exists_{\tau_B\geq 0}\ T(t)B\subset\mathcal{O}\quad\text{for } t\geq\tau_B. \tag{2.174}$$

Let $\mathcal{A}:=\omega(\mathcal{O})$. Using again Proposition 2.3.12, we find that $\mathcal{A}$ is compact, is invariant and attracts $\mathcal{O}$. Furthermore, $\mathcal{A}$ attracts bounded subsets of V since, as shown in (2.174), bounded sets are absorbed by $\mathcal{O}$. The proof is complete. □

Remark 2.3.16. As a result of Theorem 2.3.15 and Observation 2.3.11, we conclude that *if* $\{T(t)\}$ is a C^0*-semigroup on a metric space* V *and if, under* $\{T(t)\}$*, the orbits of bounded sets are bounded, then* $\{T(t)\}$ *has a global attractor in* V *if and only if* $\{T(t)\}$ *is point dissipative and asymptotically smooth.* In general, the boundedness of orbits for bounded sets is not necessary for the existence of a global attractor, although this is true for a large number of systems (e. g., compact semigroups corresponding to sectorial equations discussed in Chapter 3 have this property). From the point of view of Theorem 2.3.15, such an assumption may be weakened since, what we actually used in the proof was the property that

$$\forall_{\substack{B\subset V\\ B\text{-bounded}}}\ \exists_{t_B\geq 0}\ \bigcup_{t\geq t_B}T(t)B\text{ is bounded in V.} \tag{2.175}$$

On the basis of this remark, we have (cf. [113, Theorem 3.4]) the following.

Corollary 2.3.17. *A C^0-semigroup $\{T(t)\}$ on a metric space V has a global attractor if and only if $\{T(t)\}$ is point dissipative, is asymptotically smooth, and satisfies the condition* (2.175).

Remark 2.3.18. Looking again at the proof of Theorem 2.3.15 let us also note, that instead of the point dissipativeness of $\{T(t)\}$ we might assume a weaker condition:

$$\underset{\substack{B\subset V\\ B\text{-bounded}}}{\exists}\ \forall_{u_0\in V}\ \exists_{t_{u_0}\geq 0}\ T(t_{u_0})u_0 \in B \tag{2.176}$$

(i. e., $\gamma^+(u_0)\cap B \neq \leftarrow$ for each $u_0 \in V$). Corollary 2.3.17 then becomes the following.

Corollary 2.3.19. *A C^0-semigroup on a metric space V has a global attractor if and only if it is asymptotically smooth and satisfies the conditions* (2.175) *and* (2.176).

2.3.3 The Lyapunov function

If it exists, the Lyapunov function provides an order in the set of trajectories. In particular, it proves to be very useful in the study of long time behavior for solutions of evolutionary equations. There are several distinct definitions of the Lyapunov function. The following definition is taken from [83].

Definition 2.3.20. Let $\{T(t)\}$ be a C^0-semigroup in a metric space V. A functional $\mathcal{L}: V \to \mathbb{R}$ is said to be a **Lyapunov function** for $\{T(t)\}$ if:
- $\mathcal{L}$ is continuous on V and bounded below,
- For each $v \in V$ function, $(0, +\infty) \ni t \to \mathcal{L}(T(t)v) \in \mathbb{R}$ is nonincreasing,
- For any $v \in V$, we have:

$$\mathcal{L}(T(t)v) = \text{const.} \quad \text{for all } t \geq 0 \text{ implies that } T(t)v = v \text{ for all } t \geq 0.$$

Proposition 2.3.21. *Let $\{T(t)\}$ be an asymptotically smooth C^0-semigroup on a metric space V and suppose that $\{T(t)\}$ admits a Lyapunov function $\mathcal{L}$. Also, suppose that for each point $v \in V$ the orbit $\gamma^+(v)$ is bounded in V. Then, for each point $v \in V$ the corresponding ω-limit set $\omega(v)$ lies within the set $\mathcal{E}$ of all equilibria for $\{T(t)\}$;*

$$\mathcal{E} := \{w \in V;\ T(t)w = w \quad \text{for all } t \geq 0\}. \tag{2.177}$$

If, moreover, the set $\mathcal{E}$ is bounded in V, then $\{T(t)\}$ is point dissipative in V.

Proof. Let $\mathcal{L}$ be a Lyapunov function for $\{T(t)\}$. Based on Proposition 2.3.12, our assumptions ensure that the ω-limit set $\omega(v)$ of any element $v \in V$ is nonempty and attracts v. For fixed $v \in V$, let us choose any $y \in \omega(v)$ and a sequence $t_n \to +\infty$, that

$$\lim_{n\to\infty} T(t_n)v = y.$$

Since $\mathcal{L}$ is bounded below then, for some $\beta \in \mathbb{R}$,

$$\inf_{t\geq 1} \mathcal{L}(T(t)v) = \beta. \tag{2.178}$$

Using the above condition and the property that $\mathcal{L}(T(t)v)$ is a nonincreasing function of t, we justify the existence of a limit

$$\lim_{t\to+\infty} \mathcal{L}(T(t)v) = \beta.$$

From the continuity of $\mathcal{L}$ and the semigroup properties of $\{T(t)\}$, we also obtain

$$\begin{aligned}\mathcal{L}(T(t)y) &= \mathcal{L}\Big(T(t)\lim_{n\to\infty} T(t_n)v\Big) = \lim_{n\to\infty} \mathcal{L}(T(t)T(t_n)v) \\ &= \lim_{n\to\infty} \mathcal{L}(T(t+t_n)v) = \beta \quad \text{for each } t \geq 0.\end{aligned}$$

This proves that y is an equilibrium point and, therefore, the union $\bigcup_{v\in V} \omega(v)$ is contained in $\mathcal{E}$. Recalling that $\bigcup_{v\in V} \omega(v)$ attracts points of V, it is now clear that $\{T(t)\}$ is point dissipative whenever $\mathcal{E}$ is bounded in V. The proof is complete. □

The following result is an immediate consequence of Proposition 2.3.21 and Corollary 2.3.17.

Corollary 2.3.22. *Let $\{T(t)\}$ be an asymptotically smooth C^0-semigroup on a metric space V, and suppose that $\{T(t)\}$ satisfies the condition* (2.175). *Furthermore, suppose that all the equilibria of $\{T(t)\}$ lie in a bounded subset of V. If $\{T(t)\}$ has a Lyapunov function $\mathcal{L}$, then $\{T(t)\}$ has a global attractor.*

2.3.4 Compact semigroups

An important class of semigroups appearing in applications, especially when those applications involve parabolic problems, is the class of compact semigroups. To show that Theorem 2.3.15 is applicable to the study of such semigroups, we shall first prove that compact semigroups are asymptotically smooth and fulfill the requirement (2.175) (cf. [84, Theorem 1.2]).

Lemma 2.3.23. *If $\{T(t)\}$ is a compact C^0-semigroup on a metric space V, then $\{T(t)\}$ is asymptotically smooth.*

Proof. In the spirit of Definition 2.3.10, we consider any nonempty, closed, bounded, positively invariant set $B \subset V$. Since $\gamma^+(B) \subset B$ and since $\{T(t)\}$ is compact, the set $\mathrm{cl}_V T(1)\gamma^+(B)$ is compact. But $\omega(B)$ is a closed subset of $\mathrm{cl}_V T(1)\gamma^+(B)$. Consequently, $\omega(B)$ is compact.

We also see that $\omega(B) \subset \mathrm{cl}_V \gamma^+(B) \subset \mathrm{cl}_V B = B$. Hence it only remains for us to show that $d(T(t)B, \omega(B)) \to 0$ as $t \to +\infty$.

For the sake of argument, suppose that there is an $\varepsilon > 0$ and a sequence $t_n \to +\infty$ such that

$$d(T(t_n)B, \omega(B)) > \varepsilon, \quad n \in \mathbb{N}. \tag{2.179}$$

Then there must exist a sequence $\{y_n\} \subset B$ for which the sequence $\{T(t_n)y_n\}$ does not have a convergent subsequence. However, almost all elements of the sequence $\{T(t_n)y_n\}$ are contained in the compact set $\mathrm{cl}_V T(1)\gamma^+(B)$ and this contradicts (2.179). Finally, recalling the *Cantor condition* in metric spaces, namely, the intersection of a decreasing family of nonempty compact sets is nonempty, we conclude that $\omega(B)$ is empty only if B is empty. The proof is complete. □

Lemma 2.3.24. *If $\{T(t)\}$ is a compact C^0-semigroup on a metric space V, then*

$$\forall_{\substack{B \subset V\\ B\text{-bounded}}} \forall_{\tau_2 > \tau_1 > 0} \bigcup_{t \in [\tau_1, \tau_2]} T(t)B \text{ is bounded in } V. \tag{2.180}$$

Proof. If (2.180) is not true, one can choose: a nonempty, bounded set $B \subset V$; two numbers $\tau_2 > \tau_1 > 0$; and two sequences $\{v_n\} \subset B$, $\{t_n\} \subset [\tau_1, \tau_2]$ such that

$$y_n := \mathrm{dist}_V(T(t_n)v_n, T(t_0)v_0) > n, \quad n = 1, 2, \ldots. \tag{2.181}$$

However, there is a subsequence $\{t_{n'}\}$ such that $t_{n'} \to \tau^* \in [\tau_1, \tau_2]$. Furthermore, by the compactness of $T(t)$, $t > 0$, there is a subsequence $\{v_{n''}\}$, for which $T(\frac{\tau^*}{2})v_{n''} \to v^* \in \mathrm{cl}_V T(\frac{\tau^*}{2})B$. Since the semigroup is continuous with respect to both its arguments we obtain

$$T(t_{n''})v_{n''} = T\left(t_{n''} - \frac{\tau^*}{2}\right)T\left(\frac{\tau^*}{2}\right)v_{n''} \to T\left(\frac{\tau^*}{2}\right)v^*.$$

But this last result implies that $\{T(t_{n''})v_{n''}\}$ is bounded, which contradicts (2.181). The proof is complete. □

Corollary 2.3.25. *Let $\{T(t)\}$ be a C^0-semigroup on a metric space V. If $\{T(t)\}$ is compact and point dissipative, then $\{T(t)\}$ has a global attractor in V.*

Proof. In view of Corollary 2.3.17, it suffices to verify the condition (2.175).

Let W_0 be a bounded subset of V which attracts points and let $\mathcal{N}_{W_0}$ be any bounded, open neighborhood of W_0. We first show that

$$\forall_{\substack{B \subset V\\ B\text{-bounded}}} \exists_{\tau_B \geq 2}\, T(\tau_B)B \subset \bigcup_{t \in [1, \tau_B]} T(t)\mathcal{N}_{W_0}. \tag{2.182}$$

Indeed, as in (2.170) we have

$$\forall_{v \in V} \exists_{\tau_v \geq 0} \exists_{B_V(v, \varepsilon_v)}\, T(\tau_v)B_V(v, \varepsilon_v) \subset \mathcal{N}_{W_0}, \tag{2.183}$$

where $B_V(v,\varepsilon_v) \subset V$ is an open ball centered at v and having radius ε_v. Since B is bounded and since $T(1)$ is compact, we can choose finitely many such balls so that

$$\mathrm{cl}_V T(1)B \subset B_V(v_1,\varepsilon_{v_1}) \cup \cdots \cup B_V(v_{k(B)},\varepsilon_{v_{k(B)}}). \tag{2.184}$$

For $\tau_B := 2 + \tau_{\max} = 2 + \max\{\tau_{v_j};\ j = 1,2,\ldots\}$, the conditions (2.184) and (2.183) lead to the inclusion

$$\begin{aligned}
T(\tau_B)B &= T(\tau_{\max}+2)B \subset T(\tau_{\max}+1)\mathrm{cl}_V T(1)B \\
&\subset T(\tau_{\max}+1)\bigcup_{j=1}^{k(B)} B_V(v_j,\varepsilon_{v_j}) \subset \bigcup_{j=1}^{k(B)} T(\tau_{\max}+1-\tau_{v_j})T(\tau_{v_j})B_V(v_j,\varepsilon_{v_j}) \\
&\subset \bigcup_{j=1}^{k(B)} T(\tau_{\max}+1-\tau_{v_j})\mathcal{N}_{W_0} \subset \bigcup_{t\in[1,\tau_B]} T(t)\mathcal{N}_{W_0}.
\end{aligned}$$

As a consequence of Lemma 2.3.24, the set $\mathcal{O} = \bigcup_{t\in[1,2]} T(t)\mathcal{N}_{W_0}$ is bounded in V. Thus, applying (2.182) to $\mathcal{O}$, we obtain

$$\bigcup_{t\in[\tau_{\mathcal{O}}+1,\tau_{\mathcal{O}}+2]} T(t)\mathcal{N}_{W_0} = T(\tau_{\mathcal{O}})\mathcal{O} \subset \bigcup_{t\in[1,\tau_{\mathcal{O}}]} T(t)\mathcal{N}_{W_0}. \tag{2.185}$$

Next, from (2.185) there follows

$$\bigcup_{t\in[1,\tau_{\mathcal{O}}+2]} T(t)\mathcal{N}_{W_0} = \bigcup_{t\in[1,\tau_{\mathcal{O}}+1]} T(t)\mathcal{N}_{W_0} \cup \bigcup_{t\in[\tau_{\mathcal{O}}+1,\tau_{\mathcal{O}}+2]} T(t)\mathcal{N}_{W_0} \subset \bigcup_{t\in[1,\tau_{\mathcal{O}}+1]} T(t)\mathcal{N}_{W_0}.$$

This last inclusion further ensures that

$$\bigcup_{t\geq 1} T(t)\mathcal{N}_{W_0} \subset \bigcup_{t\in[1,\tau_{\mathcal{O}}+1]} T(t)\mathcal{N}_{W_0},$$

and now (2.182) yields the condition

$$\forall_{\substack{B\subset V\\ B\text{-bounded}}}\ \exists_{\tau_B\geq 2} \bigcup_{t\geq\tau_B} T(t)B \subset \bigcup_{t\geq 1} T(t)\mathcal{N}_{W_0} \subset \bigcup_{t\in[1,\tau_{\mathcal{O}}+1]} T(t)\mathcal{N}_{W_0}.$$

By virtue of Lemma 2.3.24, the set $\bigcup_{t\in[1,\tau_{\mathcal{O}}+1]} T(t)\mathcal{N}_{W_0}$ is bounded. The proof is complete. □

The next statement relates the notion of a Lyapunov function to Corollary 2.3.25.

Corollary 2.3.26. *Let $\{T(t)\}$ be a compact C^0-semigroup on a metric space V. Furthermore, suppose that all the equilibria of $\{T(t)\}$ lie in a bounded subset of V. If there exists a Lyapunov function for $\{T(t)\}$, then $\{T(t)\}$ has a global attractor.*

Remark 2.3.27. Note that, for the case in which the semigroup $\{T(t)\}$ is compact, Proposition 2.3.21 (and hence also Corollary 2.3.26) can be proved without the assumption that the Lyapunov function is bounded below. Indeed, the only place in the proof of Proposition 2.3.21 where we specifically used this assumption was formula (2.178). But, if the semigroup $\{T(t)\}$ is known to be compact and if the orbit $\gamma^+(v)$ is known to be bounded, then clearly the image $T(1)\gamma^+(v)$ is precompact in V and (2.178) immediately follows. Of course, the preceding argument in no way refers to an assumed boundedness of $\mathcal{L}$.

In addition to all this, we want to point out that the examples which we shall present in Chapter 5 yield semigroups which are compact in a certain natural way. In fact, as we shall see, even if the semigroup $\{T(t)\}$ is generically noncompact, as, for example, in the case of Cauchy problems considered on the whole of $\mathbb{R}^N$, we are able to slightly alter the original phase space so as to bring about the required compactness of $\{T(t)\}$. In order when the semigroup is compact, we have much better control over the behavior of the given system.

In general, we shall not again use Theorem 2.3.15 (or Corollary 2.3.22), which explicitly require asymptotic smoothness for $\{T(t)\}$. Rather, we will employ Corollaries 2.3.25 and 2.3.26, which require compactness for $\{T(t)\}$.

Remarks on completely continuous semigroups

We want to formulate some assertions concerning the special case in which the semigroup $\{T(t)\}$ is *completely continuous for $t > 0$* in the sense of [83, p. 36], which means

$$\begin{gathered}\textit{the semigroup is compact and for each bounded set } B \subset V \\ \textit{and each number } \tau > 0 \textit{ the set } \bigcup_{t\in[0,\tau]} T(t)B \textit{ is bounded in } V.\end{gathered} \tag{2.186}$$

For brevity's sake, we shall henceforth refer to such semigroups $\{T(t)\}$ as *completely continuous*, omitting explicit reference to the inequality $t > 0$.

Note that the second requirement in (2.186) can be significantly weakened inasmuch as, for a compact C^0-semigroup $\{T(t)\}$, Theorem 2.3.25 implies that the boundedness of $\bigcup_{t\in[0,\tau]} T(t)B$ for a single number $\tau > 0$ is equivalent to the boundedness of $\bigcup_{t\in[0,\tau]} T(t)B$ for every number $\tau > 0$.

Regarding complete continuity, we also have the following.

Corollary 2.3.28. *If $\{T(t)\}$ is a C^0-semigroup on a metric space V and if $\{T(t)\}$ is completely continuous and point dissipative on V, then $\{T(t)\}$ has a global attractor $\mathcal{A}$ in V.*

Remark 2.3.29. Note that in both Corollaries 2.3.25 and 2.3.28 the assumption of point dissipativeness for $\{T(t)\}$ can be replaced by the weaker condition (2.176) (cf. Remark 2.3.18).

Asymptotic compactness of $\{T(t)\}$

It is sometimes convenient to have available criteria for asymptotic smoothness distinct from those stipulated in the definition. With this in mind, we introduce the following condition (cf. [113, Chapter 3]):

$$\begin{aligned}&\textit{for any nonempty set } B \textit{ having the property that there}\\ &\textit{exists a number } t_B \geq 0 \textit{ such that } \bigcup_{t \geq t_B} T(t)B \textit{ is bounded,}\\ &\textit{each sequence } \{T(t_n)v_n\}, \textit{ where } t_n \nearrow +\infty \textit{ and}\\ &\{v_n\} \subset B, \textit{ has a convergent subsequence.}\end{aligned} \tag{2.187}$$

We shall say that $\{T(t)\}$ is *asymptotically compact* in V if (2.187) holds.

The above two notions are equivalent.

Proposition 2.3.30. *Let $\{T(t)\}$ be a C^0-semigroup on a metric space V. Then $\{T(t)\}$ is asymptotically smooth if and only if $\{T(t)\}$ is asymptotically compact.*

Bibliographical notes

A nice review of the elementary inequalities of Section 2.1.3 is given in [78]. The uniform Gronwall inequality is taken from [172] whereas Volterra-type inequalities may be found in [86]. Results reported in Sections 2.1.4, 2.1.5 are treated mostly in [173]; we like the presentation of [87]. Section 2.2 is based on the monographs [66], [86] and [126]. We remark that the most complete description concerning the powers of positive operators may be found in [173] and in [130] in case of nonnegative operators; see also the source papers [106], [107]. The monograph [166] contains the full proof of the Nirenberg–Gagliardo-type inequalities as well as the description of the properties of elliptic operators. The papers [76], [77] contain the estimates of imaginary powers of the Stokes operator studied under natural smoothness assumptions on $\partial\Omega$; see also the elegant monograph [158]. Stability concepts of Section 2.3 are based on the monograph [83]. We also refer to the nice description given in [112]. Another approach can be found in [172]; in particular, an alternative approach to existence of a global attractor is presented in [172, Theorem 1.1].

3 Solvability of the abstract Cauchy problem

This book is devoted to the studies of evolutionary equations that can be written in a form of an abstract semilinear Cauchy's problem with sectorial operator in the main part. Many important initial boundary value problems having their origin in the applied sciences fall into such class which was chosen for the studies in the famous monograph [86] by D. Henry. Examples like semilinear heat equation, Navier–Stokes system, or the quasi-geostrophic equation together with their generalizations including fractional power operators will be discussed further in the text.

The present chapter is devoted to solvability of an abstract Cauchy problem with sectorial operator. We introduce an integral equation equivalent to this problem, prove existence of its local solution, and study regularity properties. The approach here is a far extension of the E. Picard theorem known in the theory of ordinary differential equations.

3.1 Semilinear evolutionary equation with sectorial operator

We are considering an abstract Cauchy problem

$$\begin{cases} u_t + Au = F(u), & t > 0, \\ u(0) = u_0, \end{cases} \tag{3.1}$$

with sectorial operator A in a Banach space X. Without lack of generality (adding a multiple of identity to both sides of (3.1), if needed), we may assume that A is *positive*, which means that the spectrum of A is contained in a positive half-line. Since the resolvent set of a closed linear operator is open in the complex plane, then

$$\mathrm{Re}(\sigma(A)) > a, \quad \text{for some positive } a,$$

so that the *power spaces* X^α, $\alpha > 0$, are defined as the ranges of an appropriate power operators $A^{-\alpha}$ (cf. Section 2.2.3). In the described situation, we need always to assume that for fixed $\alpha \in [0, 1)$ the nonlinear term $F : X^\alpha \to X$ is Lipschitz continuous on bounded subsets of X^α. Equivalently, there exists a nondecreasing function $L : \mathbb{R}^+ \to \mathbb{R}^+$, such that the condition

$$\|F(v) - F(w)\|_X \leq L(r)\|v - w\|_{X^\alpha}$$

holds for each $v, w \in B_{X^\alpha}(r)$, where $B_{X^\alpha}(r)$ denotes an open ball in X^α centered at zero with radius r.

To simplify further references, we introduce the following assumption:

Assumption A. *Let X be a Banach space, $A : D(A) \to X$ sectorial and positive operator in X and let, for some $\alpha \in [0, 1)$, $F : X^\alpha \to X$ be Lipschitz continuous on bounded subsets of X^α.*

https://doi.org/10.1515/9783110599831-003

Then we introduce a *local X^α-solution* of (3.1).

Definition 3.1.1. Let X be a Banach space, $\alpha \in [0,1)$ and $u_0 \in X^\alpha$. A function u is called local X^α-solution of (3.1) if, for some $\tau > 0$, u belongs to $C([0,\tau); X^\alpha)$ and satisfies:

$$u(0) = u_0,$$
$$u \in C^1((0,\tau); X)$$
$$u(t) \text{ belongs to } D(A) \text{ for each } t \in (0,\tau),$$
$$\text{the first equation in (3.1) holds in } X \text{ for all } t \in (0,\tau).$$

It is common in the literature that searching for solutions of the abstract Cauchy problem (3.1) we are solving instead corresponding to it *integral equation* (3.5) (the idea inspired by the proof of the Picard theorem). The solutions obtained that way are called *mild solutions* to (3.1) (see, e. g., [138]). Sometimes in the text we will just use the name *mild solution*, instead of calling precisely the X^α-solution.

The rest of this section will be devoted to the following local existence theorem (compare [83, Section 4.2], [86, Chapter 3]).

Theorem 3.1.2. *Under the Assumption* A, *for each $u_0 \in X^\alpha$, there exists a unique X^α-solution $u = u(t,u_0)$ of* (3.1) *defined on its maximal interval of existence $[0,\tau_{u_0})$, which means that*

$$\begin{aligned} &\textit{either } \tau_{u_0} = +\infty, \textit{ or} \\ &\textit{if } \tau_{u_0} < +\infty, \textit{ then } \limsup_{t \to \tau_{u_0}^-} \|u(t,u_0)\|_{X^\alpha} = +\infty. \end{aligned} \tag{3.2}$$

As a consequence of the embeddings between fractional power spaces (see (2.121)), if the function F is Lipschitz continuous on bounded subsets of X^α ($\alpha \in [0,1)$), then it has similar property as a map from X^β into X for each $\beta \in [\alpha,1)$. This allows to extend many results concerning the X^α-solutions onto X^β-solutions with arbitrary $\beta \in [\alpha,1)$. In particular, the conclusion below holds.

Corollary 3.1.3. Under the Assumption A, for each $\beta \in [\alpha,1)$ and $u_0 \in X^\beta$, there exists a unique X^β-solution $u = u(t,u_0)$ of (3.1) defined on its maximal interval of existence $[0,\tau_{u_0})$.

Bochner integral in Banach spaces

When h is a function acting from a measure space $(Y,\mathcal{M},m)$ into a Banach space X, then h is known to be m-integrable on $\mathcal{D} \in \mathcal{M}$ if and only if there is a sequence of finitely valued functions $h_n : \mathcal{D} \to X$, such that

$$(\alpha) \quad \lim_{n\to\infty} \|h_n(y) - h(y)\|_X = 0 \quad \text{for a. e. } y \in \mathcal{D},$$

$$(\beta) \quad \lim_{n\to\infty} \int_{\mathcal{D}} \|h_n(y) - h(y)\|_X dy = 0.$$

Conditions (α) implies that h is a strongly measurable function. The latter property is crucial to show that f is Bochner integrable if and only if $\|f\|_X$ is Lebesgue integrable (cf. [190, Chapter V, Section 5]).

There is no place here to consider the notion of the *Bochner integral* in its whole generality since we will deal mostly with continuous functions on the real line. Therefore, let us focus on less general situation when m is the Lebesgue measure and $\mathcal{D}$ denotes a Lebesgue measurable subset of $Y = \mathbb{R}^n$. In that case, whenever $h : \mathcal{D} \to X$ is continuous, functions h and $\|h\|_X$ are known to be simultaneously integrable or not. Furthermore, if h is also bounded, the notion of the integral of h over a closed interval in $\mathbb{R}^n$ may be introduced with the use of the classical Riemann approximations. General notion of the Bochner integral may thus coincide in this case with the usual Riemann-type integral.

When $Y = \mathbb{R}$, the following results of elementary analysis will be further of great importance. It concerns the *improper integrals* which will appear in considerations of the next section.

Proposition 3.1.4. *Let $h : (t_1, t_2) \to X$ be continuous function. Then the Bochner integral $\int_{t_1}^{t_2} h(y)dy$ exists if and only if $\int_{t_1}^{t_2} \|h(y)\|_X dy < \infty$. Moreover, if h is integrable on (t_1, t_2), then*

$$\int_{t_1}^{t_2} h(y)dy = \lim_{(t,\tau)\to(t_1,t_2)} \int_t^\tau h(y)dy \tag{3.3}$$

and, if in addition $h \in C([t_1, t_2]; X) \cap C^1((t_1, t_2); X)$, then

$$h(t_2) - h(t_1) = \int_{t_1}^{t_2} h'(y)dy. \tag{3.4}$$

Remark 3.1.5. Note, that the integral in the equality (3.4) may have only the improper sense, that is, $\int_{t_1}^{t_2} h'(y)dy = \lim_{(t,\tau)\to(t_1,t_2)} \int_t^\tau h'(y)dy$. Also $\int_{t_1}^{t_2} h(y)dy$ in (3.3) has only improper sense unless (as in Proposition 3.1.4) it is assumed that $\int_{t_1}^{t_2} h(y)dy$ exists. Nevertheless, (3.3) points that if $\int_{t_1}^{t_2} h(y)dy$ exists, then it may be considered as an improper integral. Therefore, there is no need to introduce any special notation for improper integrals.

The following commutativity property is often used in the calculations.

Proposition 3.1.6. *Let $B : D(B) \to X$ be a closed linear operator in a Banach space X whereas $h : [0, t) \to X$ and $B \circ h : [0, t) \to X$ be continuous functions. Assume further that $h(y) \in D(B)$ for each $y \in [0, t)$ and there exist improper integrals $\int_0^t h(y)dy$ and*

$\int_0^t Bh(y)dy$. *Then* $\int_0^t h(y)dy \in D(B)$ *and*

$$B\int_0^t h(y)dy = \int_0^t Bh(y)dy.$$

For further properties of the Bochner integral, see [66], [88], [158], [190].

3.2 Variation of constants formula

Lemma 3.2.1 (Integral Cauchy formula, Duhamel formula). *Let the Assumption A hold and $u \in C([0,\tau);X^\alpha)$. Then u is a local X^α-solution of (3.1) if and only if u is a solution in X of the integral equation*

$$u(t) = e^{-At}u_0 + \int_0^t e^{-A(t-s)}F(u(s))ds, \quad \text{for } t \in [0,\tau). \tag{3.5}$$

Proof. We first prove that $u \in C([0,\tau);X^\alpha)$ satisfying (3.5) is a local X^α-solution of (3.1).

Step 1. We will need the following estimate:

$$\|(e^{-At} - I)v\|_X \leq \frac{c_{1-\delta}}{\delta}t^\delta\|v\|_{X^\delta}, \quad \delta \in (0,1],\ v \in X^\delta. \tag{3.6}$$

As a consequence of the equality,

$$\begin{aligned}(e^{-At} - I)v &= \int_0^t \frac{d}{ds}(e^{-As}v)ds \\ &= -\int_0^t Ae^{-As}vds = -\int_0^t A^{1-\delta}e^{-As}A^\delta vds,\end{aligned}$$

we find that

$$\begin{aligned}\|(e^{-At} - I)v\|_X &\leq \int_0^t \|A^{1-\delta}e^{-As}\|_{\mathcal{L}(X,X)}\|A^\delta v\|_X ds \\ &\leq c_{1-\delta}\|v\|_{X^\delta}\int_0^t \frac{e^{-as}}{s^{1-\delta}}ds \leq \frac{c_{1-\delta}}{\delta}t^\delta\|v\|_{X^\delta}.\end{aligned}$$

Step 2. Using (3.6), we will show the following local Hölder continuity of solution:

$$\begin{gathered}\forall_{0<t_0<t_1<\tau}\ \forall_{\delta\in(0,1)}\ \exists_{\text{const.}>0}\ \forall_{t,\bar{t}\in(t_0,t_1)} \\ \|u(t) - u(\bar{t})\|_{X^\alpha} \leq \text{const.}|t - \bar{t}|^\delta,\end{gathered} \tag{3.7}$$

with const. depending on α, δ, t_0, t_1.

Let $0 < h < t_1 < \tau$ be fixed. From integral representation (3.5), we obtain

$$
\begin{aligned}
u(t+h) - u(t) &= (e^{-Ah} - I)e^{-At}u_0 \\
&\quad + \left(\int_0^h + \int_h^{t+h}\right) e^{-A(t+h-s)}F(u(s))ds - \int_0^t e^{-A(t-s)}F(u(s))ds \\
&= (e^{-Ah} - I)e^{-At}u_0 + \int_0^h e^{-A(t+h-s)}F(u(s))ds \\
&\quad + \int_0^t e^{-A(t-s)}(F(u(s+h)) - F(u(s)))ds, \quad t \in (0, t_1 - h).
\end{aligned}
$$

Using next the estimate (3.6) with $0 < \delta < 1$ and Lipschitz continuity of F, we have

$$
\begin{aligned}
\|u(t+h) - u(t)\|_{X^\alpha} &\le \|(e^{-Ah} - I)e^{-At}A^\alpha u_0\|_X \\
&\quad + \int_0^h c_\alpha \frac{e^{-a(t+h-s)}}{(t+h-s)^\alpha}\|F(u(s))\|_X ds \\
&\quad + \int_0^t c_\alpha \frac{e^{-a(t-s)}}{(t-s)^\alpha}\|F(u(s+h)) - F(u(s))\|_X ds \\
&\le h^\delta \frac{c_{1-\delta}c_\delta}{\delta t^\delta}\|u_0\|_{X^\alpha} + \frac{c_\alpha \sup_{s\in[0,t_1]}\|F(u(s))\|_X}{t^\alpha}h \\
&\quad + c_\alpha L_{t_1}\int_0^t \frac{1}{(t-s)^\alpha}\|u(s+h) - u(s)\|_{X^\alpha} ds \\
&\le \frac{h^\delta}{t^{\delta_{\max}}}\left(\frac{c_{1-\delta}c_\delta}{\delta}t_1^{\delta_{\max}-\delta}\|u_0\|_{X^\alpha} + c_\alpha t_1^{1-\delta}t_1^{\delta_{\max}-\alpha}\sup_{s\in[0,t_1]}\|F(u(s))\|_X\right) \\
&\quad + c_\alpha L_{t_1}\int_0^t \frac{1}{(t-s)^\alpha}\|u(s+h) - u(s)\|_{X^\alpha} ds, \quad t \in (0, t_1 - h),
\end{aligned}
$$

where $\delta_{\max} = \max\{\alpha, \delta\}$.

Lemma 2.1.11 now implies an estimate

$$
t^{\delta_{\max}}\|u(t+h) - u(t)\|_{X^\alpha} \le \text{const.}_{t_1,\delta,\alpha}h^\delta, \quad t \in (0, t_1 - h),
$$

(with $\text{const.}_{t_1,\delta,\alpha}$ independent of h), so that

$$
\|u(t+h) - u(t)\|_{X^\alpha} \le \frac{\text{const.}_{t_1,\delta,\alpha}}{t_0^{\delta_{\max}}}h^\delta, \quad t \in (t_0, t_1 - h), \tag{3.8}
$$

which shows (3.7).

Step 3. Condition (3.7) will be used to justify that

$$\int_0^t e^{-A(t-s)}F(u(s))ds \in D(A), \quad \text{for } t \in (0,\tau), \tag{3.9}$$

$$u(t) \in D(A), \text{ for } t \in (0,\tau), \tag{3.10}$$

and also to show commutativity

$$\int_0^t Ae^{-A(t-s)}F(u(s))ds = A\int_0^t e^{-A(t-s)}F(u(s))ds, \quad t \in (0,\tau), \tag{3.11}$$

where the left integral of (3.11) is understood in an improper sense.

Indeed, since $e^{-A(t-s)}F(u(s)) \in D(A)$ for $s \in [0,t)$, from (3.7) and local Lipschitz continuity of F we have

$$\begin{aligned}
&\int_{t_0}^t \|Ae^{-A(t-s)}[F(u(s)) - F(u(t))]\|_X ds \\
&\quad \le \int_{t_0}^t c_1 \frac{e^{-a(t-s)}}{t-s}\|F(u(t)) - F(u(s))\|_X ds \\
&\quad \le c_1 \int_{t_0}^t \frac{1}{t-s} L\|u(t) - u(s)\|_{X^\alpha} ds \\
&\quad \le c_1 L \int_{t_0}^t \frac{1}{t-s}\text{const.}(t-s)^\delta ds \\
&\quad = \frac{c_1 \text{const.} L}{\delta}(t-t_0)^\delta,
\end{aligned} \tag{3.12}$$

whereas

$$\begin{aligned}
\int_{t_0}^t Ae^{-A(t-s)}F(u(t))ds &= \int_{t_0}^t \frac{d}{ds}(e^{-A(t-s)}F(u(t)))ds \\
&= (I - e^{-A(t-t_0)})F(u(t)).
\end{aligned} \tag{3.13}$$

This justifies existence of the improper integral $\int_0^t Ae^{-A(t-s)}f(u(s))ds$. Since there exists $\int_0^t e^{-A(t-s)}F(u(s))ds$ as the integral of continuous function, consequently (3.9) and (3.11) follow directly from Proposition 3.1.6 (note continuity of function $[0,t) \ni s \to Ae^{-A(t-s)}F(u(s)) \in X$). Therefore, (3.10) is a consequence of (3.9) and equation (3.5).

Step 4. Denoting

$$u_1(t) = \int_0^t e^{-A(t-s)}(F(u(s)) - F(u(t)))ds, \tag{3.14}$$

we will prove that

$$Au_1 \in C((0,\tau); X^\gamma) \quad \text{for each } \gamma \in (0,1) \tag{3.15}$$

(note that $Au_1(t) \in D(A)$ for $t \in [0,\tau)$ as follows from (3.12) and Proposition 3.1.6). More precisely, we will justify the stronger condition:

$$\begin{aligned} &\forall_{0<\gamma<\delta<1} \forall_{0<\varepsilon<\bar{\varepsilon}<\tau} \exists_{\text{const.}>0} \forall_{t_1,t_2 \in (\varepsilon,\bar{\varepsilon})} \\ &\|Au_1(t_2) - Au_1(t_1)\|_{X^\gamma} \le \text{const.}|t_2 - t_1|^{\delta-\gamma}, \end{aligned} \tag{3.16}$$

with const. depending on $\varepsilon, \bar{\varepsilon}, \gamma, \delta$.

For $0 < \gamma < \delta < 1$, $0 < \varepsilon \le t_1 < t_2 < \bar{\varepsilon} < \tau$, consider the difference

$$\begin{aligned} &Au_1(t_2) - Au_1(t_1) \\ &\quad = \int_0^{t_1} A(e^{-A(t_2-s)} - e^{-A(t_1-s)})(F(u(s)) - F(u(t_1)))ds \\ &\quad + (e^{-A(t_2-t_1)} - e^{-At_2})(F(u(t_1)) - F(u(t_2))) \\ &\quad + \int_{t_1}^{t_2} Ae^{-A(t_2-s)}(F(u(s)) - F(u(t_2)))ds \\ &\quad := \mathcal{J}_1 + \mathcal{J}_2 + \mathcal{J}_3. \end{aligned}$$

It suffices to show that $\|\mathcal{J}_k\|_{X^\gamma}$, $k = 1, 2, 3$ are proportional to $(t_2 - t_1)^{\delta-\gamma}$. In calculations below conditions (3.6), (3.7), Lipschitz continuity of F and the usual estimates for analytic semigroups will be used several times. The following estimates hold:

$$\begin{aligned} \|\mathcal{J}_1\|_{X^\gamma} &\le \left(\int_0^{\frac{\varepsilon}{2}} + \int_{\frac{\varepsilon}{2}}^{t_1} \right) \|A^\gamma A(e^{-A(t_2-t_1)} - I)e^{-A(t_1-s)}(F(u(s)) - F(u(t_1)))\|_X ds \\ &:= \mathcal{J}_{11} + \mathcal{J}_{12}, \end{aligned} \tag{3.17}$$

$$\begin{aligned} \mathcal{J}_{11} &\le \int_0^{\frac{\varepsilon}{2}} \|(e^{-A(t_2-t_1)} - I)A^\gamma Ae^{-A(t_1-s)}(F(u(s)) - F(u(t_1)))\|_X ds \\ &\le \int_0^{\frac{\varepsilon}{2}} \frac{c_{1-(\delta-\gamma)}}{\delta-\gamma}(t_2 - t_1)^{\delta-\gamma} \|A^{\delta-\gamma}A^\gamma Ae^{-A(t_1-s)}(F(u(s)) - F(u(t_1)))\|_X ds \end{aligned}$$

$$\leq \left(\int_0^{\frac{\varepsilon}{2}} \frac{C_{1-\delta+\gamma}}{\delta-\gamma} \frac{C_{1+\delta}}{(t_1-s)^{1+\delta}} 2 \sup_{s\in[0,\bar{\varepsilon}]} \|F(u(s))\|_X ds \right) (t_2-t_1)^{\delta-\gamma}$$

$$\leq 2 \sup_{s\in[0,\bar{\varepsilon}]} \|F(u(s))\|_X \frac{C_{1-\delta+\gamma}}{\delta-\gamma} \frac{C_{1+\delta}}{\delta} \left(\frac{2}{\varepsilon}\right)^{\delta} (t_2-t_1)^{\delta-\gamma},$$

$$\mathcal{J}_{12} \leq \left(\int_{\frac{\varepsilon}{2}}^{t_1} \frac{C_{1-\delta+\gamma}}{\delta-\gamma} \frac{C_{1+\delta}}{(t_1-s)^{1+\delta}} L_{\bar{\varepsilon}} \text{const.}_{\bar{\varepsilon},\varepsilon,\delta_1} (t_1-s)^{\delta_1} ds \right) (t_2-t_1)^{\delta-\gamma},$$

(since δ_1 appearing above is chosen between δ and 1 based on (3.7), the final integral above is bounded),

$$\begin{aligned} \|\mathcal{J}_2\|_{X^\gamma} &\leq \|A^\gamma e^{-A(t_2-t_1)}(e^{-At_1}-I)(F(u(t_1))-F(u(t_2)))\|_X \\ &\leq (C_0+1)\frac{C_\gamma}{(t_2-t_1)^\gamma} L_{\bar{\varepsilon}} \text{const.}_{\bar{\varepsilon},\varepsilon,\delta} (t_2-t_1)^\delta, \end{aligned} \tag{3.18}$$

$$\begin{aligned} \|\mathcal{J}_3\|_{X^\gamma} &\leq \int_{t_1}^{t_2} \|A^\gamma A e^{-A(t_2-s)}(F(u(s))-F(u(t_2)))\|_X ds \\ &\leq \int_{t_1}^{t_2} \frac{C_{1+\gamma}}{(t_2-s)^{1+\gamma}} L_{\bar{\varepsilon}} \text{const.}_{\bar{\varepsilon},\varepsilon,\delta} (t_2-s)^\delta ds \\ &= C_{1+\gamma} L_{\bar{\varepsilon}} \text{const.}_{\bar{\varepsilon},\varepsilon,\delta} \frac{(t_2-t_1)^{\delta-\gamma}}{\delta-\gamma}. \end{aligned} \tag{3.19}$$

From (3.17)–(3.19), each component $\|\mathcal{J}_k\|_{X^\gamma}$ $(k = 1,2,3)$ is proportional to $(t_2-t_1)^{\delta-\gamma}$. Therefore, (3.16) is proved.

Step 5. Due to (3.9), it will be seen that the right side derivative $\dot{u}_+(t)$ satisfies for $t \in (0,\tau)$ the equation

$$\dot{u}_+(t) + A\left(e^{-At}u_0 + \int_0^t e^{-A(t-s)}F(u(s))ds \right) = F(u(t)). \tag{3.20}$$

Indeed, letting $h \to 0^+$ in the formula

$$\begin{aligned} \frac{1}{h}(u(t+h)-u(t)) &= \frac{1}{h}(e^{-Ah}-I)e^{-At}u_0 \\ &+ \frac{1}{h}(e^{-Ah}-I)\int_0^t e^{-A(t-s)}F(u(s))ds + \frac{1}{h}\int_t^{t+h} e^{-A(t+h-s)}F(u(s))ds \end{aligned}$$

we obtain (3.20).

Step 6. Noting the equality

$$\begin{aligned} Au_1(t) &= A\int_0^t e^{-A(t-s)}F(u(s))ds - \int_0^t Ae^{-A(t-s)}F(u(t))ds \\ &= A\int_0^t e^{-A(t-s)}F(u(s))ds - F(u(t)) + e^{-At}F(u(t)), \end{aligned} \tag{3.21}$$

and connecting (3.20) and (3.21), we find a formula for $\dot{u}_+$:

$$\dot{u} = -Ae^{-At}u_0 + e^{-At}F(u(t)) - Au_1(t). \tag{3.22}$$

The functions $(0,\tau) \ni t \to Ae^{-At}u_0 \in X^\gamma$ and $(0,\tau) \ni t \to e^{-At}F(u(t)) \in X^\gamma$ are continuous, and from (3.15) also $Au_1 \in C((0,\tau);X^\gamma)$. Therefore, the right side derivative $\dot{u}_+\colon (0,\tau) \to X$ is continuous. Thus, u must be continuously differentiable on $(0,\tau)$; that means

$$\dot{u}(t) = \dot{u}_+(t) \quad \text{and} \quad u \in C^1((0,\tau);X). \tag{3.23}$$

To justify (3.23), it suffices to consider an auxiliary function $\omega : [t_0,\tau) \to X$ ($t_0 \in (0,\tau)$ arbitrary):

$$\omega(t) = \int_{t_0}^t \dot{u}_+(s)ds + u(t_0), \quad t \in [t_0,\tau).$$

Certainly, $\omega \in C^1([t_0,\tau);X)$ and $\dot{\omega}_+ \equiv \dot{u}_+$, so that for any bounded linear functional $x^* \in X^*$, the right side derivative $[x^*(\omega(t)) - x^*(u(t))]'_+$ is identically equal to zero on $[t_0,\tau)$. As a consequence of the equality $\omega(t_0) = u(t_0)$, we get further

$$x^*(\omega(t) - u(t)) = 0, \quad t \in [t_0,\tau),\ x^* \in X^*.$$

Therefore, by the standard Hahn–Banach theorem, ω needs to coincide with u. In Steps 3, 5, and 6, all of the requirements of Definition 3.1.1 were verified.

An opposite implication will be next proved, which means $u \in C([0,\tau);X^\alpha)$ being a local X^α-solution of (3.1) satisfies the Cauchy formula (3.5).

For $0 < s < t < \tau$, we obtain from (3.1) that

$$e^{-A(t-s)}\dot{u}(s) + e^{-A(t-s)}Au(s) = e^{-A(t-s)}F(u(s)). \tag{3.24}$$

Noting that the function $\mathbb{R}^+ \times \mathbb{R}^+ \ni (t,s) \to e^{-At}u(s) \in X$ has continuous first-order partial derivatives, consider the composite function $(0,t) \ni s \to e^{-A(t-s)}u(s) \in X$. Differentiating, we find that

$$\begin{aligned} \frac{d}{ds}(e^{-A(t-s)}u(s)) &= Ae^{-A(t-s)}u(s) + e^{-A(t-s)}\dot{u}(s) \\ &= e^{-A(t-s)}\dot{u}(s) + e^{-A(t-s)}Au(s), \end{aligned} \tag{3.25}$$

and due to (3.24) and (3.25) we have

$$\frac{d}{ds}(e^{-A(t-s)}u(s)) = e^{-A(t-s)}F(u(s)), \quad \text{for } 0 < s < t < \tau, \tag{3.26}$$

so that (3.5) follows by integration of (3.26) over $(0,t)$ (cf. Proposition 3.1.4).

The proof of Lemma 3.2.1 is now completed. □

3.3 Existence of the local X^α-solutions

This section is devoted to the proof of Theorem 3.1.2. The standard approach, based on considerations of the integral equation (3.5) in a Banach space $C([0,\delta];X^\alpha)$ with the supremum norm:

$$\|v\|_{C([0,\delta];X^\alpha)} := \sup_{t\in[0,\delta]} \|v(t)\|_{X^\alpha},$$

will be used. The proof of existence uses *Banach fixed-point theorem* applied to a contraction Ψ acting on the suitably chosen closed ball $B_{C([0,\delta];X^\alpha)}(u_0,r) \subset C([0,\delta];X^\alpha)$. Additionally, the solutions of (3.1) will be shown to depend continuously on the initial data.

Proof. **Step 1.** Uniqueness will be shown first. Let $u_1,u_2 \in C([0,\delta];X^\alpha)$ with $u_1(0) = u_2(0) = u_0$ be such that

$$u_i(t) = e^{-At}u_0 + \int_0^t e^{-A(t-s)}F(u_i(s))ds, \quad t \in [0,\delta],\ i = 1,2.$$

Then, since F is Lipschitz continuous on bounded sets, we obtain

$$\begin{aligned}\|u_1(t) - u_2(t)\|_{X^\alpha} &\le \int_0^t \|A^\alpha e^{-A(t-s)}\|_{\mathcal{L}(X,X)}\|F(u_1(s)) - F(u_2(s))\|_X ds \\ &\le L_\delta c_\alpha \int_0^t \frac{1}{(t-s)^\alpha}\|u_1(s) - u_2(s)\|_{X^\alpha} ds.\end{aligned}$$

Therefore, from Lemma 2.1.11, u_1, u_2 are equal on $[0,\delta]$, which implies uniqueness of the solution.

Step 2. It is clear that each initial data $u_0 \in X^\alpha$ may be identified with a constant (in time) function $u_0 \in C([0,\delta];X^\alpha)$. Fix $u_0 \in X^\alpha$, $r > 0$, and consider the *integral transformation* $\Psi : B_{C([0,\delta];X^\alpha)}(u_0,r) \to C([0,\delta];X^\alpha)$;

$$B_{C([0,\delta];X^\alpha)}(u_0,r) \ni v \xrightarrow{\Psi} \Psi(v)(t) := e^{-At}u_0 + \int_0^t e^{-A(t-s)}F(v(s))ds \quad \text{for } t \in [0,\delta],$$

where $B_{C([0,\delta];X^\alpha)}(u_0, r)$ denotes a closed ball in a Banach space $C([0,\delta]; X^\alpha)$ centered in u_0 with radius $r > 0$ (the value of $\delta > 0$ will be restricted further in (3.32)). Analogously, as in the proof of (3.7) in Lemma 3.2.1, we see that $\Psi(v)(t)$ is a continuous function of $t > 0$ with values in X^α. To justify the right side continuity at $t = 0$, it suffices to note that since F is bounded on bounded sets of X^α and $v \in B_{C([0,\delta];X^\alpha)}(u_0, r)$, then

$$\begin{aligned}\|\Psi(v)(t) - u_0\|_{X^\alpha} &\le \|A^\alpha(e^{-At} - I)u_0\|_X + \int_0^t \|A^\alpha e^{-A(t-s)}F(v(s))\|_X ds \\ &\le \|(e^{-At} - I)A^\alpha u_0\|_X + \sup_{\|v\|_{X^\alpha} \le r + \|u_0\|_{X^\alpha}} \|F(v)\|_X c_\alpha \int_0^t \frac{e^{-a(t-s)}}{(t-s)^\alpha} ds \\ &\le \|(e^{-At} - I)A^\alpha u_0\|_X + \sup_{\|v\|_{X^\alpha} \le r + \|u_0\|_{X^\alpha}} \|F(v)\|_X c_\alpha \frac{t^{1-\alpha}}{1-\alpha}, \quad t \in [0,\delta]. \end{aligned} \tag{3.27}$$

Transformation Ψ is thus properly defined.

Step 3. Now, we shall choose the parameter $\delta > 0$ sufficiently small, for which the range of Ψ will be contained in $B_{C([0,\delta];X^\alpha)}(u_0, r)$, that is,

$$\Psi : B_{C([0,\delta];X^\alpha)}(u_0, r) \to B_{C([0,\delta];X^\alpha)}(u_0, r). \tag{3.28}$$

Recalling (3.27) we find that

$$\begin{aligned}\|\Psi(v)(t) - u_0\|_{X^\alpha} &\le \varepsilon \|u_0\|_{X^\alpha} + \sup_{\|v\|_{X^\alpha} \le r + \|u_0\|_{X^\alpha}} \|F(v)\|_X c_\alpha \frac{\delta^{1-\alpha}}{1-\alpha} \\ &\le \frac{r}{2} + \frac{r}{2}, \quad \text{for } t \in [0,\delta], \end{aligned} \tag{3.29}$$

provided that $\delta \le \delta_1$, where δ_1 is chosen to $\varepsilon := \frac{r}{2\|u_0\|_{X^\alpha}}$ from continuity of the semigroup $e^{-At} : X \to X$ at $t = 0$ and, in addition, it satisfies the condition

$$\sup_{\|v\|_{X^\alpha} \le r + \|u_0\|_{X^\alpha}} \|F(v)\|_X c_\alpha \frac{\delta_1^{1-\alpha}}{1-\alpha} = \frac{r}{2}. \tag{3.30}$$

Step 4. We will impose next the second restriction on δ, making Ψ to be a contraction:

$$\exists_{\omega \in [0,1)} \forall_{v_1, v_2 \in B_{C([0,\delta];X^\alpha)}(u_0,r)} \|\Psi(v_1) - \Psi(v_2)\|_{C([0,\delta];X^\alpha)} \le \omega \|v_1 - v_2\|_{C([0,\delta];X^\alpha)}.$$

Take arbitrary $v_1, v_2 \in B_{C([0,\delta];X^\alpha)}(u_0, r)$ and fix any $\omega \in (0,1)$. Due to Lipschitz continuity of F on bounded sets, we have

$$\begin{aligned}\|\Psi(v_1)(t) - \Psi(v_2)(t)\|_{X^\alpha} &\le \int_0^t \|A^\alpha e^{-A(t-s)}\|_{\mathcal{L}(X,X)} \|F(v_1(s)) - F(v_2(s))\|_X ds \\ &\le L_{2r} \sup_{s\in[0,t]} \|v_1(s) - v_2(s)\|_{X^\alpha} c_\alpha \int_0^t \frac{e^{-a(t-s)}}{(t-s)^\alpha} ds \\ &\le L_{2r} \sup_{s\in[0,t]} \|v_1(s) - v_2(s)\|_{X^\alpha} c_\alpha \frac{\delta^{1-\alpha}}{1-\alpha} \\ &\le \omega \sup_{s\in[0,\delta]} \|v_1(s) - v_2(s)\|_{X^\alpha},\end{aligned}$$

provided that δ does not exceed δ_2 given by

$$L_{2r} c_\alpha \frac{\delta_2^{1-\alpha}}{1-\alpha} = \omega. \tag{3.31}$$

Step 5. Evidently, for

$$\delta_0 := \min\{\delta_1; \delta_2\}, \tag{3.32}$$

where δ_1, δ_2 were defined in (3.30) and (3.31), Ψ is a contraction on a complete metric space $B_{C([0,\delta_0];X^\alpha)}(u_0, r)$. As a consequence of the Banach fixed-point theorem there exists in $B_{C([0,\delta_0];X^\alpha)}(u_0, r)$ a unique fixed-point $\Psi(u) = u$, which simultaneously satisfies the integral equation

$$u(t) = e^{-At} u_0 + \int_0^t e^{-A(t-s)} F(u(s)) ds \quad \text{for } t \in [0, \delta_0]. \tag{3.33}$$

Step 6. We will characterize now maximal interval of existence of the solution to (3.33). Introduce the set $\mathcal{I}_{u_0}$:

$$\mathcal{I}_{u_0} := \left\{\delta > 0;\ \exists_{u^\delta \in C([0,\delta];X^\alpha)}\ \forall_{t\in[0,\delta]}\ u^\delta(t) = e^{-At}u_0 + \int_0^t e^{-A(t-s)} F(u^\delta(s)) ds\right\}.$$

Considerations of the previous step show that $\mathcal{I}_{u_0}$ is nonempty. Furthermore, if $\delta_1, \delta_2 \in \mathcal{I}_{u_0}$ and $\delta_1 < \delta_2$ then from the uniqueness of the local solution (Step 1):

$$u^{\delta_1} = u^{\delta_2} \quad \text{on } [0, \delta_1]. \tag{3.34}$$

Define further

$$\tau_{u_0} := \sup \mathcal{I}_{u_0} \tag{3.35}$$

and observe that

$$\tau_{u_0} \notin \mathcal{I}_{u_0}. \tag{3.36}$$

Indeed, assume that $\tau_{u_0} \in \mathcal{I}_{u_0}$. Then there is $u^{\tau_{u_0}} \in C([0,\tau_{u_0}];X^\alpha)$ satisfying

$$u^{\tau_{u_0}}(t) = e^{-At}u_0 + \int_0^t e^{-A(t-s)}F(u^{\tau_{u_0}}(s))ds, \quad t \in [0,\tau_{u_0}]. \tag{3.37}$$

On the other hand, the Banach fixed-point theorem ensures (cf. Steps 2–5) the existence of certain $w \in C([0,\varepsilon];X^\alpha)$, $\varepsilon > 0$, which fulfills

$$w(t) = e^{-At}u^{\tau_{u_0}}(\tau_{u_0}) + \int_0^t e^{-A(t-s)}F(w(s))ds, \quad t \in [0,\varepsilon]. \tag{3.38}$$

Introducing

$$u(t) = \begin{cases} u^{\tau_{u_0}}(t), & t \in [0,\tau_{u_0}], \\ w(t-\tau_{u_0}), & t \in [\tau_{u_0}, \tau_{u_0}+\varepsilon], \end{cases}$$

we see that $u \in C([0,\tau_{u_0}+\varepsilon];X^\alpha)$ and $\tau_{u_0}+\varepsilon \in \mathcal{I}_{u_0}$, which contradicts (3.35).

Step 7. The results of the previous step ensure that $\mathcal{I}_{u_0} = (0,\tau_{u_0})$ and the function $u : [0,\tau_{u_0}) \to X^\alpha$ given by

$$u(t,u_0) := u^\delta(t) \quad \text{whenever } \delta \in \mathcal{I}_{u_0} \text{ and } t \in [0,\delta]$$

is well-defined for any $u_0 \in X^\alpha$, $u(\cdot,u_0) \in C([0,\tau_{u_0});X^\alpha)$ and satisfies the integral equation

$$u(t,u_0) = e^{-At}u_0 + \int_0^t e^{-A(t-s)}F(u(s,u_0))ds, \quad \text{for each } t \in [0,\tau_{u_0}).$$

It then follows from Lemma 3.2.1 that $u(\cdot,u_0)$ is a local X^α solution of (3.1) on $[0,\tau_{u_0})$ which, as a result of Step 1, is unique in the considered class.

It remains to show that $[0,\tau_{u_0})$ is a *maximal interval of existence*, which means that

$$\text{if } \tau_{u_0} < +\infty \quad \text{then } \limsup_{t\to\tau_{u_0}^-}\|u(t,u_0)\|_{X^\alpha} = +\infty.$$

Assume at contrary that

$$\sup_{t\in[0,\tau_{u_0})}\|u(t,u_0)\|_{X^\alpha} \le \text{const} \quad \text{for some } \tau_{u_0} < +\infty,$$

and set

$$M := \sup_{t\in[0,\tau_{u_0})} \|F(u(t,u_0))\|_X.$$

Then calculations of Lemma 3.2.1 may be adapted and (3.7) (with $\delta = \alpha$ and $t_0 = \frac{\tau_{u_0}}{2}$, $t_1 = \tau_{u_0}$) gives

$$\begin{aligned}&\|u(t,u_0) - u(\bar{t},u_0)\|_{X^\alpha}\\&\quad\le \text{const.}(\alpha,\tau_0,\|u_0\|_{X^\alpha},M)(t-\bar{t})^\alpha, \quad \frac{\tau_{u_0}}{2} < \bar{t} < t < \tau_{u_0}.\end{aligned} \tag{3.39}$$

It is seen from (3.39), that there exists $\lim_{t\to\tau_{u_0}^-} u(t,u_0)$. Thus, $u : [0,\tau_{u_0}) \to X^\alpha$ may be extended to a function $\bar{u} \in C([0,\tau_{u_0}];X^\alpha)$ and $\bar{u}$ satisfies (3.37). The latter implies however that $\tau_{u_0} \in \mathcal{I}_{u_0}$ which contradicts (3.36). The proof of Theorem 3.1.2 is now completed. □

Additional regularity of X^α-solutions

From considerations of Lemma 3.2.1, extra smoothness of the time derivative of u will be concluded.

Corollary 3.3.1. Let the Assumption A hold and $u \in C([0,\tau);X^\alpha)$. If u satisfies the integral equation (3.5) in X, then

$$u \in C((0,\tau);X^1), \tag{3.40}$$

$$\dot{u} \in C((0,\tau);X^\gamma) \quad \text{for each } \gamma \in [0,1). \tag{3.41}$$

Proof. Since $u \in C((0,\tau);X^\alpha) \cap C^1((0,\tau);X)$ and $F : X^\alpha \to X$ is continuous, $Au = F(u) - \dot{u}$ belongs to $C((0,\tau);X)$ as written in (3.40).

Recalling the definition of u_1 in (3.14),

$$u_1(t) = \int_0^t e^{-A(t-s)}(F(u(s)) - F(u(t)))ds,$$

equation (3.22) reads

$$\dot{u} = -Ae^{-At}u_0 + e^{-At}F(u(t)) - Au_1(t),$$

and (3.41) follows from continuity of the maps

$$(0,\tau) \ni t \to Ae^{-At}u_0 \in X^\gamma, \quad (0,\tau) \ni t \to e^{-At}F(u(t)) \in X^\gamma \quad (\gamma \in [0,1))$$

and condition (3.15). □

The next proposition shows that solutions of (3.1) originating in a bounded set $B \subset X^\alpha$ exist at least until some common positive time $\tau_B > 0$.

Proposition 3.3.2. *Under the Assumption* A, *for any bounded set $B \subset X^\alpha$ there is a time $\tau_B > 0$ that solutions $u(t, u_0)$ of* (3.1) *having $u_0 \in B$ exist and are bounded in X^α (uniformly for $t \in [0, \tau_B]$ and $u_0 \in B$).*

Proof. Set $\rho = \sup_{\phi \in B} \|\phi\|_{X^\alpha}$, let $B_1 = B_{X^\alpha}(0, c_0\rho + 1)$ be a ball in X^α, where c_0 is a constant in the estimate

$$\|e^{-At}\|_{\mathcal{L}(X,X)} \le c_0 e^{-at}, \quad t \ge 0,$$

and define $b := \sup_{v \in B_1} \|F(v)\|_X$.

Using integral equation corresponding to (3.1), for $u_0 \in B$ and as long as $u(t, u_0) \in B_1$, we have an estimate

$$\begin{aligned}
\|u(t,u_0)\|_{X^\alpha} &\le c_0 e^{-at}\|u_0\|_{X^\alpha} + \int_0^t \|A^\alpha e^{-A(t-s)}\|_{\mathcal{L}(X,X)} \|F(u(s,u_0))\|_X ds \\
&\le c_0\rho + b\int_0^t c_\alpha \frac{e^{-a(t-s)}}{(t-s)^\alpha} ds \le c_0\rho + bc_\alpha \int_0^t \frac{dy}{y^\alpha} \\
&= c_0\rho + \frac{bc_\alpha}{1-\alpha} t^{1-\alpha}.
\end{aligned} \tag{3.42}$$

Hence the solutions starting from B will not leave B_1 until the second term in the final line of estimate (3.42) reaches the value 1. Such a time τ_B is given by the condition

$$\frac{bc_\alpha}{1-\alpha}\tau_B^{1-\alpha} = 1. \tag{3.43}$$

For $t \in [0, \tau_B]$, we thus have

$$\sup_{u_0 \in B} \|u(t,u_0)\|_{X^\alpha} \le c_0\rho + 1, \tag{3.44}$$

which completes the proof. □

The latter result will be useful in the proof that the solutions of (3.1) depend continuously on initial data.

Proposition 3.3.3. *Let the Assumption* A *be satisfied $\{u_n\} \subset X^\alpha$ and $u_n \to u_0$ in X^α. Then $T_0 := \inf\{\tau_{u_n}, n = 1, 2, \ldots\}$ is positive, with $T_0 \le \tau_{u_0}$ (τ_{u_n} being the lifetime of $u(t, u_n)$). Moreover,*

$$\forall_{T \in (0,T_0)} \sup_{t \in [0,T]} \|u(t,u_n) - u(t,u_0)\|_{X^\alpha} \to 0 \quad \text{when } n \to \infty. \tag{3.45}$$

Proof. Since $\{u_n\} \subset X^\alpha$ is convergent to u_0 in X^α, there is $\rho > 0$ such that

$$\|u_n - u_0\|_{X^\alpha} < \rho \quad \text{for all } n \in \mathbb{N} \tag{3.46}$$

and hence due to Proposition 3.3.2, $T_0 := \inf\{\tau_{u_n}, n = 1, 2, \ldots\}$ is a positive number. For any $T \in (0, T_0)$ and $n = 1, 2, \ldots$, we will show validity of the following implication:

$$\text{if } T' \in [0, T] \text{ and } \sup_{t\in[0,T']} \|u(t, u_n) - u(t, u_0)\|_{X^\alpha} \le \rho \text{ then}$$
$$\sup_{t\in[0,T']} \|u(t, u_n) - u(t, u_0)\|_{X^\alpha} \le \|u_n - u_0\|_{X^\alpha}\text{const.}(\alpha, \rho, T). \tag{3.47}$$

To justify this condition, based on the integral equation (3.5), we estimate the difference of the solutions $(u(t, u_n) - u(t, u_0))$. Lipschitz condition for F ensures that

$$\|F(u(t, u_n)) - F(u(t, u_0))\|_X \le L_\rho \|u(t, u_n) - u(t, u_0)\|_{X^\alpha}, \quad t \in [0, T'],$$

and we obtain that

$$\begin{aligned}
&\|u(t, u_n) - u(t, u_0)\|_{X^\alpha} \\
&\quad \le \left\| e^{-At}(u_n - u_0) + \int_0^t e^{-A(t-s)}[F(u(s, u_n)) - F(u(s, u_0))]ds \right\|_{X^\alpha} \\
&\quad \le c_0\|u_n - u_0\|_{X^\alpha} + c_\alpha L_\rho \int_0^t \frac{e^{-a(t-s)}}{(t-s)^\alpha} \|u(s, u_n) - u(s, u_0)\|_{X^\alpha} ds, \quad t \in [0, T'].
\end{aligned} \tag{3.48}$$

From (3.48), using Lemma 2.1.11, we get

$$\sup_{t\in[0,T']} \|u(t, u_n) - u(t, u_0)\|_{X^\alpha} \le c_0\|u_n - u_0\|_{X^\alpha}\text{const.}(c_\alpha L_\rho, \alpha, T'). \tag{3.49}$$

It follows from Lemma 2.1.11, that the const. appearing at the right-hand side in (3.49) is increasing with respect to its third argument, so that we can increase it to the value const.$(c_\alpha L_\rho, \alpha, T)$, which proves the claim.

Fix any $T \in (0, T_0)$, $n = 1, 2 \ldots$, and consider the solution $u(t, u_n)$. From (3.46) and continuity of the X^α-solution at $t = 0$, there exists $t_n > 0$ such that

$$\|u(t, u_n) - u(t, u_0)\|_{X^\alpha} < \rho \quad \text{for } t \in [0, t_n].$$

Let T_n' denote the (hypothetical) first positive time in which

$$\|u(t, u_n) - u(t, u_0)\|_{X^\alpha} = \rho.$$

If $T_n' \le T$ then, using (3.47), we find that

$$\sup_{t\in[0,T_n']} \|u(t, u_n) - u(t, u_0)\|_{X^\alpha} \le \|u_n - u_0\|_{X^\alpha}\text{const.}(\alpha, \rho, T).$$

However, since

$$\|u_n - u_0\|_{X^\alpha}\text{const.}(\alpha,\rho,T) \le \frac{\rho}{2} \quad \text{for } n \ge n_T,$$

we get, for each $n \ge n_T$, a contradiction with the hypothesis that $T_n' \le T$.

Thus $\tau_{u_0} \ge T_0$ and the following condition must be satisfied:

$$\forall_{T\in(0,T_0)}\, \exists_{n_T\in\mathbb{N}}\, \forall_{n\ge n_T} \sup_{t\in[0,T]} \|u(t,u_n) - u(t,u_0)\|_{X^\alpha} < \rho. \tag{3.50}$$

As a consequence of (3.50), (3.47), we get further

$$\forall_{T\in(0,T_0)}\, \exists_{n_T\in\mathbb{N}}\, \forall_{n\ge n_T} \sup_{t\in[0,T]} \|u(t,u_n) - u(t,u_0)\|_{X^\alpha} \le \|u_n - u_0\|_{X^\alpha}\text{const.}(\alpha,\rho,T)$$

and, therefore,

$$\lim_{n\to\infty} \sup_{t\in[0,T]} \|u(t,u_n) - u(t,u_0)\|_{X^\alpha} = 0 \quad \text{for each } T \in (0,T_0).$$

The proof is complete. □

Bibliographical notes

Formulation of the Cauchy problem was given in [86] with the correction of [134] and was remarked also in [83]. Sections 3.2, 3.3 are based on [86] and [126].

The definition of fractional powers was proposed by A. V. Balakrishnan in 1960 [11] and studied in a series of 6 papers by H. Komatsu through the years 1966–1972 ([106, 107] are among them). Application of that notion to abstract parabolic equation in Hilbert or Banach spaces was given by T. Kato and P. E. Sobolevskii around 1960; see the Comments on the results in Chapter XIV of K. Yosida's famous monograph [190]. An early reference is also the monograph [108] with Russian edition from 1967. Formulation of the Navier–Stokes equation as an integral equation in a Banach space connected with fractional powers of the Stokes operator was used in particular by the Japan mathematicians; among them T. Kato and H. Fujita [95, 67], Y. Giga and T. Miyakawa (e. g., [73–75]), and also by P. E. Sobolevskii [155, 156]. The 1981 monograph [86] by D. Henry provides a unified abstract approach, using the semigroup technique, to general parabolic equation and systems. That approach was extended in later monographs [32, 187].

4 Global in time continuation of solutions

In this chapter, we continue our study of an abstract Cauchy problem (3.1):

$$\begin{cases} u_t + Au = F(u), & t > 0, \\ u(0) = u_0, \end{cases}$$

under the Assumption A. Our concern now is the existence of the global in time solutions of (3.1) and conditions guaranteeing existence of the corresponding semigroup of global X^α-solutions.

Definition 4.0.1. A function $u = u(t)$ is called a global X^α-solution of (3.1) if it fulfills requirements of the Definition 3.1.1 with $\tau = +\infty$.

According to the condition (3.2) of Theorem 3.1.2, for the global X^α-solvability of (3.1) it suffices to check boundedness of the norm $\|u(t, u_0)\|_{X^\alpha}$ for all positive times. More precisely, the property,

$$\forall_{T>0} \limsup_{t \to T^-} \|u(t)\|_{X^\alpha} < +\infty.$$

Popular criteria for existence of the global solutions are connected with suitable limitation of the growth of the nonlinear term F (valid at least when F is operating on an arbitrary solution inside the considered class).

Sublinear growth restriction

It is in particular easy to observe that global solvability of (3.1) follows whenever the growth of F is not faster than linear function:

$$\|F(v)\|_X \leq \text{const.}(1 + \|v\|_{X^\alpha}), \quad \text{for } v \in X^\alpha. \tag{4.1}$$

Proof. For $u_0 \in X^\alpha$, denote by $u = u(t, u_0)$ the corresponding local X^α-solution of (3.1) obtained in Theorem 3.1.2. Such solution is known to exist on certain interval $[0, \tau_{u_0})$, maximal in the sense of (3.2). Assume that τ_{u_0} is finite. Simultaneously, u satisfies the integral equation

$$u(t, u_0) = e^{-At}u_0 + \int_0^t e^{-A(t-s)}F(u(s, u_0))ds \quad \text{for } t \in [0, \tau_{u_0}).$$

Applying (4.1) together with the usual estimates for analytic semigroups, we get

$$\begin{aligned} \|u(t, u_0)\|_{X^\alpha} &\leq \|e^{-At}\|_{\mathcal{L}(X,X)}\|u_0\|_{X^\alpha} + \int_0^t \|A^\alpha e^{-A(t-s)}\|_{\mathcal{L}(X,X)}\|F(u(s, u_0))\|_X ds \\ &\leq c_0\|u_0\|_{X^\alpha} + c_\alpha \,\text{const.} \int_0^t \frac{e^{-a(t-s)}}{(t-s)^\alpha}(1 + \|u(t, u_0)\|_{X^\alpha})ds, \quad t \in [0, \tau_{u_0}). \end{aligned}$$

https://doi.org/10.1515/9783110599831-004

Consider further the function $y(t) := \|u(t,u_0)\|_{X^\alpha}$. Due to Theorem 3.1.2, $y : [0,\tau_{u_0}) \to \mathbb{R}^+$ is a continuous function. From the estimate above, it is also seen that y satisfies a Volterra-type integral inequality

$$y(t) \le a + b\int_0^t \frac{1}{(t-s)^\beta} y(s)ds, \quad \text{for } t \in [0,\tau),$$

with $a = c_0\|u_0\|_{X^\alpha} + c_\alpha \,\text{const.} \int_0^{\tau_{u_0}} \frac{e^{-a(t-s)}}{(t-s)^\alpha}$ and $b = c_\alpha$const. Therefore, as a result of Lemma 2.1.11, an estimate holds

$$\sup_{t\in[0,\tau)} y(t) \le a\,\text{const.}(b,\beta,\tau),$$

which shows boundedness of $y(t)$. According to Theorem 3.1.2, all the solutions of (3.1) are then global in time. □

Unfortunately condition (4.1) is very restrictive. Moreover, it cannot be significantly weakened, which can be seen solving explicitly simple ordinary differential equations

$$z'(t) = z^{1+\epsilon}(t), \quad z(0) = z_0 > 0, \tag{4.2}$$

with $\epsilon > 0$. Hopefully, the restriction (4.1) will be improved provided that we have an additional knowledge concerning solutions of (3.1).

For this purpose, we shall introduce an additional Banach space Y and assume that all possible solutions of (3.1) are estimated in the norm of this space (known as the *"a priori estimate"*). This assumption allows to replace (4.1) through a more general condition (4.6), which is applicable to a large class of equations in the applied sciences.

4.1 Generation of nonlinear semigroups

If to any $u_0 \in X^\alpha$ corresponds a global X^α-solution $u(t,u_0)$ of (3.1), then a C^0-semigroup $\{T(t)\}$ corresponding to (3.1) will be defined on X^α through the relation

$$T(t)u_0 = u(t,u_0), \quad t \ge 0. \tag{4.3}$$

Remark 4.1.1. Recalling Definition 2.3.1, we discuss briefly properties of the family of maps $\{T(t)\}$ defined in (4.3). From the definition of the X^α-solution to (3.1), it follows immediately that $T(0)u_0 = u(0,u_0) = u_0$. Semigroup property:

$$T(t_1)T(t_2)u_0 = T(t_1+t_2)u_0, \quad t_1,t_2 \ge 0, \tag{4.4}$$

is a consequence of the *uniqueness* of solutions. By (4.3), $T(t_1)T(t_2)u_0 = T(t_1)u(t_2, u_0)$ and, since $u(t+t_2, u_0)$ is a unique X^α-solution of (3.1) which is equal to $u(t_2, u_0)$ at $t = 0$, then $T(t_1)u(t_2, u_0) = u(t_1 + t_2, u_0)$ and (4.4) holds. Continuity of the map

$$[0, \infty) \times X^\alpha \ni (t, u_0) \to T(t)u_0 = u(t, u_0) \in X^\alpha$$

follows from Proposition 3.3.3 and the property $u(\cdot, u_0) \in C([0, \infty); X^\alpha)$ (note that Proposition 3.3.3 reads that $T(t)u_0$ is continuous as a function of u_0 uniformly with respect to t varying in compact subsets of $[0, \infty)$).

We will formulate next an effective condition guaranteeing global X^α-solvability of (3.1). Furthermore, we will find for (3.1) such a set of requirements which, besides the existence of $\{T(t)\}$, would guarantee an additional property that orbits of bounded sets would be bounded.

4.1.1 Semigroups on X^α

We introduce the following two conditions:

(A_1) *Formula* (4.3) *defines on* X^α *a* C^0*-semigroup* $\{T(t)\}$ *of global* X^α*-solutions to* (3.1), *having orbits of bounded sets bounded.*

(A_2) *Given Banach space* Y *with* $D(A) \subset Y$, *a locally bounded function* $c : \mathbb{R}^+ \to \mathbb{R}^+$, *a nondecreasing function* $g : \mathbb{R}^+ \longrightarrow \mathbb{R}^+$, *and a certain number* $\theta \in [0, 1)$, *such that, for each* $u_0 \in X^\alpha$, *the conditions:*

$$\|u(t, u_0)\|_Y \le c(\|u_0\|_{X^\alpha}), \quad t \in (0, \tau_{u_0}), \tag{4.5}$$

and

$$\|F(u(t, u_0))\|_X \le g(\|u(t, u_0)\|_Y)(1 + \|u(t, u_0)\|_{X^\alpha}^\theta), \quad t \in (0, \tau_{u_0}), \tag{4.6}$$

are satisfied.

Condition (A_2) is often referred as the *subordination condition*. For many particular equations, it will be obtained using specific a priori estimates together with a suitable version of the Nirenberg–Gagliardo inequality.

Remark 4.1.2. The subordination condition (4.6) was written in another form in [29]:

$$\|F(u(t, u_0))\|_X \le g(\|u(t, u_0)\|_Y)(1 + \|u(t, u_0)\|_{X^{\alpha'}}), \quad t \in (0, \tau_{u_0}), \tag{4.7}$$

with some $\alpha' \in [0, 1)$, when $Y \subset X$. Equivalence of the two formulations follows through Proposition 2.2.34 and the estimate

$$\|u(t, u_0)\|_{X^{\alpha'}} \le c_{\frac{\alpha'}{\alpha}} \|u(t, u_0)\|_X^{1-\frac{\alpha'}{\alpha}} \|u(t, u_0)\|_{X^\alpha}^{\frac{\alpha'}{\alpha}}, \tag{4.8}$$

where $\alpha \in (\alpha', 1)$. The first right-hand side factor in (4.8) may be thus incorporated to g in (4.7) without changing its properties. Implication from (4.6) to (4.7) follows through the Young inequality.

We are able to formulate the following sufficient condition of the global solvability.

Theorem 4.1.3. *Under the Assumption* A, *conditions* (A_1) *and* (A_2) *are equivalent.*

Proof. Let $u_0 \in X^\alpha$ and $t > 0$. Using the integral equation,

$$u(t, u_0) = e^{-At}u_0 + \int_0^t e^{-A(t-s)}F(u(s, u_0))ds, \tag{4.9}$$

we obtain

$$\begin{aligned}\|u(t, u_0)\|_{X^\alpha} &\le \|A^\alpha e^{-At}u_0\|_X + \int_0^t \|A^\alpha e^{-A(t-s)}\|_{\mathcal{L}(X,X)}\|F(u(s, u_0))\|_X ds \\ &\le c_0 e^{-at}\|u_0\|_{X^\alpha} + \int_0^t c_\alpha \frac{e^{-a(t-s)}}{(t-s)^\alpha} g(\|u(s)\|_Y)(1 + \|u(s, u_0)\|^\theta_{X^\alpha})ds.\end{aligned} \tag{4.10}$$

Using further condition (4.5) and recalling the equality $\int_0^\infty \frac{e^{-ay}}{y^\alpha}dy = \frac{\Gamma(1-\alpha)}{a^{1-\alpha}}$, we get from the latter estimate that

$$\|u(t, u_0)\|_{X^\alpha} \le c_0\|u_0\|_{X^\alpha} + c_\alpha g(c(\|u_0\|_{X^\alpha}))\frac{\Gamma(1-\alpha)}{a^{1-\alpha}}\Big(1 + \sup_{\tau\in[0,t]}\|u(\tau)\|^\theta_{X^\alpha}\Big). \tag{4.11}$$

Since the right-hand side above is nondecreasing in t, (4.11) may be strengthened to

$$\sup_{\tau\in[0,t]}\|u(\tau, u_0)\|_{X^\alpha} \le c_0\|u_0\|_{X^\alpha} + g(c(\|u_0\|_{X^\alpha}))c_\alpha\frac{\Gamma(1-\alpha)}{a^{1-\alpha}}\Big(1 + \sup_{\tau\in[0,t]}\|u(\tau)\|^\theta_{X^\alpha}\Big).$$

Consequently, the real number $z := \sup_{\tau\in[0,t]}\|u(\tau, u_0)\|_{X^\alpha}$ fulfills the relation

$$0 \le z \le b(u_0)(1 + z^\theta), \tag{4.12}$$

where $b(u_0) := \max\{c_0\|u_0\|_{X^\alpha}, g(c(\|u_0\|_{X^\alpha}))c_\alpha\frac{\Gamma(1-\alpha)}{a^{1-\alpha}}\}$. However, any number z satisfying (4.12) is bounded by a positive root $z_1(u_0)$ of an algebraic equation

$$b(u_0)(1 + z^\theta) - z = 0,$$

so that $\sup_{\tau\in[0,t]}\|u(\tau, u_0)\|_{X^\alpha}$ remains bounded by $z_1(u_0)$ for all $t > 0$.

Hence, by Theorem 3.1.2, for each $u_0 \in X^\alpha$, the problem (3.1) has a global solution $u(t, u_0)$ and, moreover,

$$\sup_{\tau\in[0,\infty)}\|u(\tau, u_0)\|_{X^\alpha} \le z_1(u_0). \tag{4.13}$$

Therefore, (3.1) generates a strongly continuous semigroup of operators $T(t) : X^\alpha \longrightarrow X^\alpha$, $t \geq 0$, through (4.3). Additionally, as follows from (4.11)–(4.13), the orbit $\gamma^+(\mathcal{B}_{X^\alpha}(R))$ is bounded in X^α whenever function $c(\cdot)$ is locally bounded on $[0, \infty)$. This holds because $b(u_0)$ depends only on the radius R of a ball $B_{X^\alpha}(R)$ and is independent of the choice of particular $u_0 \in B_{X^\alpha}(R)$. The implication $A_2 \Rightarrow A_1$ is justified.

The proof of the opposite implication is formal. If (4.3) defines a C^0-semigroup $\{T(t)\}$ of global X^α-solutions to (3.1), having orbits of bounded sets bounded, then for each $r > 0$ there exists $m_r > 0$ such that

$$\sup_{u_0 \in B_{X^\alpha}(r)} \sup_{t \geq 0} \|u(t, u_0)\|_{X^\alpha} = m_r,$$

whereas $F : X^\alpha \to X$ is bounded on bounded sets as a consequence of the basic assumption. Then (A_2) holds with $Y = X^\alpha$, $c(s) = m_s$, $g(s) = \sup_{\|v\|_{X^\alpha} \leq s} \|F(v)\|_X$ and $\theta = 0$. The proof is complete. □

Remark 4.1.4. An implicit assumption of the above theorem is the inclusion $X^\alpha \subset Y$ (cf. the estimate (4.5)). Nevertheless, we skipped expressing it directly since, as seen further in Theorem 4.2.3, condition (4.5) may be also formulated in a weaker form (4.18), in which validity of this inclusion is not needed.

Remark 4.1.5. It should be also recalled (cf. Corollary 3.1.3) that, under the Assumption A, the nonlinear term F is Lipschitz continuous on bounded subsets of X^β whenever $\beta \in [\alpha, 1)$. Therefore, if condition (A_1) holds then, simultaneously, (3.1) generates C^0-semigroup $\{T_\beta(t)\}$ having orbits of bounded sets bounded on each of the spaces X^β, $\beta \in [\alpha, 1)$.

4.2 Smoothing properties of the semigroup

We will discuss further the smoothing action of the semigroup generated by (3.1) on X^α.

Smoothing effects

Assume the Assumption A and denote by $T(t)u_0$ a local X^α-solution of (3.1) originated at $u_0 \in X^\alpha$. Then, thanks to Proposition 3.3.2, for any $t \in (0, \tau_B)$ the image $T(t)B$ of a bounded set $B \subset X^\alpha$ will be well-defined.

Below, we shall prove that for each $t \in (0, \tau_B)$ the image $T(t)B$ (which, by the Definition 3.1.1, is contained in $D(A) = X^1$) remains bounded in $X^{\alpha+\varepsilon}$ whenever $\alpha+\varepsilon < 1$.

Proposition 4.2.1. *Under the Assumption A, let $T(t)u_0$ be a local X^α-solution of (3.1) corresponding to $u_0 \in X^\alpha$. Then*

$$\forall_{\substack{B \subset X^\alpha \\ B\text{-bounded}}} \forall_{t \in (0, \tau_B)} \; T(t)B \text{ is bounded in } X^{\alpha+\varepsilon} \text{ whenever } \alpha + \varepsilon < 1.$$

Proof. For bounded set $B \subset X^\alpha$, sufficiently large $\rho > 0$ such that $B \subset B_{X^\alpha}(\rho)$, $u_0 \in B$ and $t \in (0, \tau_B)$ (where τ_B is chosen to B in Proposition 3.3.2), we have

$$\|A^\varepsilon u(t,u_0)\|_{X^\alpha} \le \|A^\varepsilon e^{-At}\|_{\mathcal{L}(X,X)}\|u_0\|_{X^\alpha} + \int_0^t \|A^{\alpha+\varepsilon} e^{-A(t-s)}\|_{\mathcal{L}(X,X)}\|F(u(s))\|_X ds$$

$$\le c_\varepsilon \frac{e^{-at}}{t^\varepsilon}\rho + b\int_0^t c_{\alpha+\varepsilon}\frac{e^{-ay}}{y^{\alpha+\varepsilon}}dy,$$

where $b := \sup_{v\in B_{X^\alpha}(0,c\rho+1)} \|F(v)\|_X$. Therefore, $T(t)B$ is bounded in $X^{\alpha+\varepsilon}$, which completes the proof. □

Below we will prove a different in spirit smoothing result, which will be particularly useful in considerations of problems where the resolvent of A is not compact.

Theorem 4.2.2. *Let the Assumption* A *be satisfied and* (4.3) *define a* C^0*-semiflow* $T(t)$: $X^\alpha \to X^\alpha$, $t > 0$, *of global* X^α*-solutions of* (3.1). *Then, for each* $t > 0$, *the* $T(t)$ *images of the sets that are bounded in* X^α *and precompact in* X *are precompact in* X^α.

Proof. We should show that if B is a bounded subset of X^α which is also precompact in X, then for each $t > 0$ the image $T(t)B$ is precompact in X^α.

Fix $t > 0$ and consider a sequence $\{u_n\} \subset B$. It suffices to show that $\{T(t)u_n\}$ has a subsequence convergent in X^α. Since $\{u_n\}$, as a subset of B, is bounded, then from compactness of $\mathrm{cl}_X B$ in X, there exists a subsequence $\{u_{n_k}\}$ which is convergent in X. In addition, since the sequence $\{u_{n_k}\} \subset B$ and B is bounded in X^α, the set $\{T(s)u_{n_k}, s \in [0,\tau_B], k \in \mathbb{N}\}$ (where $\tau_B \in (0,t)$ is chosen to B according to Proposition 3.3.2) is bounded in X^α. Therefore, through the Assumption A, there exists Lipschitz constant $L(\tau_B, B)$ such that

$$\|F(T(s)u_{n_r}) - F(T(s)u_{n_k})\|_X \le L(\tau_B, B)\|T(s)u_{n_r} - T(s)u_{n_k}\|_{X^\alpha}, \tag{4.14}$$

for all $s \in [0,\tau_B]$ and $r, k \in \mathbb{N}$.

Writing integral equation for the difference $[T(\tau)u_{n_r} - T(\tau)u_{n_k}]$:

$$T(\tau)u_{n_r} - T(\tau)u_{n_k} = e^{-A\tau}[u_{n_r} - u_{n_k}] + \int_0^\tau e^{-A(\tau-s)}[F(T(s)u_{n_r}) - F(T(s)u_{n_k})]ds, \quad \tau \in (0,\tau_B), \tag{4.15}$$

and estimating in a standard way, applying condition (4.14), we obtain

$$\|[T(\tau)u_{n_r} - T(\tau)u_{n_k}]\|_{X^\alpha} \le \|A^\alpha e^{-A\tau}\|_{\mathcal{L}(X,X)}\|u_{n_r} - u_{n_k}\|_X + \int_0^\tau \|A^\alpha e^{-A(\tau-s)}\|_{\mathcal{L}(X,X)}\|F(T(s)u_{n_r}) - F(T(s)u_{n_k})\|_X ds$$

$$\leq c_0 \frac{e^{-a\tau}}{\tau^\alpha} \|u_{n_r} - u_{n_k}\|_X + \int_0^\tau c_\alpha \frac{e^{-a(\tau-s)}}{(\tau-s)^\alpha} L(\tau_B, B) \|A^\alpha [T(s)u_{n_r} - T(s)u_{n_k}]\|_{X^\alpha} ds$$

$$\leq \frac{c_0 \|u_{n_r} - u_{n_k}\|_X}{\tau^\alpha} + c_\alpha L(\tau_B, B) \int_0^\tau \frac{1}{(\tau-s)^\alpha} \|T(s)u_{n_r} - T(s)u_{n_k}\|_{X^\alpha} ds, \quad \tau \in (0, \tau_B). \tag{4.16}$$

Let us choose any $0 < t^* < \min\{\tau_B, t\}$ satisfying

$$\frac{c_\alpha L(\tau_B, B) t^{*^{1-\alpha}}}{2^{1-2\alpha}} \frac{2}{1-\alpha} \leq \frac{1}{2}.$$

As a consequence of Lemma 2.1.11, inequality (4.16) extends to

$$\|[T(t^*)u_{n_r} - T(t^*)u_{n_k}]\|_{X^\alpha} \leq \frac{2c_0 \|u_{n_r} - u_{n_k}\|_X}{t^{*\alpha}}. \tag{4.17}$$

Since $\{u_{n_k}\}$ was a Cauchy sequence in X, it follows from (4.17) that $\{T(t^*)u_{n_k}\}$ is a Cauchy sequence in X^α, so that $\{T(t^*)u_{n_k}\}$ is convergent in the topology of X^α. Therefore, $\{T(t)u_{n_k}\} = \{T(t - t^*)T(t^*)u_{n_k}\}$ is convergent in X^α as a consequence of continuity of $T(t)$. The proof is complete. □

Further existence result

As a consequence of the smoothing action described above, the existence result of Theorem 4.1.3 will be strengthened. Let us introduce the following set of requirements (A_2') being a counterpart of (A_2).

(A_2') *Given a Banach space Y with $D(A) \subset Y$, a function $c : [0,\infty) \times (0,\infty) \to \mathbb{R}$ for which $c(\cdot, s)$ is locally bounded for each fixed $s > 0$, a nondecreasing function $g : \mathbb{R}^+ \longrightarrow \mathbb{R}^+$, and a number $\theta \in [0,1)$, such that, for each $u_0 \in X^\alpha$, both conditions*

$$\|u(t, u_0)\|_Y \leq c(\|u_0\|_{X^\alpha}, s), \quad 0 < s < t < \tau_{u_0} \tag{4.18}$$

and

$$\|F(u(t, u_0))\|_X \leq g(\|u(t, u_0)\|_Y)(1 + \|u(t, u_0)\|_{X^\alpha}^\theta), \quad t \in (0, \tau_{u_0}) \tag{4.19}$$

are satisfied.

Such extended formulation of the requirements shows that the knowledge of the Y-estimate of solutions is important only for t strictly positive. In particular, the Y-estimate may be singular at $t = 0$ and this, in fact, does not influence the behavior of the system in the phase space X^α.

We proceed now to the proof of the following.

Theorem 4.2.3. *Under the Assumption* A, *conditions* (A_1) *and* (A_2') *are equivalent.*

Proof. Let us take $R > 0$, $\|u_0\|_{X^\alpha} < R$, quantity $\tau_{B_{X^\alpha}(R)}$ satisfying condition (3.43) of Proposition 3.3.2 and any $t \in [\tau_{B_{X^\alpha}(R)}, \tau_{u_0})$. Writing an integral equation equivalent to (3.1) in the form

$$u(t,u_0) = e^{-At}u_0 + \left(\int_0^{\tau_{B_{X^\alpha}(R)}} + \int_{\tau_{B_{X^\alpha}(R)}}^t \right) e^{-A(t-s)}F(u(s,u_0))ds, \tag{4.20}$$

we obtain that

$$\begin{aligned} \|u(t,u_0)\|_{X^\alpha} &\le c_0\|u_0\|_{X^\alpha} + 1 \\ &+ \int_{\tau_{B_{X^\alpha}(R)}}^t c_\alpha \frac{e^{-a(t-s)}}{(t-s)^\alpha} g(\|u(s)\|_Y)(1 + \|u(s,u_0)\|_{X^\alpha}^\theta)ds. \end{aligned} \tag{4.21}$$

Application of (4.18) to the right-hand side of (4.21) gives

$$\begin{aligned} \|u(t,u_0)\|_{X^\alpha} &\le c_0R + 1 \\ &+ g(c(\|u_0\|_{X^\alpha}, \tau_{B_{X^\alpha}(R)}))c_\alpha \frac{\Gamma(1-\alpha)}{a^{1-\alpha}} \left(1 + \sup_{\tau\in[0,t]} \|u(\tau)\|_{X^\alpha}^\theta\right). \end{aligned} \tag{4.22}$$

Therefore, $z := \sup_{\tau\in[0,t]} \|u(\tau,u_0)\|_{X^\alpha}$ fulfills condition (4.12) of Theorem 4.1.3 with $b(u_0)$ defined now as

$$b(u_0) := \max\left\{c_0R + 1, g(c(\|u_0\|_{X^\alpha}, \tau_{B_{X^\alpha}(R)}))c_\alpha \frac{\Gamma(1-\alpha)}{a^{1-\alpha}}\right\}.$$

The proof can be finished analogously to that of Theorem 4.1.3. □

4.3 Compactness results

One of the advantages of the parabolic initial boundary value problems in bounded domain is that the corresponding semigroups are usually compact. The latter provides the strong control over the trajectories of a system and, essentially, is related to the compactness of the resolvent of the operator appearing in the main part of the equation.

We start with the necessary and sufficient condition for compactness of a C^0-semigroup of bounded linear operators which is due to A. Pazy [138].

Proposition 4.3.1. *A* C^0*-semigroup* $\{T(t)\}$ *of bounded linear operators is compact for* $t > 0$ *if and only if it is continuous in the uniform operator topology for* $t > 0$ *and the resolvent of its infinitesimal generator A is a compact linear operator for every* λ *in the resolvent set* $\rho(A)$ *of A.*

In this vein, we continue next the studies of the problem (3.1).

Theorem 4.3.2. *Let the Assumption A hold, the resolvent of A be compact, and the relation* (4.3) *define a* C^0*-semiflow* $T(t) : X^\alpha \to X^\alpha$, $t > 0$, *of global* X^α*-solutions of* (3.1). *Then, for each* $t > 0$, $T(t)$ *is a compact map on* X^α.

Proof. Let us fix $t > 0$ and bounded subset B of X^α. Then choose τ_B according to Proposition 3.3.2 and take $t_1 \in (0, \max\{t_1, \tau_B\})$. From Proposition 3.3.2, it may be seen that $T(t_1)B$ is bounded in $X^{\alpha+\varepsilon}$ for $\alpha + \varepsilon < 1$. Since the resolvent of A is compact, we know that the inclusion $X^{\alpha+\varepsilon} \subset X^\alpha$ is compact and $T(t_1)B$ is thus precompact in X^α. Continuity of the map $T(t - t_1)$ in X^α ensures then that the image $T(t)B = T(t - t_1)T(t_1)B$ will be precompact in X^α. The proof is complete. □

Remark 4.3.3. Such theorem was proved in [83, Theorem 4.2.2]. The proof given there was based on compactness of the maps e^{-At} for $t > 0$ which in order follows from compactness of the embeddings between the domains of fractional powers of A (implied by the compactness of the resolvent of A). Theorem 4.3.2 may be also concluded from Theorem 4.2.2 based on compactness of the inclusion $X^\alpha \subset X$, while the property of $T(t)$ described in Theorem 4.2.2 is more general. As shown below, it will be possible to prove compactness of $T(t)$ on a suitably chosen subspace of X^α, despite the generic noncompactness of the resolvent of A.

Corollary 4.3.4. *Under the assumptions of Theorem* 4.3.2, *the semigroup* $\{T(t)\}$ *generated by* (3.1) *on* X^α *is completely continuous. More precisely, for each bounded subset* B *of* X^α *and each* $T \geq 0$, *the following conditions hold:*

$$\text{the union } \bigcup_{0\le t\le T} T(t)B \text{ is bounded in } X^\alpha, \tag{4.23}$$

$$\text{the union } \bigcup_{t_1\le t\le T} T(t)B \text{ is a precompact set in } X^\alpha \text{ for each } t_1 \in (0, T). \tag{4.24}$$

Proof. Since, by Theorem 4.3.2, $\{T(t)\}$ is compact semigroup on X^α, (4.24) is a consequence of Lemma 2.3.24. Condition (4.23) follows then from (4.24) and Lemma 3.3.2. □

The compactness result of Theorem 4.3.2 was proved under the essential assumption that the resolvent of A was compact. Nevertheless, as shown below, Theorem 4.2.2 allows to make $T(t)$ compact on a suitable metric subspace of X^α even in the case when this latter assumption is not satisfied.

Assumption. *Consider a Banach space* Z *and the product Banach space* $X^\alpha \cap Z$ *with the usual norm*

$$\|\cdot\|_{X^\alpha\cap Z} := \|\cdot\|_{X^\alpha} + \|\cdot\|_Z.$$

Let $X^\alpha \cap Z$ *be compactly embedded in* X, *and assume that there is a subset* V_α *of the product* $Z \cap X^\alpha$ *such that:*

(i) *V_α is a bounded set in Z,*
(ii) *$T(t)V_\alpha \subset V_\alpha$, for $t \geq 0$ (i. e., V_α is $T(t)$ positively invariant).*

Now, the following generalization of Theorem 4.2.2 can be proved.

Theorem 4.3.5. *Let the Assumption* A *be satisfied and* (4.3) *define a C^0-semiflow $T(t) : X^\alpha \to X^\alpha$, $t > 0$, of global X^α-solutions of* (3.1). *If, in addition, the Assumption above is satisfied, then for each $t > 0$, a map $T(t) : \mathrm{cl}_{X^\alpha} V_\alpha \to \mathrm{cl}_{X^\alpha} V_\alpha$ is compact.*

Proof. From (ii) in the Assumption above, C^0-semigroup $\{T(t)\}$ defined by (3.1) can be restricted to a complete metric space $\mathrm{cl}_{X^\alpha} V_\alpha$, so that we have a family of mappings

$$T(t) : \mathrm{cl}_{X^\alpha} V_\alpha \to \mathrm{cl}_{X^\alpha} V_\alpha, \quad t \geq 0.$$

Let $B \subset \mathrm{cl}_{X^\alpha} V_\alpha$ be bounded, nonempty, and $d(B)$ denote the diameter of B. For fixed $v_0 \in B$, the following inclusions hold:

$$B \subset \{v \in \mathrm{cl}_{X^\alpha} V_\alpha : \|v - v_0\|_{X^\alpha} < d(B) + 1\} := K, \quad B \subset \mathrm{cl}_{X^\alpha}(K \cap V_\alpha).$$

Since $T(t)$ is continuous in the topology of X^α, we obtain further that

$$T(t)B \subset T(t)(\mathrm{cl}_{X^\alpha}[K \cap V_\alpha]) \subset \mathrm{cl}_{X^\alpha} T(t)(K \cap V_\alpha). \tag{4.25}$$

From (i) in the Assumption and the definition of K, the product $K \cap V_\alpha$ is seen to be bounded both in X^α and Z. Therefore, $K \cap V_\alpha$ is precompact in X and, as a consequence of Theorem 4.2.2, the image $T(t)(K \cap V_\alpha)$ is precompact in X^α. From inclusion (4.25), the set $T(t)B$ is precompact in X^α, which completes the proof of Theorem 4.3.5. □

Corollary 4.3.6. *Let $\{T(t)\}$ be a C^0-semigroup in a metric space X and W be a dense subset of X. Then the following two conditions are equivalent:*
(i) *for each open ball $B_X(r)$, the set $T(t)[W \cap B_X(r)]$ is precompact in X,*
(ii) *for each bounded subset B of X, the set $T(t)B$ is precompact in X.*

4.4 Equation (3.1) in fractional power scale

All of the reported above results concerning local and global X^α-solvability of semilinear equation (3.1) considered in the base space X can be generalized to the case when the problem is shifted from X onto another level $X^\beta = D(A^\beta)$, $\beta \in \mathbb{R}$, of the *fractional power scale* connected with the positive sectorial operator A. We will sketch the corresponding generalization recalling however, [6, Chapter V], for a more formal presentation.

Studying semilinear equation with sectorial operator (3.1), it is often convenient to consider it on a particular level of the fractional powers scale X^β, $\beta \in \mathbb{R}$, instead of X. As well known (e. g., [118, (3.9)]),

$$A^\mu : D(A^\alpha) \to D(A^{\alpha-\mu}), \quad \alpha, \mu \in \mathbb{R},$$

as an isometry on the fractional powers scale.

Observe next, following [145, p. 17], that when dealing with the Cauchy problem (3.1) in the base space X^β, where $F : X^\gamma \to X^\beta$ is *Lipschitz continuous on bounded sets* of X^γ and $0 \le \gamma - \beta < 1$, we can set $v(t) := A^\beta u(t)$, $t \ge 0$, and $v_0 = A^\beta u_0 \in X^{\gamma-\beta}$. Then v fulfills formally the equation in X,

$$\begin{aligned} &v_t + Av = \tilde{F}(v) = A^\beta \circ F(A^{-\beta} v), \\ &v(0) = A^\beta u_0 \in X^{\gamma-\beta}. \end{aligned} \tag{4.26}$$

The latter equation is obtained applying the operator A^β to (3.1).

Bibliographical notes

Subordination condition (A_2) proves to be a very effective tool in verification of the global solvability of various equations known in the applied sciences. First, the condition for global solvability of (3.1) of that type can be found in [66, p. 180]. It was later generalized in [5], [176] to the form similar to (A_2), again in connection with global solvability of semilinear parabolic problems. Finally, as was observed in [30], this condition slightly modified is also sufficient for the existence of a global attractor for such type problems.

Systematic study of evolution equations in fractional power scale was presented in [6, Chapter V]. A classical reference will serve [145], followed by its recent extension [35].

5 Definitions, properties, estimates, and inequalities

The present chapter contains information necessary for studying examples reported in Chapters 6–11. We recall first a collection of definitions of fractional power of operator. Various authors are using various definitions of fractional power which are most pleasant for them in their studies. Fortunately, in case of the basic elliptic operators like the negative Laplacian considered in $L^p(\mathbb{R}^N)$ (or some other spaces over $\mathbb{R}^N$), the popular definitions of fractional powers are equivalent (see [109]). The situation is not that good when we study negative Dirichlet Laplacian in a bounded regular domain. Only some definitions of its fractional powers are known to be equivalent in that case.

Next, we compare the notions of *positive* and *nonnegative* operators in a Banach spaces. An example of the negative Laplace operator in $\mathbb{R}^N$ forced us to formulate certain properties and estimates for the general class of nonnegative operators (compare [130]). While some useful properties of operators are valid only in a smaller class of positive operators (like the negative Dirichlet Laplacian in bounded regular domain $\Omega \subset \mathbb{R}^N$). We compare the two classes here.

When estimating solutions of critical problems, we need often to use the strongest versions of the standard estimates and inequalities. We quote here in particular several versions of the *commutator estimates*, and some borderline versions of the Sobolev-type embeddings. We also discuss several *moment inequalities* (called also *interpolation inequalities*), controlling carefully the constants appearing in it. Finally, we include a version of the Kato–Beurling–Deny inequality needed when estimating solutions of the Dirichlet problem for equations with fractional operator in the main part.

5.1 Various definitions of the fractional Laplace operator

This section is based on a review article [109]. We recall several definitions of the fractional Laplace operator $L = -(-\Delta)^{\alpha/2}$ in $\mathbb{R}^N$, with $\alpha \in (0,2)$ and $N \in \{1,2,\ldots\}$. Several such definitions can be found in literature: L as a Fourier multiplier with symbol $-|\xi|^\alpha$, as a fractional power in the sense of Bochner or Balakrishnan, as the inverse of the Riesz potential operator, as a singular integral operator, as an operator associated to an appropriate Dirichlet form, as an infinitesimal generator of an appropriate semigroup of contractions, or as the Dirichlet-to-Neumann operator for an appropriate harmonic extension problem. $L^p = L^p(\mathbb{R}^N)$ denotes the Lebesgue space for $p \in [0,\infty)$, $C_0 = C_0(\mathbb{R}^N)$ denotes the space of continuous functions vanishing at infinity, and $C_{bu} = C_{bu}(\mathbb{R}^N)$ denotes the spaces of bounded uniformly continuous functions. Note that we restrict α to $(0,2)$. Let X be any of spaces L^p, $p \in [0,\infty)$, C_0 or C_{bu}, and let $f \in X$. The following definitions of $Lf \in X$ are equivalent; see [109] for the complete proofs.

https://doi.org/10.1515/9783110599831-005

5.1.1 Fourier transform definition

Definition 5.1.1. The fractional Laplace operator in $\mathbb{R}^N$ is given by

$$\mathcal{F}(L_F f)(\xi) = -|\xi|^\alpha \mathcal{F}f(\xi). \tag{5.1}$$

More formally, let $X = L^p(\mathbb{R}^N)$ is the Lebesgue space with $p \in [1,2]$. We say that $f \in D(L_F, X)$ whenever $f \in X$ and there is $L_F f \in X$ such that (5.1) holds.

Observe that the Laplace operator Δ takes diagonal form in the Fourier variable: it is a Fourier multiplier with symbol $-|\xi|^2$, namely $\mathcal{F}(\Delta f)(\xi) = -|\xi|^2 \mathcal{F}f(\xi)$ for $f \in \mathcal{S}$, where $\mathcal{S}$ denotes the class of Schwartz functions, and $\mathcal{S}'$ is space of Schwartz distributions. By the spectral theorem, the operator $L = -(-\Delta)^{\alpha/2}$ also takes diagonal form in the Fourier variable: it is a Fourier multiplier with symbol $-|\xi|^\alpha$.

5.1.2 Distributional definition

The Fourier transform of a convolution of two functions is the product of their Fourier transforms, and $-|\xi|^\alpha$ is a Schwartz distribution, so it is the Fourier transform of some $\tilde{L} \in \mathcal{S}'$. Therefore, the fractional Laplace operator L is the convolution operator with kernel $\tilde{L}$. Unfortunately, the convolution of two Schwartz distributions is not always well-defined. Nevertheless, the following definition is a very general one.

Definition 5.1.2. Let $\tilde{L}$ be the distribution in $\mathcal{S}'$ with Fourier transform $-|\xi|^\alpha$. The distributional fractional Laplace operator is given by the convolution

$$L_W f = \tilde{L} * f. \tag{5.2}$$

We write $f \in D(L_W, \mathcal{S}')$ whenever $f \in \mathcal{S}'$ and the convolution of $\tilde{L}$ and f is well-defined in $\mathcal{S}'$, that is, for all $\varphi, \psi \in \mathcal{S}$ the functions $\tilde{L} * \varphi$ and $f * \psi$ are convolvable in the usual sense, and

$$L_W f * (\varphi * \psi) = (\tilde{L} * \varphi) * (f * \psi). \tag{5.3}$$

We write $f \in D(L_W, X)$, when both f and $L_W f$ belong to X.

Note that $f \in D(L_W, X)$ if and only if $f \in X$ and there is $L_W f \in X$ such that

$$L_W f * \varphi = (\tilde{L} * \varphi) * f.$$

The last equality follows from (5.3) with $\varphi = \varphi_n$ to be an approximate identity, and passing to the limit. Conversely, a function $L_W f \in X$ with the above property satisfies (5.3).

5.1.3 Bochner's definition

If an operator generates a strongly continuous semigroup on a Banach space, its fractional power can be defined using Bochner's subordination. By integrating by parts the integral defining $\Gamma(1-\frac{\alpha}{2})$, we have

$$\lambda^{\alpha/2} = \frac{1}{|\Gamma(-\frac{\alpha}{2})|}\int_0^\infty (1-e^{-\lambda t})t^{-1-\alpha/2}dt.$$

Therefore, at least in the sense of spectral theory on L^2,

$$L = -(-\Delta)^{\alpha/2} = \frac{1}{|\Gamma(-\frac{\alpha}{2})|}\int_0^\infty (e^{t\Delta}-1)t^{-1-\alpha/2}dt.$$

Here, $e^{t\Delta}$ is the convolution operator with the Gauss–Weierstrass kernel $k_t(z)$.

Definition 5.1.3. Bochner's definition of the fractional Laplace operator is given by

$$\begin{aligned} L_{\mathcal{B}}f(x) &= \frac{1}{|\Gamma(-\frac{\alpha}{2})|}\int_0^\infty (f * k_t(x) - f(x))t^{-1-\alpha/2}dt \\ &= \frac{1}{|\Gamma(-\frac{\alpha}{2})|}\int_0^\infty \left(\int_{\mathbb{R}^d} (f(x+z)-f(x))k_t(z)dz\right)t^{-1-\alpha/2}dt. \end{aligned} \tag{5.4}$$

More formally, we say that $f \in D(L_B, x)$ if the integrals converge for a given $x \in \mathbb{R}^N$. If X is a Banach space, $f \in X$ and $\|f * k_t - f\|_X t^{-1-\alpha/2}$ is integrable in $t \in (0,\infty)$, then the first expression for $L_{\mathcal{B}}f(x)$ in (5.4) can be understood as the Bochner's integral of a function with values in X, and in this case we write $f \in D(L_B, X)$.

5.1.4 Balakrishnan's definition

A similar approach uses the representation of $\lambda^{\alpha/2}$ as a complete Bernstein function,

$$\lambda^{\alpha/2} = \frac{\sin\frac{\alpha\pi}{2}}{\pi}\int_0^\infty \frac{\lambda}{s+\lambda}s^{\alpha/2-1}ds$$

(which, through a substitution $s = \lambda(1-t)/t$, reduces to a beta integral). This definition of fractional power of a dissipative operator was introduced by Balakrishnan.

Definition 5.1.4. The Balakrishnan definition of the fractional Laplace operator is given by

$$L_{\hat{\mathcal{B}}}f(x) = \frac{\sin\frac{\alpha\pi}{2}}{\pi}\int_0^\infty \Delta(sI-\Delta)^{-1}f(x)s^{\alpha/2-1}ds. \tag{5.5}$$

More formally, if X is a Banach space on which Δ has a strongly continuous family of resolvent operators $(sI - \Delta)^{-1}$, $s > 0$, $f \in X$ and $\|\Delta(sI + \Delta)^{-1} f\|_X s^{\alpha/2-1}$ is integrable in $s \in (0, \infty)$, then the integral (5.5) can be understood as Bochner's integral of a function with values in X, and in this case we write $f \in D(L_{\widehat{B}}, X)$.

A complete treatment of the theory of fractional powers of operators is given in the monograph [130].

5.1.5 Singular integral definition

If the order of integration in (5.4) could be reversed, we would have $L_B f(x) = \int_{\mathbb{R}^N} (f(x + z) - f(x))\nu(z)dz$, where using a substitution $t = |z|^2/(4s)$,

$$\begin{aligned} \nu(z) &= \frac{1}{|\Gamma(-\frac{\alpha}{2})|} \int_0^\infty k_t(z) t^{-1-\alpha/2} dt \\ &= \frac{1}{2^N \pi^{N/2} |\Gamma(-\frac{\alpha}{2})|} \int_0^\infty t^{-1-(N+\alpha)/2} e^{-|z|^2/(4t)} dt \\ &= \frac{2^\alpha}{\pi^{N/2} |\Gamma(-\frac{\alpha}{2})| |z|^{N+\alpha}} \int_0^\infty s^{-1+(N+\alpha)/2} e^{-s} ds \\ &= \frac{2^\alpha \Gamma(\frac{N+\alpha}{2})}{\pi^{N/2} |\Gamma(-\frac{\alpha}{2})| |z|^{N+\alpha}}. \end{aligned} \tag{5.6}$$

This leads to the classical pointwise definition of the fractional Laplace operator.

Definition 5.1.5. The fractional Laplace operator is given by the Cauchy principal value integral

$$\begin{aligned} L_I f(x) &= \lim_{r \to 0^+} c_{N,\alpha} \int_{\mathbb{R}^N \setminus B_r} (f(x + z) - f(x)) \frac{1}{|z|^{N+\alpha}} dz \\ &= \lim_{r \to 0^+} \int_{\mathbb{R}^N} (f(x + z) - f(x)) \nu_r(z) dz, \end{aligned} \tag{5.7}$$

where

$$\nu_r(z) = \frac{c_{N,\alpha}}{|z|^{N+\alpha}} \mathbf{1}_{\mathbb{R}^N \setminus B_r}, \quad c_{N,\alpha} = \frac{2^\alpha \Gamma(\frac{N+\alpha}{2})}{\pi^{N/2} |\Gamma(-\frac{\alpha}{2})|}. \tag{5.8}$$

We write $f \in D(L_I, x)$ if the limit in (5.7) exists for a given $x \in \mathbb{R}^N$. We write $f \in D(L_I, X)$ if $f \in X$ and the limit in (5.7) exists in X.

The following variant of (5.7) is used in probability theory.

Definition 5.1.6. The fractional Laplace operator is given by

$$L_{\hat{I}}f(x) = \int_{\mathbb{R}^d} (f(x+z) - f(x) - \nabla f(x) \cdot z\mathbf{1}_B(z))\nu(z)dz. \tag{5.9}$$

We write $f \in D(L_{\hat{I}}, x)$ if $\nabla f(x)$ exists and the above integral converges absolutely. Another regularization of the singular integral (5.7) can be found in analysis.

Definition 5.1.7. The fractional Laplace operator is given by

$$L_{\check{I}}f(x) = \frac{1}{2}\int_{\mathbb{R}^d} (f(x+z) + f(x-z) - 2f(x))\nu(z)dz. \tag{5.10}$$

We write $f \in D(L_{\check{I}}, x)$ if the above integral converges absolutely. We quote from [109] in the following.

Proposition 5.1.8. *If $f \in D(L_{\hat{I}}, x)$, then $f \in D(L_I, x)$, and $L_{\hat{I}}f(x) = L_If(x)$. Similarly, if $f \in D(L_{\check{I}}, x)$, then $f \in D(L_I, x)$, and $L_{\check{I}}f(x) = L_If(x)$.*

See [109] for the proof.

5.1.6 Dynkin's definition

The following definition introduces L as the Dynkin characteristic operator of the isotropic α-stable Lévy process.

Definition 5.1.9. The definition of the fractional Laplace operator as the Dynkin characteristic operator is given by

$$\begin{aligned} L_Df(x) &= \lim_{r\to 0^+} c_{N,\alpha} \int_{\mathbb{R}^N\setminus \overline{B}_r} (f(x+z) - f(x)) \frac{1}{|z|^N(|z|^2 - r^2)^{\alpha/2}} dz \\ &= \lim_{r\to 0^+} \int_{\mathbb{R}^N} (f(x+z) - f(x))\tilde{\nu}_r(z)dz, \end{aligned} \tag{5.11}$$

where

$$\tilde{\nu}_r(z) = \frac{c_{N,\alpha}}{|z|^N(|z|^2 - r^2)^{\alpha/2}} \mathbf{1}_{\mathbb{R}^N\setminus \overline{B}_r}(z),$$

and $c_{N,\alpha}$ is given by (5.8). We write $f \in D(L_D, x)$ if the limit in (5.11) exists for a given $x \in \mathbb{R}^N$, and $f \in D(L_D, X)$ if $f \in X$ and the limit in (5.11) exists in X.

Following [109], we quote the following.

Proposition 5.1.10. *If $f \in D(L_D, x)$, then $f \in D(L_I, x)$, and $L_Df(x) = L_If(x)$.*

See [109] for the proof.

It is easy to construct $f \in D(L_I, x)$ which is not in $D(L_D, x)$.

Example. Let

$$f(x+z) = \sum_{n=1}^{\infty} \frac{\epsilon_n}{(|z|^2 - r_n^2)^{1-\alpha/2}} \mathbf{1}_{[r_n, 2r_n]}(|z|),$$

where, for example, $r_n = 2^{-n}$ and $\epsilon_n = 5^{-n}$. Then

$$\int_{\mathbb{R}^d} f(x+z)\nu(z)dz = \sum_{n=1}^{\infty} \frac{c_{N,\alpha}\epsilon_n}{r_n^2} \int_{B_2 \setminus B_1} \frac{1}{|y|^{N+\alpha}(|y|^2 - 1)^{1-\alpha/2}} dy < \infty,$$

so $f \in D(L_I, x)$, but

$$\int_{\mathbb{R}^d} f(x+z)\tilde{\nu}_{r_n}(z)dz = \infty,$$

so that $f \notin D(L_D, x)$.

5.1.7 Quadratic form definition

A self-adjoint operator on L^2 is completely described by its quadratic form. By Fubini, for all $r > 0$ and all integrable f, g,

$$\begin{aligned}
&\int_{\mathbb{R}^N} \left(\int_{\mathbb{R}^N} (f(x+z) - f(x))\nu_r(z)dz \right) \overline{g(x)}dx \\
&= \int_{\mathbb{R}^N} \int_{\mathbb{R}^N} (f(y) - f(x))\overline{g(x)}\nu_r(y-x)dydx \\
&= \frac{1}{2} \int_{\mathbb{R}^N} \int_{\mathbb{R}^N} (f(y) - f(x))\overline{g(x)}\nu_r(y-x)dydx \\
&\quad + \frac{1}{2} \int_{\mathbb{R}^N} \int_{\mathbb{R}^N} (f(x) - f(y))\overline{g(y)}\nu_r(y-x)dydx \\
&= \frac{1}{2} \int_{\mathbb{R}^N} \int_{\mathbb{R}^N} (f(y) - f(x))(\overline{g(x)} - \overline{g(y)})\nu_r(y-x)dydx.
\end{aligned}$$

A formal limit as $r \to 0^+$ leads to the following definition.

Definition 5.1.11. Let

$$\mathcal{E}(f, g) = \frac{c_{N,\alpha}}{2} \int_{\mathbb{R}^N} \int_{\mathbb{R}^N} \frac{(f(y) - f(x))(\overline{g(y)} - \overline{g(x)})}{|x-y|^{N+\alpha}} dxdy$$

$$= \frac{1}{2} \int_{\mathbb{R}^N} \int_{\mathbb{R}^N} (f(y) - f(x))(\overline{g(y)} - \overline{g(x)})\nu(x - y)dxdy,$$

where $\nu(z) = c_{N,\alpha}|z|^{-N-\alpha}$ and $c_{N,\alpha}$ is given by (5.8). We write $f \in D(\mathcal{E})$ if $f \in L^2$ and $\mathcal{E}(f,f)$ is finite. We write $f \in D(L_Q, L^2)$ if $f \in L^2$ and there is $L_Q f \in L^2$ such that for all $g \in D(\mathcal{E})$,

$$\int_{\mathbb{R}^N} L_Q f(x)\overline{g(x)}dx = -\mathcal{E}(f,g). \tag{5.12}$$

We note that $D(\mathcal{E})$ is the Sobolev space $H^{\alpha/2}$, which consists functions f in L^2 such that $|\xi|^\alpha |\mathcal{F}f(\xi)|^2$ is integrable. Furthermore,

$$\mathcal{E}(f,g) = (2\pi)^{-d} \int_{\mathbb{R}^N} \mathcal{F}|\xi|^\alpha f(\xi)\overline{\mathcal{F}g(\xi)}d\xi.$$

The quadratic form $\mathcal{E}$ is positive definite and it is an important example of a (nonlocal) Dirichlet form.

5.1.8 Semigroup definition

By the spectral theorem, the fractional Laplace operator L generates a strongly continuous semigroup of operators $P_t = e^{tL}$ on L^2, and P_t is the Fourier multiplier with symbol $e^{-t|\xi|^\alpha}$. Hence, P_t is the convolution operator with a symmetric kernel function $p_t(z)$, given by $\mathcal{F}p_t(\xi) = e^{-t|\xi|^\alpha}$.

We note some well-known properties of the kernel $p_t(z)$. For $\alpha = 1$, $p_t(z)$ is the Poisson kernel of the half-space in $\mathbb{R}^{N+1}$,

$$p_t(z) = \frac{\Gamma(\frac{N+1}{2})}{\pi^{(N+1)/2}} \frac{t}{(t^2 + |z|^2)^{(N+1)/2}}. \tag{5.13}$$

For arbitrary $\alpha \in (0,2)$, $\mathcal{F}p_t$ is rapidly decreasing and, therefore, $p_t(z)$ is infinitely smooth. We also have $\mathcal{F}p_t(\xi) = \mathcal{F}p_1(t^{1/\alpha}\xi)$, and so

$$p_t(z) = t^{-N/\alpha} p_1(t^{-1/\alpha} z).$$

Definition 5.1.12. The fractional Laplace operator is given by

$$\begin{aligned} L_S f(x) &= \lim_{t \to 0^+} \frac{P_t f(x) - f(x)}{t} \\ &= \lim_{t \to 0^+} \int_{\mathbb{R}^N} (f(x+z) - f(x)) \frac{p_t(z)}{t} dz. \end{aligned} \tag{5.14}$$

Here, $\mathcal{F}p_t(\xi) = e^{-t|\xi|^\alpha}$ and $P_t f(x) = f * p_t(x)$ (note that $p_t(z) = p_t(-z)$). More precisely, we write $f \in D(L_S, x)$ if the limit exists for a given $x \in \mathbb{R}^N$. If $f \in X$ and the limit (5.14) exists in X, then we write $f \in D(L_S, X)$.

The above approach is distinguished due to the general theory of strongly continuous semigroups of operators on Banach spaces. The operators L form a semigroup of contractions on every L^p, $p \in [1, \infty)$, and on C_0, C_{bu}, and C_b. This semigroup is strongly continuous, except on L^∞ and C_b.

Proposition 5.1.13. *If $f \in D(L_I, x)$ and f is locally integrable near x, then $f \in D(L_S, x)$, and $L_I f(x) = L_S f(x)$.*

See [109] for the proof.

5.1.9 Riesz potential definition

Suppose that $\alpha < N$, that is, $N \geq 2$ and $\alpha \in (0, 2)$ or $N = 1$ and $\alpha \in (0, 1)$. Then the function $|\xi|^{-\alpha}$ is integrable in the unit ball and bounded in its complement, and hence it is the Fourier transform of a tempered distribution. Using the identity

$$\lambda^{-\alpha/2} = \frac{1}{\Gamma(\frac{\alpha}{2})} \int_0^\infty e^{-t\lambda} t^{-1+\alpha/2} dt$$

and following the argument used in Section 5.1.5 and in (5.6), one easily shows that, at least in the sense of spectral theory on L^2,

$$(-L)^{-1} f(x) = c_{N,-\alpha} \int_{\mathbb{R}^N} f(x+z)|z|^{-d+\alpha} dz,$$

with $c_{N,-\alpha}$ defined as in (5.15).

The same expression can be obtained using the general theory of strongly continuous semigroups, which tells that, at least formally, $(-L)^{-1}$ is the 0-resolvent operator

$$U_0 f = \int_0^\infty P_t f dt.$$

Indeed, using the Bochner's formula,

$$p_t(z) = t^{-2/\alpha} \int_0^\infty k_s(z) \eta(t^{-2/\alpha}) ds, \tag{5.15}$$

where k_s is the Gauss–Weierstrass kernel with respect to $s \in (0, \infty)$, and η is a function on $(0, \infty)$ with Fourier–Laplace transform equal

$$\int_0^\infty e^{-\xi s} \eta(s) ds = \exp(-\xi^{\alpha/2}),$$

one easily shows that

$$\int_0^\infty p_t(z)dt = \frac{1}{\Gamma(\frac{\alpha}{2})}\int_0^\infty t^{-1+\alpha/2}K_t(z)dt = c_{N,-\alpha}|z|^{-N+\alpha}.$$

The above considerations motivate the following definition.

Definition 5.1.14. If $\alpha < N$, then the fractional Laplace operator is the inverse of the Riesz potential, namely $L_R f = -g$ whenever

$$f(x) = I_\alpha g(x) = c_{N,-\alpha}\int_{\mathbb{R}^N} g(x+z)|z|^{-N+\alpha}dz, \tag{5.16}$$

with $c_{N,-\alpha}$ defined as in (5.8).

More precisely, if $f, g \in X$, the integral in (5.16) is finite and the equality therein holds for all $x \in \mathbb{R}^N$ (when $X = C_0$) or for almost all $x \in \mathbb{R}^N$ (when $X = L^p$ with $p \in [1,\infty)$), then we write $f \in D(L_R, X)$.

For more information concerning the Riesz operators I_α, we refer to [129, 130, 148, 160].

5.1.10 Harmonic extension definition

For $\lambda \geq 0$, there is exactly one solution φ_λ of the second-order ordinary differential equation:

$$\begin{aligned} &\alpha^2 c_\alpha^{2/\alpha} y^{2-2/\alpha}\partial_y^2\varphi_\lambda(y) = \lambda\varphi_\lambda(y), \\ &\varphi_\lambda(0) = 1, \end{aligned} \tag{5.17}$$

with $c_\alpha = 2^{-\alpha}|\Gamma(-\frac{\alpha}{2})|/\Gamma(\frac{\alpha}{2})$, which is nonnegative, continuous, and bounded on $[0,\infty)$. This solution can be written as

$$\varphi_\lambda(y) = \frac{2^{1-\alpha/2}}{\Gamma(\frac{\alpha}{2})}\left(\frac{\lambda^{\alpha/2}y}{c_\alpha}\right)^{1/2} K_{\alpha/2}\left(\left(\frac{\lambda^{\alpha/2}y}{c_\alpha}\right)^{1/\alpha}\right),$$

where $K_{\alpha/2}$ is the modified Bessel function of the second kind (by continuity, we let $\varphi_\lambda(0) = \varphi_0(y) = 1$). By a simple calculation, we have $\varphi'_\lambda(0) = -\lambda^{\alpha/2}$. This property is crucial for the extension technique, described below.

In particular, for $\alpha = 1$ the equation reduces to $\partial_y^2\varphi_\lambda(y) = \lambda\varphi_\lambda(y)$ having the solution $\varphi_\lambda(y) = \exp(\sqrt{\lambda}y)$. There is another solution of (5.17), linearly independent from φ_λ, given by a similar formula, with $K_{\alpha/2}$ replaced by the modified Bessel function of the first kind $I_{\alpha/2}$.

For $f \in L^2$, consider next the elliptic partial differential equation:

$$\begin{cases} \Delta_x u(x,y) + \alpha^2 c_\alpha^{2/\alpha} y^{2-2/\alpha} \partial_y^2 u(x,y) = 0, & y > 0, \\ u(x,0) = f(x), \end{cases}$$

together with the regularity conditions:
- $u(x,y)$ as a function of $x \in \mathbb{R}^N$, is in L^2 for each $y \in [0,\infty)$, with norm bounded uniformly in $y \in [0,\infty)$,
- $u(x,y)$ depends continuously on $y \in [0,\infty)$ with respect to the L^2 norm.

Then for each $\xi \in \mathbb{R}^N$ the Fourier transform $\mathcal{F}u(\xi,y)$ of $u(x,y)$ with respect to x is bounded in $y \in [0,\infty)$, satisfies (5.17) with $\lambda = |\xi|^2$, and is equal to $\mathcal{F}f(\xi)$ for $y = 0$. Consequently,

$$\mathcal{F}u(\xi,y) = \varphi_{|\xi|^2}(y)\mathcal{F}f(\xi) \tag{5.18}$$

and, therefore,

$$\partial_y \mathcal{F}u(\xi,0) = \varphi'_{|\xi|^2}(0)\mathcal{F}f(\xi) = -|\xi|^\alpha \mathcal{F}f(\xi),$$

that is, $\partial_y u(x,0)$ is equal to $L_F f(x)$, the fractional Laplace operator applied to f.

The same method applies not only to L^2. Similar construction is possible in the spaces L^p, $p \in [1,\infty)$, C_0 or C_{bu}.

The fractional Laplace operator L on a Banach space X, defined using harmonic extensions, is *the Dirichlet-to-Neumann operator* for the weighted Dirichlet problem in the half-space:

$$\begin{cases} \Delta_x u(x,y) + \alpha^2 c_\alpha^{2/\alpha} y^{2-2/\alpha} \partial_y^2 u(x,y) = 0, & y > 0, \\ u(x,0) = f(x), \\ \partial_y u(x,0) = L_H f(x), \end{cases} \tag{5.19}$$

where

$$c_\alpha = \frac{|\Gamma(-\frac{\alpha}{2})|}{2^\alpha \Gamma(\frac{\alpha}{2})}. \tag{5.20}$$

The problem (5.19) requires a regularity condition on u, which asserts that the (distributional) Fourier transform of u has the desired form (5.18).

Equation (5.19) leads to the following definition. The property (5.18) of the distributional Fourier transform is equivalent to the condition $u(x,y) = f * q_y(x)$, where $\mathcal{F}q_y(\xi) = \varphi_{|\xi|^2}(y)$. The definition can be given in the same way as in (5.14), using the kernel q_y instead of p_t.

Definition 5.1.15. The fractional Laplace operator is given by

$$\begin{aligned} L_H f(x) &= \lim_{y\to 0^+} \frac{f * q_y(x) - f(x)}{y} \\ &= \lim_{y\to 0^+} \int\limits_{\mathbb{R}^N} (f(x+z) - f(x)) \frac{q_y(z)}{y} dz, \end{aligned} \tag{5.21}$$

where

$$\mathcal{F}q_y(\xi) = \frac{2^{1-\alpha/2}}{\Gamma(\frac{\alpha}{2})} \left(\frac{|\xi|^\alpha y}{c_\alpha} \right)^{1/2} K_{\alpha/2}\left(\left(\frac{|\xi|^\alpha y}{c_\alpha} \right)^{1/\alpha} \right),$$

$K_{\alpha/2}$ is the modified Bessel function of the second kind and c_α is given by (5.20).

We write $f \in D(L_H, x)$ if the limit exists for a given $x \in \mathbb{R}^N$. If $f \in X$ and the limit exists in X, then we write $f \in D(L_H, X)$.

Following [109], we quote the following.

Proposition 5.1.16. *If $f \in D(L_I, x)$ and f is locally integrable near x, then $f \in D(L_H, x)$, and $L_I f(x) = L_H f(x)$.*

See [109] for the proof.

5.2 Fractional powers of abstract operators

We extend here the discussion of Subsection 5.1 to the case of general sectorial operator. We compare first the two classes of such operators, for which the fractional powers will be defined; the *nonnegative* and *positive* operators in Banach spaces. We recall also the notion of the *spectral resolution*, which is used to define fractional powers of self-adjoint operators in Hilbert spaces.

For the corresponding abstract operator, basically the *Balakrishnan definition* of fractional powers is used. We will recall it below.

Definition 5.2.1. Let A be a closed linear densely defined operator in a Banach space X, such that its resolvent set contains $(-\infty, 0)$ and the resolvent satisfies

$$\|\lambda(\lambda + A)^{-1}\| \le M, \quad \lambda > 0.$$

Then, for $\eta \in (0, 1)$, we set

$$A^\eta \phi = \frac{\sin(\pi\eta)}{\pi} \int\limits_0^\infty s^{\eta-1} A(s+A)^{-1} \phi ds, \tag{5.22}$$

where the original constant $\frac{\Gamma(1)}{\Gamma(\alpha)\Gamma(1-\alpha)}$ was transformed using the known properties of the Γ function (e. g., [130, p. 60]):

$$\Gamma(1+\alpha) = \alpha\Gamma(\alpha), \quad \Gamma(1) = 1,$$
$$\Gamma(\alpha)\Gamma(1-\alpha) = \frac{\pi}{\sin(\pi\alpha)} \quad \text{for } \alpha \in (0,1). \tag{5.23}$$

We will formulate next a simple observation concerning commutativity property of the nonnegative operator with its resolvent.

Remark 5.2.2. The nonnegative operator A (eventually unbounded) commutes with its resolvent on $D(A)$. Indeed, for $\lambda > 0$ and $\phi \in D(A)$ we have

$$\begin{aligned} A(\lambda + A)^{-1}\phi &= (\lambda + A - \lambda)(\lambda + A)^{-1}\phi = (I - \lambda(\lambda + A)^{-1})\phi \\ &= \phi - (\lambda + A)^{-1}\lambda\phi = \phi - (\lambda + A)^{-1}(\lambda + A - A)\phi \\ &= \phi - \phi + (\lambda + A)^{-1}A\phi = (\lambda + A)^{-1}A\phi. \end{aligned} \tag{5.24}$$

Consequently, the order of operators in the Balakrishnan definition will be reversed,

$$A^{\eta}\phi = \frac{\sin(\pi\eta)}{\pi}\int_0^{\infty} s^{\eta-1}A(s+A)^{-1}\phi ds = \frac{\sin(\pi\eta)}{\pi}\int_0^{\infty} s^{\eta-1}(s+A)^{-1}A\phi ds, \tag{5.25}$$

whenever $\eta \in (0,1)$, $\phi \in D(A)$.

5.2.1 Nonnegative versus positive operators

Two classes of operators are used when defining the fractional powers: *nonnegative operators* and *positive operators*. More precisely, we have (see, e. g., [130, p. 1]) the following.

Definition 5.2.3. Let $A : D(A) \subset X \to X$ be a closed linear operator in a Banach space X. A is called *nonnegative* if $(-\infty, 0)$ is contained in the resolvent set $\rho(A)$ of A and

$$M = \sup_{\lambda>0} \|\lambda(\lambda + A)^{-1}\| < \infty. \tag{5.26}$$

The constant M is called the *nonnegativity constant* of A.

A closed linear operator A is called *positive* if it is nonnegative and $0 \in \rho(A)$.

Note the equality $A(\lambda + A)^{-1} = I - \lambda(\lambda + A)^{-1}$, so that for nonnegative operators $A(\lambda + A)^{-1}$ is uniformly bounded for $\lambda > 0$. Moreover,

$$\left|\sup_{\lambda>0}\|A(\lambda + A)^{-1}\| - \sup_{\lambda>0}\|\lambda(\lambda + A)^{-1}\|\right| \le 1. \tag{5.27}$$

Another characterization of positive operators in the class of nonnegative operators is common in the literature. We have namely the following.

Proposition 5.2.4. *A closed linear operator A is positive if and only if it is nonnegative and*

$$\sup_{\lambda>0}(1+\lambda)\|(\lambda+A)^{-1}\| < \infty. \tag{5.28}$$

Proof. When A is positive, A^{-1} is bounded and $A(\lambda+A)^{-1}$ is uniformly bounded with respect to $\lambda > 0$, so that

$$\sup_{\lambda>0}\|(\lambda+A)^{-1}\| \leq \|A^{-1}\| \sup_{\lambda>0}\|A(\lambda+A)^{-1}\| < \infty. \tag{5.29}$$

Therefore, (5.28) is satisfied due to (5.27) and (5.26).

In order, if (5.28) holds then in particular

$$\sup_{\lambda>0}\|(\lambda+A)^{-1}\| < \infty.$$

By the *resolvent identity*, it can be deduced that $\{(\lambda+A)^{-1}; \lambda > 0\}$ is a Cauchy net in the space $\mathcal{L}(X)$ of bounded linear operators on X, therefore, there is a limit

$$\lim_{\lambda\to 0}(\lambda+A)^{-1} = A^{-1},$$

since A is a closed operator. □

5.2.2 Fractional powers of positive self-adjoint operators in Hilbert spaces

As was already mentioned in Proposition 2.2.13, the linear, densely defined, self-adjoint and bounded below operator in a Hilbert space is sectorial in the sense considered in this book. Moreover, fractional powers of such operators will be defined in a way specific for Hilbert spaces using the, so called, *spectral resolution*. Such approach was presented in the classical monograph by S. G. Krein [108, Section I].

We recall next briefly basic facts concerning the spectral resolution of self-adjoint operators in a Hilbert space H; see [158, Chapter II], [190, Chapter XI, 5–7 and 12]. Consider a positive self-adjoint operator B, that means a closed linear operator with dense domain $D(B) \subset H$ and range $R(B) = D(B') \subset H$ (here B' denotes the adjoint operator), such that $Bv = B'v$ for all $v \in D(B)$, and that B is symmetric

$$\langle w, Bv\rangle = \langle Bw, v\rangle \quad \forall_{w,v\in D(B)},$$

and positive

$$\langle w, Bw\rangle \geq 0 \quad \forall_{w\in D(B)}.$$

For each $\lambda \in [0, \infty)$, let E_λ be a projection operator from H onto a subspace $D(\lambda) \subset H$, and let $\{E_\lambda; \lambda > 0\}$ denotes a family of such projections. For arbitrary $0 \le \lambda_0 \le \infty$, we define

$$E_{\lambda_0} = \lim_{\lambda \to \lambda_0} E_\lambda,$$

strongly in the operator topology.

If the family of operators, $\{E_\lambda; \lambda \ge 0\}$ has the following three properties:

- $E_\lambda E_\mu = E_\mu E_\lambda = E_\lambda$, $0 \le \lambda \le \mu < \infty$,
- $E_\lambda = s - \lim_{\mu \to \lambda} E_\mu$, $0 < \mu < \lambda < \infty$,
- $E_0 = 0$, $s - \lim_{\lambda \to \infty} E_\lambda = I$,

then $\{E_\lambda; \lambda \ge 0\}$ is called a *(spectral) resolution of the identity* on $[0, \infty)$.

Note also, that for each continuous function $g : [0, \infty) \to \mathbb{R}$, the Stieltjes integral can be defined

$$\int_0^b g(\lambda) d\|E_\lambda u\|^2,$$

as a limit of the Riemann–Stieltjes sums

$$\sum_{j=1}^m g(\lambda_j) \left\| (E_{\lambda_j} - E\lambda_{j-1}) v \right\|^2,$$

where $0 = \lambda_0 < \lambda_1 < \cdots < \lambda_m = b$, $\max |\lambda_j - \lambda_{j-1}| \to 0$. If $g(\lambda) \ge 0$ for $\lambda \ge 0$, and if

$$\int_0^\infty g(\lambda) d\|E_\lambda v\|^2 = \lim_{b \to \infty} \int_b^\infty g(\lambda) d\|E_\lambda\|^2$$

exists for some $v \in H$, we will write $\int_0^\infty g(\lambda) d\|E_\lambda v\|^2 < \infty$.

It is possible to define a *functional calculus* corresponding to the resolution of the identity $\{E_\lambda; \lambda \ge 0\}$. More precisely, for arbitrary continuous real function $g : [0, \infty) \to \mathbb{R}$, the integral

$$\int_0^b g(\lambda) dE_\lambda v \in H, \quad 0 < b < \infty, v \in H, \tag{5.30}$$

is well-defined as the strong limit of the Riemann sums. If $\int_0^\infty g^2(\lambda) d\|E_\lambda v\|^2 < \infty$ for some $v \in H$, then the integral

$$\int_0^\infty g(\lambda) dE_\lambda v := s - \lim_{b \to \infty} \int_0^b g(\lambda) dE_\lambda v$$

exists. We thus obtain an operator

$$\int_0^\infty g(\lambda)dE_\lambda : v \to \int_0^\infty g(\lambda)dE_\lambda v$$

being self-adjoint and having dense domain in H

$$D\left(\int_0^\infty g(\lambda)dE_\lambda\right) = \left\{v \in H; \int_0^\infty g^2(\lambda)d\|E_\lambda v\|^2 < \infty\right\}.$$

Moreover, the following properties hold:

$$\left\|\int_0^\infty g(\lambda)dE_\lambda v\right\|^2 = \int_0^\infty g^2(\lambda)d\|E_\lambda v\|^2,$$

and

$$\left\langle\left(\int_0^\infty g(\lambda)dE_\lambda\right)v, v\right\rangle = \int_0^\infty g(\lambda)d\|E_\lambda v\|^2$$

for all $v \in D(\int_0^\infty g(\lambda)dE_\lambda)$. In particular for all $v \in H$ the equality holds

$$v = \int_0^\infty dE_\lambda v, \quad \|v\|^2 = \int_0^\infty d\|E_\lambda v\|^2. \tag{5.31}$$

With that knowledge, it is possible to introduce a *functional calculus* for positive self-adjoint operators in H. Let $B : D(B) \to H$ be any positive self-adjoint operator with dense domain $D(B) \subset H$. Then there exists a unique resolution of identity such that

$$B = \int_0^\infty \lambda dE_\lambda, \quad D(B) = \left\{v \in H; \int_0^\infty \lambda^2 d\|E_\lambda v\|^2 < \infty\right\}, \tag{5.32}$$

known as the *spectral representation* of the operator B; see [190, Chapter XI, 5].

Further, for each continuous real function $g : [0, \infty) \to \mathbb{R}$, we will define a self-adjoint operator

$$g(B) = \int_0^\infty g(\lambda)dE_\lambda$$

having the domain

$$D(g(B)) = \left\{v \in H; \int_0^\infty g^2(\lambda)d\|E_\lambda v\|^2 < \infty\right\}.$$

As a particular case *fractional powers* of B will be defined

$$B^\alpha = \int_0^\infty \lambda^\alpha dE_\lambda, \quad D(B^\alpha) = \left\{ v \in H; \int_0^\infty \lambda^{2\alpha} d\|E_\lambda v\|^2 < \infty \right\} \tag{5.33}$$

for all $\alpha \geq 0$; note that $B^0 = I$. Further, the *resolvent operator* is defined for all $\mu > 0$ as

$$(I + B)^{-1} = \int_0^\infty (\mu + \lambda)^{-1} dE_\lambda.$$

It is bounded with the norm satisfying

$$\|(\mu + B)^{-1}\| \leq \sup_{\lambda \geq 0} (\mu + \lambda)^{-1} \leq \mu^{-1}.$$

If $E_\lambda = 0$ for $0 \leq \lambda \leq \delta$ with certain $\delta > 0$, then evidently B is invertible with bounded inverse operator

$$B^{-1} = \int_\delta^\infty \lambda^{-1} dE_\lambda, \quad \|B^{-1}\| \leq \sup_{\lambda \geq \delta} \lambda^{-1}. \tag{5.34}$$

When $N(B) = \{0\}$, we can define the operator $B^{-\alpha} : D(B^{-\alpha}) \to H$ for $\alpha \geq 0$ as

$$B^{-\alpha} v = \int_0^\infty \lambda^{-\alpha} dE_\lambda v = s - \lim_{\delta \to 0} \lambda^{-\alpha} dE_\lambda v, \quad v \in D(B^{-\alpha}), \tag{5.35}$$

on the domain

$$D(B^{-\alpha}) = \left\{ v \in H; \|B^{-\alpha} v\|^2 = \int_0^\infty \lambda^{-2\alpha} d\|E_\lambda v\|^2 < \infty \right\}.$$

Thus, if $N(B) = \{0\}$, then $N(B^\alpha) = \{0\}$, $D(B^{-\alpha}) \subset H$ is dense, and $B^{-\alpha}$ is a positive self-adjoint operator satisfying $B^{-\alpha} = (B^{-1})^\alpha = (B^\alpha)^{-1}$. Further, $D(B^\alpha) = R(B^{-\alpha})$ and $D(B^{-\alpha}) = R(B^\alpha)$. It is seen from integral representation of fractional powers that $D(B) \subset D(B^\alpha)$, $D(B^{-1}) \subset D(B^{-\alpha})$ when $0 \leq \alpha \leq 1$.

In case of strictly positive self-adjoint operators A in a *separable* Hilbert space H having compact inverse a particularly simple characterization of fractional powers of A is available (see, e. g., [145]). This is also the simplest case for the described above spectral resolution when the spectrum of A is pointwise, containing only eigenvalues (see, e. g., [190, Chapter XI, 8]). We will use it for verification of a version of the Poincaré inequality. A particular example of such operator is given by the negative Dirichlet Laplacian in a bounded C^2 domain densely defined in $L^2(\Omega)$.

Assume that A is a strictly positive self-adjoint operator in H (see [145, 158, 190]) and let A^{-1} be compact. Then A in H has an increasing to infinity sequence of positive eigenvalues

$$0 < \lambda_1 \le \lambda_2 \le \cdots \le \lambda_n \le \cdots$$

and the corresponding sequence of ortho-normalized eigenfunctions, denoted

$$e_1, e_2, \dots, e_n, \dots,$$

forming an orthonormal basis in H (so called *"Hilbert base"*). In particular, fractional powers of A can be expressed as

$$A^\alpha \phi = \sum_{n=1}^{\infty} \lambda_n^\alpha \langle \phi, e_n \rangle e_n, \quad \phi \in H^\alpha, \tag{5.36}$$

where $\langle \cdot, \cdot \rangle$ stands for a scalar product in H (see [144, Sections 3.9 and 3.10]). As a simple consequence of (5.36) and Parseval's identity:

$$\|\phi\|_{H^\alpha}^2 = \sum_{n=1}^{\infty} \lambda_n^{2\alpha} |\langle \phi, e_n \rangle|^2,$$

a version of the Poincaré inequality holds

$$\lambda_1^{\beta-\alpha} \|\phi\|_{H^\alpha} \le \|\phi\|_{H^\beta} \quad \text{for } \beta > \alpha, \tag{5.37}$$

since $\lambda_1 \le \lambda_n$ (here H^α, $\alpha \in \mathbb{R}$, stands for fractional power space connected with H).

5.2.3 Fractional powers of operators in bounded regular domains

Unfortunately, in case of differential operators considered in bounded regular domains then mentioned in Section 5.1 definitions of fractional powers of the negative Laplace operator are in general *not equivalent* (see [159] for an example). Only certain of the above definitions will be used equivalently, while such equivalence is valid for general class of operators, not limited to the Dirichlet Laplacian itself. We recall here a result taken from [130, Section 3.2, Section 6.1.1], where A denotes a nonnegative operator and the standard Balakrishnan Definition 5.2.1 is used to introduce its fractional powers.

Assume the operator A generates an equibounded C^0-semigroup $\{P_t : t > 0\}$, and a positive integer $m > \Re\alpha > 0$ is given. Assume further that real numbers α and m satisfy the condition

$$K_{\alpha,m} = \int_0^{\infty} t^{-\alpha-1}(1 - e^{-t})^m \, dt \neq 0. \tag{5.38}$$

Then the following property was proved in ([130, Theorem 6.1.6]).

Proposition 5.2.5. *Let $m > \Re\alpha$ be a positive integer. If $\phi \in D(A^\alpha)$, then there exists*

$$\lim_{\epsilon\to 0}\int_\epsilon^\infty t^{-\alpha-1}(I-P_t)^m\phi dt.$$

If also $K_{\alpha,m} \neq 0$ (so, in particular when $\alpha > 0$), then

$$A^\alpha\phi = \frac{1}{K_{\alpha,m}}\lim_{\epsilon\to 0}\int_\epsilon^\infty t^{-\alpha-1}(I-P_t)^m\phi dt. \tag{5.39}$$

Consequently, under the assumptions of the above proposition, the Balakrishnan and Bochner definitions of fractional powers are equivalent.

Under the assumption of the previous proposition, we recall next from [130, p. 66] the formula defining fractional powers of higher order (not limited to $\alpha \in (0,1)$).

Proposition 5.2.6. *Let $\alpha \in \mathbb{C}^+ \setminus \mathbb{N}$, $n \in \mathbb{N}$, $n > \Re\alpha$. If $\phi \in D(A^n)$, then*

$$\begin{aligned} A^\alpha\phi = \frac{1}{\Gamma(-\alpha)}\Bigg[\int_0^a t^{-\alpha-1}\Big(P_t - \sum_{0\le k\le n-1}(-1)^k\frac{t^k}{k!}A^k\Big)\phi dt \\ + \int_a^\infty t^{-\alpha-1}P_t\phi dt + \sum_{0\le k\le n-1}(-1)^k\frac{a^{k-\alpha}}{k!(k-\alpha)}A^k\phi\Bigg]. \end{aligned} \tag{5.40}$$

In particular, if $n-1 < \Re\alpha < n$, this formula can be rewritten as

$$A^\alpha\phi = \frac{1}{\Gamma(-\alpha)}\int_0^\infty\Big[P_t - \sum_{0\le k\le n-1}(-1)^k\frac{t^k}{k!}A^k\Big]\phi dt. \tag{5.41}$$

Higher powers of generators of equibounded C^0-semigroups will be also defined equivalently (see [130, p. 67]) through the Bochner type definition.

Definition 5.2.7. Let $\alpha \in \mathbb{C}^+$, $m, n \in \mathbb{N}$, $m \ge n > \Re\alpha$. If $\phi \in D(A^n)$, then

$$\int_0^\infty t^{-\alpha-1}(I-P_t)^m\phi dt = K_{\alpha,m}A^\alpha\phi, \tag{5.42}$$

where $K_{\alpha,m} = \int_0^\infty t^{-\alpha-l}(1-e^{-t})^m dt$.

Remark 5.2.8. Positive self-adjoint operators in a Hilbert space form a subclass of general sectorial positive operators. For such operators in a Hilbert space equivalently with Balakrishnan ([108, Chapter I, 9]), we will also use the definition of fractional powers introduced through spectral resolution, as described in Subsection 5.2.2. This covers an example of negative Dirichlet Laplacian in a bounded regular domain Ω considered on fractional power scale built on $L^2(\Omega)$.

Indeed, let A be a positive self-adjoint operator in a Hilbert space H with $\sigma(A) \subset [\epsilon, \infty)$, $\epsilon > 0$. Choose an integral contour Γ surrounding $\sigma(A)$ counterclockwise in $\mathbb{C} \setminus (-\infty, 0] \cap \rho(A)$, and let $\{E_\lambda, \lambda \geq \epsilon\}$ be the spectral resolution of the operator A. When $\Re z > 0$, we then have a formula expressing bounded linear operator A^{-z} through Dunford integral (see [106, p. 285])

$$A^{-z} = \frac{1}{2\pi i} \int_\Gamma \lambda^{-z} (\lambda - A)^{-1} d\lambda = \frac{1}{2\pi i} \int_\Gamma \int_\epsilon^\infty \lambda^{-z} (\lambda - \mu)^{-1} dE_\mu d\lambda,$$

$$\int_\epsilon^\infty \frac{1}{2\pi i} \int_\Gamma \lambda^{-z} (\lambda - \mu)^{-1} d\lambda dE_\mu = \int_\epsilon^\infty \mu^{-z} dE_\mu, \tag{5.43}$$

which proves the two definitions coincide for negative powers. Note that the contour in Dunford integral above will be, for real z, $0 < z < 1$, shifted to reduplicated half-lines $(\infty e^{i\pi}, 0] \cup [0, \infty e^{-i\pi})$, without changing the value of the integral, and we obtain

$$\begin{aligned} A^{-z} &= \frac{1}{2\pi i} \int_\Gamma \lambda^{-z} (\lambda - A)^{-1} d\lambda \\ &= \frac{1}{2\pi i} \int_0^\infty (se^{\pi i})^{-z} (s + A)^{-1} ds - \frac{1}{2\pi i} \int_0^\infty (se^{-\pi i})^{-z} (s + A)^{-1} ds. \end{aligned} \tag{5.44}$$

Consequently,

$$\begin{aligned} A^{-z} &= \frac{e^{-i\pi z}}{2\pi i} \int_0^\infty s^{-z} (s + A)^{-1} ds - \frac{e^{i\pi z}}{2\pi i} \int_0^\infty s^{-z} (s + A)^{-1} ds \\ &= -\frac{\sin(\pi z)}{\pi} \int_0^\infty s^{-z} (s + A)^{-1} ds, \quad 0 < z < 1, \end{aligned} \tag{5.45}$$

(see [187, p. 93]). Thus the Balakrishnan definition coincides with the one through spectral resolution in case of negative exponents $\alpha \in (-1, 0)$.

Since in both compared definitions the formula $(A^{-\alpha})^{-1} = A^{-\alpha}$, $\alpha > 0$, $D(A^\alpha) = R(A^{-\alpha})$ is valid (e. g., [86, p. 25], (5.35)), the two considered definitions coincide also for positive exponents $\alpha \in (0, 1)$.

5.3 Some properties of fractional powers of operators

In the present subsections, we collect several technical estimates and properties of the fractional powers that are needed in the following chapters. Some of that technical results, in particular Proposition 5.3.7, commutator estimates, moment inequalities, or the Kato–Beurling–Deny inequality are vital in the studies of the examples presented further in the text.

5.3.1 The continuity property of fractional powers

We note first (e. g., [130, p. 61]) that the dependence

$$\alpha \to A^\alpha \phi, \quad n \in \mathbb{N},\ \phi \in D(A^n),$$

is analytic for $\alpha \in \mathbb{C}$, $0 < \Re\alpha < n$. In particular, it is continuous (for fixed ϕ).

Passing to the limit in Chapter 11, it is important that the constants in the estimates can be taken uniformly with respect to α. Therefore, in the technical lemmas below, we need to take care of that uniformity. While such estimates can be found in the literature, usually the uniformity is not clear from the presentation.

Lemma 5.3.1. *When $(\frac{1}{2}, 1] \ni \alpha \to \frac{1}{2}^+$, then $A^{-\alpha} \to A^{-\frac{1}{2}}$ pointwise in the domain of A^{-1}, where A is a sectorial nonnegative operator in a reflexive Banach space X.*

Proof. Using the formula (e. g., [130, p. 175]) for negative powers of nonnegative operators:

$$A^{-\alpha}\phi = \frac{\sin(\pi\alpha)}{\pi} \int_0^\infty \lambda^{-\alpha}(A + \lambda I)^{-1}\phi\, d\lambda,$$

we find that, for $\phi \in D(A^{-1})$ and any $L > 1$,

$$\begin{aligned} \|A^{-\alpha}\phi - A^{-\frac{1}{2}}\phi\|_X &= \frac{1}{\pi}\left\| \int_0^\infty \left[\frac{\sin(\pi\alpha)}{\lambda^\alpha} - \frac{1}{\lambda^{\frac{1}{2}}}\right](A + \lambda I)^{-1}\phi\, d\lambda \right\|_X \\ &\le \frac{1}{\pi}\left\| \int_0^L + \int_L^\infty \right\|_X. \end{aligned} \tag{5.46}$$

Further, we will need the following two estimates:

$$\begin{aligned} \|(A + \lambda I)^{-1}\| &\le \frac{M}{|\lambda|}, \\ \|(A + \lambda I)^{-1}\phi\|_X &= \|A(A + \lambda I)^{-1}A^{-1}\phi\|_X \\ &= \|[I - \lambda(A + \lambda I)^{-1}]A^{-1}\phi\|_X \le (M + 1)\|A^{-1}\phi\|_X. \end{aligned} \tag{5.47}$$

Consequently, we will estimate the right-hand side terms in (5.46) as follows:

$$\begin{aligned} &\frac{1}{\pi}\int_0^L \left|\frac{\sin(\pi\alpha)}{\lambda^\alpha} - \frac{1}{\lambda^{\frac{1}{2}}}\right| \|(A + \lambda I)^{-1}\phi\|_X d\lambda \\ &\quad \le \frac{M+1}{\pi}\|A^{-1}\phi\|_X \int_0^L \lambda^{-\alpha}|\sin(\pi\alpha) - \lambda^{\alpha - \frac{1}{2}}|\, d\lambda, \end{aligned}$$

$$\frac{1}{\pi}\int_L^\infty \left|\frac{\sin(\pi\alpha)}{\lambda^\alpha} - \frac{1}{\lambda^{\frac{1}{2}}}\right| \|(A+\lambda I)^{-1}\phi\|_X d\lambda$$
$$\le \frac{M\|\phi\|_X}{\pi}\int_L^\infty \left(\frac{1}{\lambda^{\alpha+1}} + \frac{1}{\lambda^{\frac{3}{2}}}\right) d\lambda \le \frac{4M\|\phi\|_X}{L^{\frac{1}{2}}\pi}.$$

We will choose now $L > 1$ so large that the second term is less than $\epsilon\|\phi\|_X$, next choose $\alpha > \frac{1}{2}$ so close to $\frac{1}{2}$ that the first term is less than $\epsilon\|A^{-1}\phi\|_X$. Consequently,

$$\|A^{-\alpha}\phi - A^{-\frac{1}{2}}\phi\|_X \le \epsilon(\|\phi\|_X + \|A^{-1}\phi\|_X). \tag{5.48}$$

This proves $A^{-\alpha} \to A^{-\frac{1}{2}}$ pointwise in the domain of A^{-1}. Indeed, recall here the general definition of the domain of linear operator in the space X applied to A^{-1}:

$$D(A^{-1}) = \{\phi \in X; A^{-1}\phi \in X\}.$$

Thus, the convergence in (5.48) holds for $\phi \in D(A^{-1})$. □

Similarly, when studying the convergence of the solutions u_α as $\alpha \to \frac{1}{2}$ in Chapter 8, we need the following continuity property of fractional powers (e. g., [130, p. 62]).

Lemma 5.3.2. *Let A be a nonnegative operator in a Banach space X. For arbitrary $\phi \in D(A)$ and $0 < \beta \to 0^+$, we have*

$$\|(I - A^\beta)\phi\| \to 0 \quad \text{as } \beta \to 0^+.$$

In fact, when passing to the limit in [53, 7.3], we need a variant of the above lemma which uses the formula (e. g., [130, p. 117]):

$$(I + A^\alpha)^{-1} = \frac{\sin(\alpha\pi)}{\pi}\int_0^\infty \frac{\lambda^\alpha}{\lambda^{2\alpha} + 2\lambda^\alpha\cos(\pi\alpha) + 1}(\lambda I + A)^{-1} d\lambda,$$

valid for $\alpha \in \mathbb{C} : \Re\alpha > |\alpha|^2$; in particular for $0 < \alpha < 1$. Applying the above expression to $A = (-\Delta)$, with α and with $\frac{1}{2}$, we have the following.

Lemma 5.3.3. *Let $p \in [2, +\infty)$ be arbitrary. When $(\frac{1}{2}, 1] \ni \alpha \to \frac{1}{2}^+$, then $A_\alpha^{-1} := \frac{1}{\kappa}[(-\Delta)^\alpha + I]^{-1} \to A_{\frac{1}{2}}^{-1}$ pointwise in*

$$D((-\Delta)^{-1}) = \{\phi \in L^p(\mathbb{R}^2); (-\Delta)^{-1}\phi \in L^p(\mathbb{R}^2)\} = \mathcal{L}^p_{-2}(\mathbb{R}^2);$$

compare (5.49) for the definition of $\mathcal{L}^p_{-2}(\mathbb{R}^2)$. The same proof works for a general sectorial nonnegative operator in a reflexive Banach space X.

Proof. We need to estimate the difference:

$$[I + (-\Delta)^{\alpha}]^{-1} - [I + (-\Delta)^{\frac{1}{2}}]^{-1}$$
$$= \frac{1}{\pi}\int_0^{\infty}\Big[\frac{\sin(\pi\alpha)\lambda^{\alpha}}{\lambda^{2\alpha} + 2\lambda^{\alpha}\cos(\pi\alpha) + 1} - \frac{\sin(\frac{\pi}{2})\lambda^{\frac{1}{2}}}{\lambda + 2\lambda^{\frac{1}{2}}\cos(\frac{\pi}{2}) + 1}\Big](\lambda I + (-\Delta))^{-1} d\lambda.$$

Now the proof goes as in the previous lemma separating the integral into two parts; over $(0, L)$ and (L, ∞), and estimating them using (5.47). More precisely, for $\phi \in D((-\Delta)^{-1})$, we have

$$\frac{1}{\pi}\int_0^{L}\Big|\frac{\sin(\pi\alpha)\lambda^{\alpha}}{\lambda^{2\alpha} + 2\lambda^{\alpha}\cos(\pi\alpha) + 1} - \frac{\lambda^{\frac{1}{2}}}{\lambda + 1}\Big| \|((-\Delta) + \lambda I)^{-1}\phi\|_{L^p(\mathbb{R}^2)} d\lambda$$
$$\le \frac{M+1}{\pi}\|(-\Delta)^{-1}\phi\|_{L^p(\mathbb{R}^2)} \int_0^{L}\Big|\frac{\sin(\pi\alpha)\lambda^{\alpha}}{\lambda^{2\alpha} + 2\lambda^{\alpha}\cos(\pi\alpha) + 1} - \frac{\lambda^{\frac{1}{2}}}{\lambda + 1}\Big| d\lambda,$$

$$\frac{1}{\pi}\int_L^{\infty}\Big|\frac{\sin(\pi\alpha)\lambda^{\alpha}}{\lambda^{2\alpha} + 2\lambda^{\alpha}\cos(\pi\alpha) + 1} - \frac{\lambda^{\frac{1}{2}}}{\lambda + 1}\Big| \|((-\Delta) + \lambda I)^{-1}\phi\|_{L^p(\mathbb{R}^2)} d\lambda$$
$$\le \frac{M\|\phi\|_{L^p(\mathbb{R}^2)}}{\pi}\int_L^{\infty}\Big(\frac{\lambda^{-1-\alpha}}{1 + 2\lambda^{-\alpha}\cos(\pi\alpha) + \lambda^{-2\alpha}} + \frac{\lambda^{-\frac{1}{2}}}{\lambda + 1}\Big) d\lambda.$$

Note that the two terms under the final integral are bounded, moreover, uniformly in $\alpha \in (\frac{1}{2}, \frac{3}{4}]$. For such a range of parameter, we have

$$\frac{\lambda^{-1-\alpha}}{1 + 2\lambda^{-\alpha}\cos(\pi\alpha) + \lambda^{-2\alpha}} \le \frac{\lambda^{-1-\alpha}}{1 - \cos^2(\frac{3\pi}{4})} = \frac{2}{\lambda^{1+\alpha}}.$$

We will choose $L > 1$ so large that the integral over (L, ∞) is less than $\kappa\epsilon\|\phi\|_{L^p(\mathbb{R}^2)}$, next choose $\alpha > \frac{1}{2}$ so close to $\frac{1}{2}$ that the integral over $(0, L)$ is less than $\kappa\epsilon\|(-\Delta)^{-1}\phi\|_{L^p(\mathbb{R}^2)}$. Consequently,

$$\frac{1}{k}\|[I + (-\Delta)^{\alpha}]^{-1}\phi - [I + (-\Delta)^{\frac{1}{2}}]^{-1}\phi\|_{L^p(\mathbb{R}^2)} \le \epsilon(\|\phi\|_{L^p(\mathbb{R}^2)} + \|(-\Delta)^{-1}\phi\|_{L^p(\mathbb{R}^2)}).$$

This proves $A_{\alpha}^{-1} \to A_{\frac{1}{2}}^{-1}$ pointwise in $\mathcal{L}^p_{-2}(\mathbb{R}^2)$. □

Passing to the limit in construction of weak solution to the critical equation [53, 7.3], we were using the "test functions" $\phi \in \mathcal{L}^q_{-2}(\mathbb{R}^2)$. In those calculations, we need to be sure that for such ϕ, $(A_{\alpha}^{-1})^*\phi$ and $(A_{\frac{1}{2}}^{-1})^*\phi$ are in $L^q(\mathbb{R}^2)$ (rather form a dense subset).

We recall next the definition of the spaces $\mathcal{L}^p_{\gamma}(\mathbb{R}^N)$ (e. g., [184, p. 1164]).

Definition 5.3.4. For $1 \le p \le \infty$ and $\gamma \in \mathbb{R}$, we define

$$\mathcal{L}^p_{\gamma}(\mathbb{R}^N) = \{f \in L^p(\mathbb{R}^N) : f = I^{\gamma}g := (-\Delta)^{-\frac{\gamma}{2}}g \text{ for certain } g \in L^p(\mathbb{R}^N)\}, \tag{5.49}$$

normed by $\|f\|_{\mathcal{L}^p_{\gamma}(\mathbb{R}^N)} = \|g\|_{L^p(\mathbb{R}^N)}$.

It is known (e. g., [130, p. 303]), that when $0 < \gamma < m$, $1 < p < \infty$, then

$$\mathcal{L}^p_\gamma(\mathbb{R}^2) \doteq \mathcal{S}^{\gamma,p}(\mathbb{R}^N) \doteq [L^p(\mathbb{R}^N), W^{m,p}(\mathbb{R}^N)]_{\frac{\gamma}{m}},$$

the last bracket stands for the complex interpolation.

Characterization of the class $I^\gamma(L^p(\mathbb{R}^N))$, $\gamma > 0$, is given in [148, Theorem 26.8].

Proposition 5.3.5. *Let f be locally integrable and $\lim_{|x|\to\infty} f(x) = 0$. Then $f \in I^\gamma(L^p(\mathbb{R}^N))$, $\gamma > 0$, $1 < p < \infty$, if and only if:*

- *When $1 < p < \frac{N}{\gamma}$, then $f \in L^q(\mathbb{R}^N)$ with $q = \frac{Np}{N-\gamma p}$ and $D^\gamma f \in L^p(\mathbb{R}^N)$,*
- *When $p \geq \frac{N}{\gamma}$, then*

$$(\Delta_h^l f) \in L^p(\mathbb{R}^N), \text{ and there exists in } L^p(\mathbb{R}^N), \lim_{\epsilon\to 0} \int_{|h|>\epsilon} |h|^{-N-\gamma}(\Delta_h^l f)(x)dh,$$

where $l > 2[\frac{\gamma}{2}]$. For the definition of the difference operators $(\Delta_h^l f)$ we refer to [148, Lemma 26.5].

Moreover, $I^\gamma(L^p(\mathbb{R}^N)) \cap L^r(\mathbb{R}^N) = L^\gamma_{p,r}(\mathbb{R}^N)$, $\gamma > 0$, $1 \leq p < \infty$, $1 \leq r < \infty$.

We are using the above description for $N = \gamma = 2$ with $I^2(\psi) = \int_{\mathbb{R}^2} \ln(\frac{1}{|x-y|})\psi(y)dy$. We also have $\mathcal{L}^p_\gamma(\mathbb{R}^2) = L^\gamma_{p,p}(\mathbb{R}^2)$.

5.3.2 Properties of Riesz operators

We present a short description of the Riesz operators R_j, $j = 1, \ldots, N$, as in [160]:

$$R_j(f)(x) = \lim_{\epsilon\to 0} c_N \int_{|y|\geq\epsilon} \frac{y_j}{|y|^{N+1}} f(x-y)dy, \quad j = 1, 2, \ldots, N, \; c_N = \frac{\Gamma(\frac{N+1}{2})}{\pi^{(N+1)/2}},$$

where $f \in L^p(\mathbb{R}^N)$, $1 \leq p < \infty$. For $f \in L^2(\mathbb{R}^N) \cap L^p(\mathbb{R}^N)$, $1 < p < \infty$, Riesz operators admit the characterization ([130, p. 299]):

$$R_j f = \left(-i\frac{x_j}{|x|}\hat{f}\right)^\vee, \quad 1 \leq j \leq N.$$

Further, if $f \in W^{1,p}(\mathbb{R}^2)$, then (e. g., [130, p. 299])

$$\frac{\partial f}{\partial x_k} = -R_k(-\Delta)^{\frac{1}{2}} f,$$

also, if $f \in C_0^2(\mathbb{R}^N)$, then

$$\frac{\partial^2 f}{\partial x_j \partial x_k} = -R_j R_k \Delta f.$$

We next quote an observation (e. g., [130, p. 299]) that the Riesz operators R_j are bounded from $L^q(\mathbb{R}^N)$ into $L^q(\mathbb{R}^N)$, $1 < q < \infty$:

$$\exists_{C>0} \forall_{\psi \in L^q(\mathbb{R}^N)} \|R_j(\psi)\|_{L^q(\mathbb{R}^N)} \leq C\|\psi\|_{L^q(\mathbb{R}^N)}, \quad j = 1, 2, \ldots, N. \tag{5.50}$$

Studying the quasi-geostrophic equation, we are using Calderon–Zygmund inequality for Riesz transforms $u = (-R_2\theta, R_1\theta)$ (see [183, p. 12]):

$$\||D^j u(t, \cdot)|\|_{L^q(\mathbb{R}^2)} \leq c\|D^j\theta(t, \cdot)\|_{L^q(\mathbb{R}^2)}, \quad q \in (1, \infty), \tag{5.51}$$

where j is a multiindex with $|j| \geq 1$.

Definition 5.3.6. In case of the Laplace operator with Dirichlet boundary condition defined in a bounded C^2 regular domain $\Omega \subset \mathbb{R}^N$, we naturally introduce the Riesz operators as

$$R_j := \frac{\partial}{\partial x_j}(-\Delta)^{-\frac{1}{2}}, \quad j = 1, 2, \ldots, N,$$

and they will be considered on $L^p(\Omega)$, $1 < p < \infty$. Fractional powers of the Laplace operator are given through the Balakrishnan definition.

In calculations of Chapters 8 and 9, we are using commutativity property of the two operators, $(-\Delta)^\beta$, $\beta > 0$, and Riesz operator $R_j := \frac{\partial}{\partial x_j}(-\Delta)^{-\frac{1}{2}}$, in $\mathbb{R}^N$ and in a bounded domain Ω for the Dirichlet Laplacian. Commutativity of these two operators was proved in case of $\mathbb{R}^N$, in [130, p. 300]:

$$(-\Delta)^\beta R_j\phi = R_j(-\Delta)^\beta\phi, \quad \beta > 0, \phi \in D((-\Delta)^\beta). \tag{5.52}$$

Such property is useful when estimating higher Sobolev norms of the solution to quasi-geostrophic equation; compare Section 8.2.

We will sketch similar property for the case of the Dirichlet Laplacian in a bounded regular domain Ω. The main point, as in [130], is to prove such property for $\beta = 1$. The generalization to positive powers given in [130] uses standard properties of fractional powers operators and is not specific for the problem in $\mathbb{R}^N$.

We will prove the property in Dirichlet case for $\beta = 1$. We need to show that, for $\psi \in D((-\Delta)) = H^2(\Omega) \cap H^1_0(\Omega)$, $R_j\psi \in D((-\Delta))$ and the equality holds

$$(-\Delta)R_j\psi = R_j(-\Delta)\psi. \tag{5.53}$$

In the Dirichlet case, when $\psi \in D((-\Delta)) = H^2(\Omega) \cap H^1_0(\Omega)$, then $(-\Delta)^{-\frac{1}{2}}\psi \in H^3(\Omega) \cap H^2_0(\Omega)$, and further $\frac{\partial}{\partial x_j}(-\Delta)^{-\frac{1}{2}}\psi \in H^2(\Omega) \cap H^1_0(\Omega)$. Indeed, according to [118, (3.9)] (see Remark 2.2.27),

$$A^\gamma : D(A^\alpha) \to D(A^{\alpha-\gamma}), \quad \alpha, \gamma \in \mathbb{R},$$

as an isometry. Thus, in our case,

$$(-\Delta)^{-\frac{1}{2}} : D((-\Delta)) \to D((-\Delta)^{\frac{3}{2}}) = H^3(\Omega) \cap H_0^2(\Omega). \tag{5.54}$$

Further, the following chain of equations holds:

$$\begin{aligned} R_j(-\Delta)\psi &= \frac{\partial}{\partial x_j}(-\Delta)^{-\frac{1}{2}}(-\Delta)\psi = \frac{\partial}{\partial x_j}(-\Delta)^{\frac{1}{2}}\psi \\ &= \frac{\partial}{\partial x_j}(-\Delta)(-\Delta)^{-\frac{1}{2}}\psi = (-\Delta)\frac{\partial}{\partial x_j}(-\Delta)^{-\frac{1}{2}}\psi = (-\Delta)R_j\psi, \end{aligned} \tag{5.55}$$

since the derivatives $\frac{\partial}{\partial x_j}$ and $(-\Delta)$ commute both for classical C^3 functions and for Sobolev derivatives in $H^3(\Omega)$. We have thus checked that for $\psi \in D((-\Delta))$ condition (5.53) is satisfied.

Following [130, p. 300], extending the property from $\beta = 1$ to $\beta > 0$, we will use the fact that having (5.53) then R_j commutes with the resolvent $(\lambda - \Delta)^{-1}$ for all $\lambda > 0$. Indeed, for $\psi \in D((-\Delta))$ evidently $R_j(\lambda - \Delta)^{-1}\psi \in D((-\Delta))$, and by (5.53),

$$(\lambda - \Delta)R_j(\lambda - \Delta)^{-1}\psi = R_j\psi, \quad \lambda > 0, \tag{5.56}$$

so that

$$R_j(\lambda - \Delta)^{-1}\psi = (\lambda - \Delta)^{-1}R_j\psi. \tag{5.57}$$

To get, in the Dirichlet case, general formula (5.52) with power $\beta > 0$ it remains to note that fractional power commutes with any bounded operator that commutes with the resolvent operator.

Using the just proved commutativity property (5.52) with power $\beta = \frac{1}{2}$, we will justify next an estimate used in the studies of the Dirichlet problem for the quasi-geostrophic equation in Chapter 9. Let $\phi \in D((-\Delta)^{-\frac{1}{2}}) = H_0^1(\Omega)$, then the following estimate holds:

$$\left\| \frac{\partial}{\partial x_k} R_j \phi \right\|_{L^2(\Omega)} \le \left\| \frac{\partial}{\partial x_k} \phi \right\|_{L^2(\Omega)}. \tag{5.58}$$

Note that, when $\phi \in D((-\Delta)^{-\frac{1}{2}})$, then $(-\Delta)^{-\frac{1}{2}}\phi \in H^2(\Omega) \cap H_0^1(\Omega)$. We have a chain of estimates:

$$\begin{aligned} \left\| \frac{\partial}{\partial x_k} R_j \phi \right\|_{L^2(\Omega)} &= \left\| \frac{\partial}{\partial x_k} \frac{\partial}{\partial x_j} (-\Delta)^{-\frac{1}{2}} \phi \right\|_{L^2(\Omega)} \\ &= \left\| \frac{\partial}{\partial x_j} \frac{\partial}{\partial x_k} (-\Delta)^{-\frac{1}{2}} \phi \right\|_{L^2(\Omega)} \quad (\text{since } (-\Delta)^{-\frac{1}{2}}\phi \in D((-\Delta))) \\ &= \left\| \frac{\partial}{\partial x_j} R_k \phi \right\|_{L^2(\Omega)} \le \left\| (-\Delta)^{\frac{1}{2}} \frac{\partial}{\partial x_k} (-\Delta)^{-\frac{1}{2}} \phi \right\|_{L^2(\Omega)} \ (\text{by } (5.52)) = \left\| \frac{\partial}{\partial x_k} \phi \right\|_{L^2(\Omega)}. \end{aligned} \tag{5.59}$$

where we have used an estimate

$$\left\| \frac{\partial}{\partial x_j} \psi \right\|_{L^2(\Omega)}^2 \le \sum_{i=1}^N \left\| \frac{\partial}{\partial x_i} \psi \right\|_{L^2(\Omega)}^2 = \int_\Omega (-\Delta)\psi\psi dx = \|(-\Delta)^{\frac{1}{2}}\psi\|_{L^2(\Omega)}^2.$$

5.3.3 Boundedness of the operators $(-\Delta_p)^{-\frac{1}{2}}\frac{\partial}{\partial x_j}$, $1 \le j \le N$

We shall recall *an important property of the square root of the negative Dirichlet Laplacian*, analogous to the one in [75] for the Stokes operator, used frequently in our considerations. Below $(-\Delta_p)$ denotes negative Dirichlet Laplacian in a bounded smooth domain Ω, densely defined in $L^p(\Omega)$.

Proposition 5.3.7. *Following the result of* [75, *Lemma 2.1*], *we recall that for each* $j = 1,\dots,N$ *and* $1 < p < +\infty$ *the operator* $(-\Delta_p)^{-\frac{1}{2}}\frac{\partial}{\partial x_j}$ *extends uniquely to a bounded linear operator from* $L^p(\Omega)$ *into itself. In particular, for the nonlinearity in the N-D quasi-geostrophic equation, an estimate holds:*

$$\forall_{p\in(1,+\infty)}\exists_{M_p>0}\ \|(-\Delta_p)^{-\frac{1}{2}}\nabla\cdot(u\theta)\|_{L^p(\Omega)} \le M_p\||u\theta|\|_{L^p(\Omega)}. \tag{5.60}$$

Proof. According to D. Fujiwara [68], $D(-\Delta_r^\alpha) \subset H_r^{2\alpha}(\Omega)$, $1 < r < \infty$, $0 \le \alpha \le 1$. Since

$$(-\Delta_r^\alpha) : \{v \in L^r(\Omega); (-\Delta_r^\alpha)v \in L^r(\Omega)\} = D(-\Delta_r^\alpha) \to L^r(\Omega)$$

is a bounded linear operator, then

$$\exists_{C>0}\forall_{v\in D(-\Delta_r^\alpha)}\|(-\Delta_r^\alpha)v\|_{L^r(\Omega)} \le C\|v\|_{H_r^{2\alpha}(\Omega)}. \tag{5.61}$$

In particular, for $\alpha = \frac{1}{2}$,

$$\exists_{C>0}\forall_{v\in D(-\Delta_r^{\frac{1}{2}})}\|(-\Delta_r^{\frac{1}{2}})v\|_{L^r(\Omega)} \le C\|v\|_{H_r^1(\Omega)}. \tag{5.62}$$

Thanks to invertibility of $(-\Delta_r^{\frac{1}{2}})$, being a consequence of the Poincaré inequality:

$$c_P\|\phi\|_{L^r(\Omega)} \le \|(-\Delta_r^{\frac{1}{2}})\phi\|_{L^r(\Omega)}, \quad \phi \in D(-\Delta_r^{\frac{1}{2}}), \tag{5.63}$$

its invert $(-\Delta_r^{\frac{1}{2}})^{-1} = (-\Delta_r^{-\frac{1}{2}})$ is bounded linear and $(-\Delta_r^{-\frac{1}{2}}) : L^r(\Omega) \to D(-\Delta_r^{\frac{1}{2}}) \subset H_r^1(\Omega)$. Therefore, the composition

$$\frac{\partial}{\partial x_j}(-\Delta_r^{-\frac{1}{2}}) : L^r(\Omega) \to L^r(\Omega),$$

is a bounded linear operator for every $1 \le j \le N$. By duality, this implies that

$$(-\Delta_{r'}^{-\frac{1}{2}})\frac{\partial}{\partial x_j} : H_{r'}^1(\Omega) \to H_{r'}^1(\Omega), \quad \frac{1}{r} + \frac{1}{r'} = 1,$$

extends uniquely to a bounded linear operator from $L^{r'}(\Omega)$ into itself for arbitrary $1 < r'(= p) < \infty$. □

It will be also seen, that the operators $\frac{\partial}{\partial x_j}(-\Delta_p)^{-\frac{1}{2}} : L^p(\Omega) \to L^p(\Omega)$, $j = 1,\dots,N$, $1 < p < \infty$, are bounded linear (see [86, p. 18] when $N = 1$, [75, p. 270] in the case of the Stokes operator, and [55, Proposition A.1]).

Remark 5.3.8. It will be observed that R_j, $j = 1, 2$, are bounded operators from $H_0^1(\Omega)$ into $H^1(\Omega)$.

The operator $(-\Delta_2)^{\frac{1}{2}} : H^2(\Omega) \cap H_0^1(\Omega) = D(-\Delta_2) \to D((-\Delta_2)^{\frac{1}{2}}) = H_0^1(\Omega)$ is bounded (see [187, p. 559] for characterization of the domains of fractional powers). Moreover, as a consequence of (9.18),

$$\lambda_1^{\frac{1}{2}} \|(-\Delta_2)^{\frac{1}{2}}\phi\|_{L^2(\Omega)} \le \|(-\Delta_2)\phi\|_{L^2(\Omega)}, \quad \phi \in D(-\Delta_2),$$

which shows the operator $(-\Delta_2)^{\frac{1}{2}} : D(-\Delta_2) \to D((-\Delta_2)^{\frac{1}{2}})$ is invertible and

$$((-\Delta_2)^{\frac{1}{2}})^{-1} = (-\Delta_2)^{-\frac{1}{2}} : D((-\Delta_2)^{\frac{1}{2}}) \to D(-\Delta_2)$$

is bounded linear. Consequently,

$$\frac{\partial}{\partial x_j}(-\Delta_2)^{-\frac{1}{2}} : H_0^1(\Omega) \to H^1(\Omega) \tag{5.64}$$

is bounded linear, and

$$\|R_j\phi\|_{H^1(\Omega)} = \left\| \frac{\partial}{\partial x_j}(-\Delta_2)^{-\frac{1}{2}}\phi \right\|_{H^1(\Omega)} \le c\|\phi\|_{H_0^1(\Omega)}.$$

5.4 The standard estimates and inequalities

5.4.1 Various commutator estimates

In this subsection, we collect several commutator estimates known in the literature.

Let $\Lambda = (-\Delta)^{1/2}$, $J = (1 - \Delta)^{1/2}$, and we set $L_s^p = L_s^p(\mathbb{R}^N) := J^{-s}L^p$, where $L^p = L^p(\mathbb{R}^N)$ is the Lebesgue space with $1 < p < \infty$ and we have used the definition $\mathcal{F}[f](\xi) = \hat{f}(\xi) = \int_{\mathbb{R}^n} e^{2\pi i x\cdot\xi} f(x)dx$, and J^s and Λ^s denote fractional derivative operators defined in terms of Fourier transforms as follows:

$$\mathcal{F}[\Lambda^s f](\xi) = |\xi|^s \hat{f}(\xi),$$
$$\mathcal{F}[J^s f](\xi) = (1 + |\xi|^2)^{s/2} \hat{f}(\xi).$$

The basic commutator estimate in the spaces L_s^p, reported in [96], has the form as in the following.

Proposition 5.4.1. *If $s > 0$, $1 < p < \infty$, $f, g \in \mathcal{S}$ (the Schwartz space), then*

$$\|J^s(fg) - fJ^s g\|_{L^p} \le c(\|\partial f\|_{L^\infty} \|J^{s-1}g\|_{L^p} + \|J^s f\|_{L^p} \|g\|_{L^\infty}), \tag{5.65}$$

where $\partial = (\partial_1,\dots,\partial_m)$, $\partial_j = \partial/\partial x_j$, and c is a constant depending only on m, p, s.

We refer to the original article [96] for the proof of Proposition 5.4.1.

With $f = u$ and $g = \partial u$, (5.65) implies

$$\|[J^s, u]\partial u\|_{L^p} \le c\|\partial u\|_{L^\infty}\|u\|_{L^p_s}, \quad 1 < p < \infty,\ s > 0, \tag{5.66}$$

where $[J^s, u] = J^s u - uJ^s$ is the commutator, in which u is regarded as a multiplication operator, and $\|u\|_{L^p_s} = \|J^s u\|_{L^p}$ is the standard norm in $L^p_s = J^{-s}L^p$. Estimate (5.65) also implies

$$|||[J^s, f]J^{1-s}|||_p \le c\|f\|_{L^p_s}, \quad 1 < p < \infty,\ s > 1 + m/p, \tag{5.67}$$

where $|||\ |||_p$ denotes the operator norm in L^p. Indeed, under the stated condition on s, the right member of (5.65) does not exceed $c\|f\|_{L^p_s}\|g\|_{L^p_{s-1}}$, because $\|\partial f\|_{L^\infty} \le c\|f\|_{L^p_s}$ and $\|g\|_{L^\infty} \le \|g\|_{L^p_{s-1}}$.

Remark 5.4.2. Actually, (5.65) is true whenever the right member is finite, more precisely, if $f \in \mathrm{Lip} \bigcap L^p_s$ and $g \in L^\infty \bigcap L^p_{s-1}$. Indeed, for such f, g it is easily seen that the expression inside $\|\ \|_{L^p}$ on the left-hand side makes sense, term by term, as a tempered distribution, with the associated mapping continuous. Since L^p is reflexive and ϕ is dense in L^p, it follows that the expression as a whole belongs to L^p and the inequality holds. Similarly, (5.66) is true whenever $u \in \mathrm{Lip} \bigcap L^p_s$, and (5.67) is true whenever $f \in L^p_s$.

As an important application of Proposition 5.4.1, we have the following commutator estimates in [96, 100].

Proposition 5.4.3. *If $s > 0, 1 < p < \infty, f, g \in \mathcal{S}$ (the Schwartz space), then*

$$\|J^s(fg) - f(J^s g)\|_{L^p} \le c\|\nabla f\|_{L^{p_1}}\|J^{s-1}g\|_{L^{p_2}} + \|J^s f\|_{L^{p_3}}\|g\|_{L^{p_4}},$$

and

$$\|J^s(fg)\|_{L^p} \le c(\|f\|_{L^{p_1}}\|J^s g\|_{L^{p_2}} + \|J^s f\|_{L^{p_3}}\|g\|_{L^{p_4}}),$$

with $p_2, p_3 \in (1, \infty)$ such that

$$\frac{1}{p} = \frac{1}{p_1} + \frac{1}{p_2} = \frac{1}{p_3} + \frac{1}{p_4}.$$

Proof. For $p_1 = p_4 = \infty$, the proof was given in Proposition 5.4.1; the general case follows easily by combining the argument used in Proposition 5.4.1 with the version of the Coifman–Meyer result found in [36, p. 154]. □

The above result is still valid not only for the operator J but also for Λ, with the proof given in [79].

Proposition 5.4.4. *Suppose that $s > 0$ and $p \in (1, \infty)$, $f, g \in \mathcal{S}$ (the Schwartz space), then*

$$\|\Lambda^s(fg) - f(\Lambda^s g)\|_{L^p} \le c\left(\|\nabla f\|_{L^{p_1}} \|\Lambda^{s-1} g\|_{L^{p_2}} + \|\Lambda^s f\|_{L^{p_3}} \|g\|_{L^{p_4}}\right).$$

and

$$\|\Lambda^s(fg)\|_{L^p} \le c(\|f\|_{L^{p_1}} \|\Lambda^s g\|_{L^{p_2}} + \|\Lambda^s f\|_{L^{p_3}} \|g\|_{L^{p_4}})$$

with $p_2, p_3 \in (1, \infty)$ such that

$$\frac{1}{p} = \frac{1}{p_1} + \frac{1}{p_2} = \frac{1}{p_3} + \frac{1}{p_4}.$$

A particular case of the Proposition 5.4.4 was proved by S. G. Resnick in [142].

As a consequence of Proposition 5.4.1, we have the following.

Corollary 5.4.5. *Given $s \ge 0$, $1 < p < \infty$, then*

$$\|J^s[(u \cdot \nabla)B] - (u \cdot \nabla)(J^s B)\|_{L^p} \le c(\|\nabla u\|_{L^\infty} \|J^{s-1}(\nabla B)\|_{L^p} + \|J^s(u)\|_{L^p} \|\nabla B\|_{L^\infty}),$$

and for $p = 2$ and $s > n/2$,

$$\|J^s[(u \cdot \nabla)B] - (u \cdot \nabla)(J^s B)\|_{L^p} \le c(|\nabla u|_{H^s} \|B\|_{H^s} + \|u\|_{H^s} \|\nabla B\|\|_{H^s}). \tag{5.68}$$

The following estimate contains only the first of the two terms on the right-hand side of (5.68); see [61] for its proof.

Proposition 5.4.6. *Given $s > n/2$, there is a constant $c = c(n, s)$ such that, for all u, B with $\nabla u, B \in H^s(\mathbb{R}^n)$,*

$$\|\Lambda^s[(u \cdot \nabla)B] - (u \cdot \nabla)(\Lambda^s B)\|_{L^2} \le c\|\nabla u\|_{H^s} ||B||_{H^s}. \tag{5.69}$$

Considering Proposition 5.4.6 for divergence-free function u,

$$\langle (u \cdot \nabla)(\Lambda^s B), \Lambda^s B \rangle = 0,$$

we immediately obtain the following corollary.

Corollary 5.4.7. *Given $s > n/2$, there is a constant $c = c(n, s)$ such that, for all u, B with $\nabla u, B \in H^s(\mathbb{R}^n)$, and $\nabla \cdot u = 0$,*

$$|\langle \Lambda^s[(u \cdot \nabla)B], \Lambda^s B \rangle| \le c\|\nabla u\|_{H^s} \|B\|_{H^s}^2.$$

Remark 5.4.8. In many applications, the following commutator estimates are used (see [103]):

$$\begin{aligned}\sum_{|\alpha|=s} \|[\partial_x^\alpha; g]f\|_{L^2} &= \sum_{|\alpha|=s} \|\partial_x^\alpha(gf) - g\partial_x^\alpha f\|_{L^2} \\ &\le C_{n,s}\Big(\|\nabla g\|_{L^\infty} \sum_{|\beta|=s-1} \|\partial_x^\beta f\|_{L^2} + \|f\|_{L^\infty} \sum_{|\beta|=s} \|\partial_x^\beta g\|_{L^2} \Big).\end{aligned}$$

Similarly, for $s \geq 1$ one has

$$\|[J^s; g]f\|_{L^2} \leq c(\|\nabla g\|_{L^\infty}\|J^{s-1}f\|_{L^2} + \|J^s g\|_{L^2}\|f\|_{L^\infty});$$

see [96].

There are several extensions of the *commutator estimates* in the recent literature; we mention here only [79, 169]. The recent papers deal mostly with extending the original version of that estimate into another function spaces (e. g., weak Lorentz spaces, [79]).

Recently in [38], using the Bochner definition of fractional power (see Subsection 5.1.3), verification of pointwise *Cordoba's estimate* [39] in a bounded domain is presented together with a version of the *commutator estimates.*

Proposition 5.4.9. *Let Φ be a C^2 convex function satisfying $\Phi(0) = 0$. Let $f \in C_0^\infty(\Omega)$ and let $0 \leq s \leq 2$. Then*

$$\Phi'(f)\Lambda_D^s(f) - \Lambda_D^s(\Phi(f)) \geq 0 \tag{5.70}$$

holds pointwise almost everywhere in Ω.

Proposition 5.4.10. *Let $a \in B(\Omega)$ where $B(\Omega) = W^{2,N}(\Omega) \cap W^{1,\infty}(\Omega)$, if $N \geq 3$, and $B(\Omega) = W^{2,p}(\Omega)$ with $p > 2$, if $N = 2$. There exists a constant C, depending only on Ω, such that*

$$\|[a, \Lambda_D]f\|_{\frac{1}{2},D} \leq C\|a\|_{B(\Omega)}\|f\|_{\frac{1}{2},D} \tag{5.71}$$

holds for any $f \in V_0(\Omega)$, with

$$\|a\|_{B(\Omega)} = \|a\|_{W^{2,N}(\Omega)} + \|a\|_{W^{1,\infty}(\Omega)} \tag{5.72}$$

if $N \geq 3$, and

$$\|a\|_{B(\Omega)} = \|a\|_{W^{2,p}(\Omega)} \tag{5.73}$$

with $p > 2$, if $N = 2$.

Remark 5.4.11. As the last estimate, we quote *a borderline version of the Nirenberg–Gagliardo inequality* (see [37]):

$$\|\psi\|_{W^{1,3}(\Omega)} \leq c\|\psi\|_{L^\infty(\Omega)}^{\frac{1}{3}}\|\psi\|_{H^{\frac{3}{2}}(\Omega)}^{\frac{2}{3}}, \quad \Omega \subset \mathbb{R}^2. \tag{5.74}$$

To show (5.74), we need to use two more elementary estimates (the first one discussed in [166, Theorem 3.5]):

$$\begin{aligned} &\|\phi\|_{W^{2,3}} \leq c\|\phi\|_{L^\infty}^{\frac{1}{3}}\|\phi\|_{H^3}^{\frac{2}{3}}, \quad \phi \in C_0^3(\mathbb{R}^2), \\ &\|\phi\|_{L^3} \leq c\|\phi\|_{L^\infty}^{\frac{1}{3}}\|\phi\|_{L^2}^{\frac{2}{3}}, \end{aligned} \tag{5.75}$$

and a simple interpolation estimate for identity map (specifying in [127, Theorem 2.6]; $T = \mathrm{Id}$, $X_1 = H^3(\Omega)$, $X_2 = W^{2,3}(\Omega)$, $Y_1 = L^2(\Omega) \cap L^\infty(\Omega)$, $Y_2 = L^3(\Omega)$), which gives

$$\|\phi\|_{[W^{2,3}(\Omega),L^3(\Omega)]_\theta} \leq c\|\phi\|^{\frac{1}{3}}_{L^\infty(\Omega)}\|\phi\|^{\frac{2}{3}}_{[H^3(\Omega),L^2(\Omega)]_\theta} \quad \text{for any } \theta \in (0,1).$$

For $\theta = \frac{1}{2}$, we obtain (5.74) ($[\cdot,\cdot]_\theta$ stays for complex interpolation functor).

5.4.2 Continuity and compactness lemmas

The next result is the, frequently used in the text, *Lions–Aubin compactness* lemma. An extended discussion can be found in [153]. The original version was formulated in [124]. We quote it first for completeness.

Lemma 5.4.12. *Let $B_0 \subset B \subset B_1$ be three Banach spaces, where B_1, B_2 are reflexive and the inclusion $B_0 \hookrightarrow B$ is compact. Denote*

$$W = \{v;\ v \in L^{p_0}(0,T;B_0),\ v' \in L^{p_1}(0,T;B_1)\},$$

where T is finite and $1 < p_i < \infty$, $i = 1,2$. The space W is normed with

$$\|v\|_{L^{p_0}(0,T;B_0)} + \|v'\|_{L^{p_1}(0,T;B_1)}.$$

Then the embedding $W \hookrightarrow L^{p_0}(0,T;B)$ is compact.

We recall next an extension taken from [153].

Lemma 5.4.13. *Let X, B and Y be Banach spaces such that $X \hookrightarrow B \subset Y$,*

$$W = \{u \in L^\infty(0,T;X),\ u_t \in L^r(0,T;Y)\} \quad \text{with} \quad r > 1,$$

then $W \hookrightarrow C([0,T];B)$ (here $\hookrightarrow$ denotes compact inclusion).

The following result from [162] is often used to obtain weak continuity of solutions.

Lemma 5.4.14. *Let V and Y be Banach spaces, V reflexive, V a dense subset of Y, and the inclusion map $V \subset Y$ is continuous. Then*

$$L^\infty(0,T;V) \cap C_w([0,T];Y) = C_w([0,T];V).$$

The latter lemma has a corollary (see [162]).

Corollary 5.4.15. *Let V and W be Banach spaces, V reflexive, both contained in some fixed linear space, and $V \cap W$ dense in V and in W. If $u \in L^\infty(0,T;V)$ and $u' \in L^1(0,T;W)$, then there exists a weakly continuous function on $[0,T]$ with values in V which is equal to u almost everywhere.*

We recall finally the classical result on strong continuity borrowed from [171, Chapter III, Section 1].

Lemma 5.4.16. *Let a triple of Hilbert spaces $V \subset H \subset V'$ be given, where V' is the dual to V. If $u \in L^2(0,T;V)$ with $u' \in L^2(0,T;V')$, then function u can be identified with a function in $C([0,T];H)$ and the following equality holds in the sense of "scalar distributions" on $(0,T)$:*

$$\frac{d}{dt}\|u\|^2 = 2\langle u', u\rangle_{\langle H', H\rangle}.$$

5.4.3 Moment inequality

5.4.3.1 Moment inequality, a brief discussion

We recall here, for completeness of presentation, the *moment inequality* estimate valid for fractional powers of nonnegative operators having zero in the resolvent. According to Proposition 5.2.4, such operators are in fact positive. Recalling [187, p. 98], we have the following version of the moment inequality, for $0 \le \alpha < \beta < \gamma \le 1$,

$$\|A^\beta \phi\| \le \frac{(\sin\frac{\pi(\beta-\alpha)}{\gamma-\alpha})(\gamma-\alpha)^2}{\pi(\gamma-\beta)(\beta-\alpha)}(M+1)\|A^\gamma\phi\|^{\frac{\beta-\alpha}{\gamma-\alpha}}\|A^\alpha\phi\|^{\frac{\gamma-\beta}{\gamma-\alpha}},$$

where $\phi \in D(A^\gamma)$.

In particular, for $\alpha = 0, \beta = \frac{1}{2} < \gamma \le 1$ this reduces to

$$\|A^{\frac{1}{2}}\phi\| \le \frac{(\sin\frac{\pi}{2\gamma})\gamma^2}{\frac{\pi}{2}(\gamma-\frac{1}{2})}(M+1)\|A^\gamma\phi\|^{\frac{1}{2\gamma}}\|\phi\|^{1-\frac{1}{2\gamma}}, \quad \phi \in D(A^\gamma). \tag{5.76}$$

Using de l'Hospital rule, we see that $\frac{(\sin\frac{\pi}{2\gamma})\gamma^2}{\frac{\pi}{2}(\gamma-\frac{1}{2})} \to 1$ as $\gamma \searrow \frac{1}{2}$, so we can let $\gamma \searrow \frac{1}{2}$ in the formula above, and the constant remains bounded.

Moment inequality for nonnegative operators

We will present now the proof of the *moment inequality* under the sole assumption, that the operator A is nonnegative. Next, we will transform that estimate, as in [187, p. 98], to the form suitable to compare the $\frac{1}{2}$ and $\frac{1}{2}+\epsilon$ ($\epsilon > 0$ small) norms of fractional powers of such operators.

Lemma 5.4.17. *For arbitrary nonnegative operator A in a Banach space X with*

$$\|(\lambda + A)^{-1}\| \le \frac{M}{\lambda}, \quad \lambda > 0,$$

the following version of the moment inequality holds for $\alpha \in (0,1)$:

$$\|A^\alpha\phi\| \le C(\alpha, 1, M)\|A\phi\|^\alpha\|\phi\|^{1-\alpha}, \quad \phi \in D(A), \tag{5.77}$$

where

$$C(\alpha, 1, M) = \frac{|\sin(\pi\alpha)|}{\pi\alpha(1-\alpha)} M^{\alpha}(1+M)^{1-\alpha}.$$

Proof. We are using an estimate

$$\|(\lambda + A)^{-1}A\|_X = \|(\lambda + A)^{-1}(-\lambda + (\lambda + A))\|_X \leq M + 1, \tag{5.78}$$

in the calculations below. Splitting the integral (5.22) over $(0, L)$ and (L, ∞), we estimate now the first integral

$$\begin{aligned}
&\left\| \frac{\sin(\pi\alpha)}{\pi} \int_0^L \lambda^{\alpha-1}(\lambda + A)^{-1}A\phi d\lambda \right\|_X \\
&\leq \frac{|\sin(\pi\alpha)|}{\pi} \int_0^L \lambda^{\alpha-1}\|(\lambda + A)^{-1}(-\lambda + (\lambda + A))\phi\|_X d\lambda \\
&\leq \frac{|\sin(\pi\alpha)|}{\pi} \int_0^L \lambda^{\alpha-1}(M+1)d\lambda\|\phi\|_X \\
&= \frac{|\sin(\pi\alpha)|}{\pi} \frac{L^{\alpha}}{\alpha}(M+1)\|\phi\|_X.
\end{aligned} \tag{5.79}$$

The second integral, over (L, ∞), is estimated next:

$$\begin{aligned}
\frac{|\sin(\pi\alpha)|}{\pi} \left\| \int_L^{\infty} \lambda^{\alpha-1}(\lambda + A)^{-1}A\phi d\lambda \right\|_X &\leq \frac{|\sin(\pi\alpha)|}{\pi} \int_L^{\infty} \lambda^{\alpha-1}\frac{M}{\lambda}\|A\phi\|_X d\lambda \\
&\leq \frac{|\sin(\pi\alpha)|}{\pi} \frac{L^{\alpha-1}}{1-\alpha} M\|A\phi\|_X.
\end{aligned} \tag{5.80}$$

Putting the estimates together, we obtain for $\phi \in D(A)$,

$$\begin{aligned}
&\left\| \frac{\sin(\pi\alpha)}{\pi} \int_0^{\infty} \lambda^{\alpha-1}(\lambda + A)^{-1}A\phi d\lambda \right\|_X \\
&\leq \frac{|\sin(\pi\alpha)|}{\pi} \left[\frac{L^{\alpha}}{\alpha}(M+1)\|\phi\|_X + \frac{L^{\alpha-1}}{1-\alpha} M\|A\phi\|_X \right].
\end{aligned} \tag{5.81}$$

Minimizing the above sum with respect to $L > 0$, we find that $L_{\min} = \frac{M\|A\phi\|_X}{(M+1)\|\phi\|_X}$, which leads to the final estimate:

$$\begin{aligned}
\|A^{\alpha}\phi\|_X \leq \frac{|\sin(\pi\alpha)|}{\pi} \Bigg(&\frac{1}{\alpha}\left(\frac{M\|A\phi\|_X}{(M+1)\|\phi\|_X}\right)^{\alpha}(M+1)\|\phi\|_X \\
&+ \frac{1}{1-\alpha}\left(\frac{M\|A\phi\|_X}{(M+1)\|\phi\|_X}\right)^{\alpha-1} M\|A\phi\|_X \Bigg)
\end{aligned}$$

$$= \frac{|\sin(\pi\alpha)|}{\pi(1-\alpha)\alpha} M^{\alpha}(M+1)^{1-\alpha}\|A\phi\|_X^{\alpha}\|\phi\|_X^{1-\alpha}. \tag{5.82}$$

This is the *moment inequality* valid for $\phi \in D(A)$. □

We will transform that estimate, as in [187, p. 98], to the form suitable to compare the norms of $\frac{1}{2}$ and $\frac{1}{2}+\epsilon$ ($\epsilon > 0$ small) fractional powers. Let $0 < \beta < \gamma \le 1$. When A is a nonnegative operator, the same is true for A^{γ}. Put first in (5.82) $D(A) \ni \phi = A^{-1}f$, so that $f \in X$ and

$$\|A^{\alpha-1}f\|_X \le \frac{|\sin(\pi\alpha)|}{\pi(1-\alpha)\alpha} M^{\alpha}(M+1)^{1-\alpha}\|f\|_X^{\alpha}\|A^{-1}f\|_X^{1-\alpha}. \tag{5.83}$$

Next, replace A in the latter estimate with A^{γ} and choose $\alpha := \frac{\beta}{\gamma} < 1$. We obtain

$$\|(A^{\gamma})^{\frac{\beta}{\gamma}-1}f\|_X = \|A^{\beta-\gamma}f\|_X \le \frac{|\sin(\pi\frac{\beta}{\gamma})|}{\pi(1-\frac{\beta}{\gamma})\frac{\beta}{\gamma}} M^{\frac{\beta}{\gamma}}(M+1)^{1-\frac{\beta}{\gamma}}\|f\|_X^{\frac{\beta}{\gamma}}\|A^{-\gamma}f\|_X^{1-\frac{\beta}{\gamma}}, \tag{5.84}$$

where $A^{-\gamma}f \in D(A^{\gamma})$. Setting finally $f = A^{\gamma}U$ in the latter estimate, we obtain

$$\|A^{\beta}U\|_X \le \frac{|\sin(\pi\frac{\beta}{\gamma})|}{\pi(1-\frac{\beta}{\gamma})\frac{\beta}{\gamma}} M^{\frac{\beta}{\gamma}}(M+1)^{1-\frac{\beta}{\gamma}}\|A^{\gamma}U\|_X^{\frac{\beta}{\gamma}}\|U\|_X^{1-\frac{\beta}{\gamma}} \tag{5.85}$$

which holds for $U \in D(A^{\gamma})$. Note also that we are able to let $\gamma \searrow \beta$ (or $\frac{\beta}{\gamma} \nearrow 1$) in the last version of the moment inequality, since by the de l'Hospital rule

$$\lim_{\gamma\searrow\beta} \frac{|\sin(\pi\frac{\beta}{\gamma})|}{\pi(1-\frac{\beta}{\gamma})\frac{\beta}{\gamma}} = \left(\frac{0}{0}\right) = 1.$$

Remark 5.4.18. We will specify now the moment inequality (5.82) in the form that is suitable in the studies of the critical Hamilton–Jacobi or quasi-geostrophic equations. We thus set: $\beta = \frac{1}{2}$, $\gamma > \frac{1}{2}$ but close, and study the possibility of passing with γ to β in the resulting additive estimate. We have

$$\|A^{\frac{1}{2}}U\|_X \le \frac{|\sin(\frac{\pi}{2\gamma})|}{(1-\frac{1}{2\gamma})\frac{\pi}{2\gamma}} M^{\frac{1}{2\gamma}}(M+1)^{1-\frac{1}{2\gamma}}\|A^{\gamma}U\|_X^{\frac{1}{2\gamma}}\|U\|_X^{1-\frac{1}{2\gamma}}. \tag{5.86}$$

We apply next the Young inequality (Corollary 2.1.3)

$$ab \le \frac{1}{m}\epsilon^m a^m + \frac{m-1}{m}\epsilon^{-\frac{m}{m-1}} b^{\frac{m}{m-1}}, \quad a, b > 0,\ m > 1,\ \epsilon > 0, \tag{5.87}$$

with $m = 2\gamma$ and $\epsilon = 1$, so that $\frac{m}{m-1} = \frac{2\gamma}{2\gamma-1}$, to the right-hand side of (5.86) to obtain

$$\|A^{\frac{1}{2}}U\|_X \le \frac{|\sin(\frac{\pi}{2\gamma})|}{(1-\frac{1}{2\gamma})\frac{\pi}{2\gamma}}\left[\frac{1}{2\gamma}M\|A^{\gamma}U\|_X + \frac{1-2\gamma}{2\gamma}(M+1)\|U\|_X\right]. \tag{5.88}$$

It is important to see that we will pass to the limit $\gamma \searrow \frac{1}{2}$ in the last estimate keeping the constants bounded, since by the de l'Hospital rule

$$\lim_{\gamma \searrow \frac{1}{2}} \frac{|\sin(\frac{\pi}{2\gamma})|}{(1-\frac{1}{2\gamma})\frac{\pi}{2\gamma}} = \left(\frac{0}{0}\right) = 1. \tag{5.89}$$

Remark 5.4.19. In case of the positive self-adjoint operators A in Hilbert spaces, the moment inequality is simpler (e. g., [108, Chapter I, Section 5]). Fractional powers of such operators will be defined using *spectral resolution* (e. g., [158, p. 130], or Subsection 5.2.2) and the corresponding *moment inequality* reduces to

$$\|A^{\beta}\phi\| \le \|A^{\gamma}\phi\|^{\frac{\beta-\alpha}{\gamma-\alpha}} \|A^{\alpha}\phi\|^{\frac{\gamma-\beta}{\gamma-\alpha}}, \tag{5.90}$$

where $\alpha < \beta < \gamma$ and $\phi \in D(A^{\gamma})$. Note that in that case the constant at the right-hand side is *precisely equal to* 1.

We will prove the basic *moment inequality* for fractional powers of positive self-adjoint operators in a Hilbert space H.

Lemma 5.4.20. *Let $B : H \supset D(B) \to H$ be a positive self-adjoint operator, and $0 \le \alpha \le 1$. Then for all $v \in D(B)$ the moment inequality holds:*

$$\|B^{\alpha}v\| \le \|Bv\|^{\alpha}\|v\|^{1-\alpha}. \tag{5.91}$$

Proof. The proof follows through a direct application of the Hölder inequality (see [190, Chapter I, 3, (5)]) to spectral representation:

$$\begin{aligned} \|B^{\alpha}v\|^2 &= \int_0^{\infty} \lambda^{2\alpha} d\|E_{\lambda}v\|^2 \\ &\le \left(\int_0^{\infty} \lambda^2 d\|E_{\lambda}v\|^2\right)^{\alpha} \left(\int_0^{\infty} d\|E_{\lambda}v\|^2\right)^{1-\alpha} = \|Bv\|^{2\alpha}\|v\|^{2(1-\alpha)}. \end{aligned} \tag{5.92}$$

It is interesting to note that the constant at the right-hand side of that inequality is precisely equal to one. □

The above estimate was reported in a more general form in [108, Chapter I, 9].

We propose next an estimate of the difference of fractional powers of positive operators.

Lemma 5.4.21. *Let A be a positive operator in a Banach space X (see Subsection 5.2.1) satisfying*

$$\|(\lambda+1)(\lambda+A)^{-1}\| \le M, \quad \lambda > 0.$$

For arbitrary $\phi \in X$, *we have*

$$\forall_{\epsilon>0}\exists_L \ \|(I - A^{-\beta})\phi\|_X \leq \sin(\pi\beta)\left(\frac{2L(1+M)}{\pi} + L^{-1}M\right)\|\phi\|_X + \epsilon.$$

Consequently, the left-hand side tends to zero as $0 < \beta \to 0^+$.

We will use that lemma to estimate the difference $(I - A_{\alpha_1}^{-1}A_{\alpha_2})$ when $\frac{1}{2} < \alpha_2 \leq \alpha_1 \to \frac{1}{2}^+$.

Proof. Our task is, for fixed $\phi \in X$ and β near 0^+, to estimate the expression:

$$\begin{aligned}(A^{-\beta} - I)\phi &= \frac{\sin(\pi\beta)}{\pi}\int_0^\infty \lambda^{-\beta}(\lambda + A)^{-1}\phi d\lambda - \frac{\sin(\pi(1-\beta))}{\pi}\int_0^\infty \frac{\lambda^{(1-\beta)-1}}{\lambda+1}d\lambda\,\phi \\ &= \frac{\sin(\pi\beta)}{\pi}\int_0^\infty \lambda^{-\beta}\left[(\lambda + A)^{-1}\phi - \frac{1}{\lambda+1}\phi\right]d\lambda. \end{aligned} \tag{5.93}$$

In the estimates, we are using the following properties taken from [130, p. 62], for $\eta \in (0,1)$,

$$\int_0^\infty \frac{\lambda^{\eta-1}}{\lambda+1}d\lambda = \frac{\pi}{\sin(\pi\eta)}, \qquad |\sin(\pi\alpha)| = |\sin(\pi(1-\alpha))|,$$

the simple formula:

$$(\lambda + A)^{-1}\phi - \frac{1}{\lambda+1}\phi = \frac{1}{\lambda+1}[\lambda(\lambda + A)^{-1}\phi - \phi + (\lambda + A)^{-1}\phi], \tag{5.94}$$

and the two asymptotic properties of *nonnegative operators* valid on functions $\phi \in X$ taken from [130, Proposition 1.1.3]:

$$\begin{aligned}&\lim_{\lambda\to\infty} \lambda(\lambda + A)^{-1}\phi = \phi, \quad \phi \in \overline{D(A)}, \\ &\lim_{\lambda\to\infty} (\lambda + A)^{-1}A\phi = 0.\end{aligned} \tag{5.95}$$

Returning to the proof, we split the integral in (5.93) into $(0, L)$ and (L, ∞) and estimate the first part,

$$\begin{aligned}&\frac{\sin(\pi\beta)}{\pi}\left\|\int_0^L \lambda^{-\beta}\left(\frac{1}{\lambda+1} - (\lambda + A)^{-1}\right)\phi d\lambda\right\|_X \\ &\leq \frac{\sin(\pi\beta)}{\pi}\int_0^L \frac{\lambda^{-\beta}}{\lambda+1}(1 + \|(\lambda+1)(\lambda + A)^{-1})\|d\lambda\|\phi\|_X \\ &\leq \frac{\sin(\pi\beta)}{\pi}\int_0^L \lambda^{-\beta}(1 + M)d\lambda\|\phi\|_X \\ &= \frac{\sin(\pi\beta)}{\pi}\frac{L^{1-\beta}}{1-\beta}(1 + M)\|\phi\|_X,\end{aligned} \tag{5.96}$$

where $L > 0$ will be chosen later. Note that letting $\beta \to 0^+$ the result of the estimate above is bounded by $|\frac{\sin(\pi\beta)}{\pi}|2L(1+M)\|\phi\|_X \to 0$, for any fixed $L > 0$.

Next using (5.94), the integral over (L, ∞) is, for $\phi \in X$, estimated as follows:

$$\frac{\sin(\pi\beta)}{\pi}\left\|\int_L^\infty \lambda^{-\beta}\left[(\lambda+A)^{-1}\phi - \frac{1}{\lambda+1}\phi\right]d\lambda\right\|_X$$
$$\leq \frac{\sin(\pi\beta)}{\pi}\int_L^\infty \frac{\lambda^{-\beta}}{\lambda+1}\|\lambda(\lambda+A)^{-1}\phi - \phi + (\lambda+A)^{-1}\phi\|_X d\lambda, \tag{5.97}$$

where due to (5.95) we see that

$$\|(\lambda+1)(\lambda+A)^{-1}\phi - \phi\|_X \leq \|\lambda(\lambda+A)^{-1}\phi - \phi\|_X + \|(\lambda+A)^{-1}\phi\|_X$$
$$\leq \epsilon + \frac{M}{1+\lambda}\|\phi\|_X \quad \text{as } \lambda \to \infty, \tag{5.98}$$

$\epsilon > 0$ arbitrary fixed. Consequently, we obtain

$$\frac{\sin(\pi\beta)}{\pi}\int_L^\infty \frac{\lambda^{-\beta}}{\lambda+1}\left(\epsilon + \frac{M}{1+\lambda}\|\phi\|_X\right)d\lambda \leq \frac{\sin(\pi\beta)}{\pi}\int_L^\infty \lambda^{-\beta-1}\left(\epsilon + \frac{M}{\lambda}\|\phi\|_X\right)d\lambda$$
$$\leq \frac{\sin(\pi\beta)}{\pi}\frac{L^{-\beta}}{\beta}\epsilon + \frac{\sin(\pi\beta)}{\pi}\frac{L^{-\beta-1}}{\beta+1}M\|\phi\|_X, \tag{5.99}$$

for sufficiently large value of $L \geq 1$, as specified in (5.98) with $\lambda \geq L$. Note that letting $\beta \to 0^+$ in the resulting estimate we have

$$\frac{\sin(\pi\beta)}{\pi\beta}L^{-\beta}\epsilon + \frac{\sin(\pi\beta)}{\pi}\frac{L^{-\beta-1}}{\beta+1}M\|\phi\|_X \leq \epsilon + \sin(\pi\beta)L^{-1}M\|\phi\|_X, \tag{5.100}$$

for chosen large value of L.

For such L, we get a final estimate of the integral in (5.93) having the form:

$$\|(A^{-\beta} - I)\phi\|_X \leq \frac{\sin(\pi\beta)}{\pi}\left\|\int_0^\infty \lambda^{-\beta}\left[(\lambda+A)^{-1}\phi - \frac{1}{\lambda+1}\phi\right]d\lambda\right\|_X$$
$$\leq \sin(\pi\beta)\left(\frac{2L(1+M)}{\pi} + L^{-1}M\right)\|\phi\|_X + \epsilon, \tag{5.101}$$

where $\epsilon > 0$ was arbitrary. The right-hand side of (5.101) will be made small when we let β near 0^+, noting ϵ was an arbitrary positive number. □

5.4.3.2 Moment inequality extended

Our task here is to extend moment inequality to the form suitable to compare the powers $1+\alpha$ and 1 of a positive operator (here $\alpha > 0$, small). We need to use a more general than (5.22) expression (5.102) for the fractional powers, taken from [130, (3.4), p. 59], which states that

$$A^{1+\alpha}\phi = \frac{\sin(\pi\alpha)}{\alpha\pi}\int_0^\infty \lambda^\alpha [A(\lambda+A)^{-1}]^2 \phi d\lambda, \quad \phi \in D(A^2), \tag{5.102}$$

where the original term $\frac{\Gamma(2)}{\Gamma(1+\alpha)\Gamma(1-\alpha)}$ has been transformed using the known properties of the Γ function,

$$\Gamma(1+\alpha) = \alpha\Gamma(\alpha), \quad \Gamma(\alpha)\Gamma(1-\alpha) = \frac{\pi}{\sin(\pi\alpha)}, \quad \Gamma(2) = 1.$$

We are using the following bounds, valid for *positive operators*, in the calculations below:

$$\|A(\lambda+A)^{-1}\|_X \le M+1, \quad \|(\lambda+A)^{-1}\|_X \le \frac{M}{1+\lambda}, \quad \lambda > 0, \tag{5.103}$$

splitting the integral over $(0, L)$ and (L, ∞), we write the first integral,

$$\frac{\sin(\pi\alpha)}{\alpha\pi}\int_0^L \lambda^\alpha [A(\lambda+A)^{-1}]^2 \phi d\lambda = \frac{\sin(\pi\alpha)}{\alpha\pi}\int_0^L \lambda^\alpha (\lambda+A)^{-1}[A(\lambda+A)^{-1}]A\phi d\lambda. \tag{5.104}$$

We thus have

$$\begin{aligned}\left\|\frac{\sin(\pi\alpha)}{\alpha\pi}\int_0^L \lambda^\alpha [A(\lambda+A)^{-1}]^2 \phi d\lambda\right\|_X &\le \frac{\sin(\pi\alpha)}{\alpha\pi} M(M+1)\|A\phi\|_X \int_0^L \lambda^\alpha d\lambda \\ &\le \frac{\sin(\pi\alpha)}{\alpha\pi} M(M+1)\frac{L^{\alpha+1}}{\alpha+1}\|A\phi\|_X.\end{aligned} \tag{5.105}$$

The second integral over (L, ∞) is estimated next:

$$\begin{aligned}\frac{\sin(\pi\alpha)}{\alpha\pi}\left\|\int_L^\infty \lambda^\alpha [(\lambda+A)^{-1}]^2 A^2 \phi d\lambda\right\|_X &\le \frac{\sin(\pi\alpha)}{\alpha\pi}\int_L^\infty \lambda^\alpha \left(\frac{M}{\lambda}\right)^2 \|A^2\phi\|_X d\lambda \\ &= \frac{\sin(\pi\alpha)}{\alpha\pi}\frac{M^2}{1-\alpha} L^{\alpha-1}\|A^2\phi\|_X.\end{aligned} \tag{5.106}$$

Minimizing with respect to $L > 0$, we get $L^2_{\min} = \frac{M\|A^2\phi\|_X}{(M+1)\|A\phi\|_X}$, which leads to the final estimate:

$$\begin{aligned}\|A^{1+\alpha}\phi\|_X &\le \frac{\sin(\pi\alpha)}{\alpha\pi} M(M+1)\frac{L^{\alpha+1}_{\min}}{\alpha+1}\|A\phi\|_X + \frac{\sin(\pi\alpha)}{\alpha\pi}\frac{M^2}{1-\alpha} L^{\alpha-1}_{\min}\|A^2\phi\|_X \\ &= \frac{2\sin(\pi\alpha)}{\alpha\pi(1+\alpha)(1-\alpha)} M^{1+\frac{\alpha+1}{2}}(M+1)^{\frac{1-\alpha}{2}}\|A\phi\|_X^{\frac{1-\alpha}{2}}\|A^2\phi\|_X^{\frac{1+\alpha}{2}} \\ &\to 2M^{\frac{3}{2}}(M+1)^{\frac{1}{2}}\|A\phi\|_X^{\frac{1}{2}}\|A^2\phi\|_X^{\frac{1}{2}},\end{aligned} \tag{5.107}$$

as $\alpha \to 0^+$.

5.4.4 Kato–Beurling–Deny inequality

We now quote a version of the famous Kato–Beurling–Deny inequality borrowed from [34]. It will be proved below for completeness of the presentation. We shall focus here on the Dirichlet Laplacian $A = -\Delta_D$ considered on $L^2(\Omega)$, $\Omega \subset \mathbb{R}^n$ being a bounded C^2 domain, which is a special case of the general theory of [44].

It is well known that $(sI+A)^{-1}$, $s \geq 0$, has a positive symmetric kernel $K_s = K_s(x,y)$, $x, y \in \Omega$, satisfying the estimate in [166, 5.168, p. 210]. The following proposition holds.

Proposition 5.4.22. *Let $\Omega \subset \mathbb{R}^N$ be a bounded domain, $\partial\Omega \in C^2$, and $A = -\Delta_D$ on $L^2(\Omega)$. Then, for $\alpha \in [0,1]$, $q \in [2,+\infty)$ and ϕ in the cone C_0^+, the following inequality holds:*

$$\int_\Omega A^\alpha \phi \phi^{q-1} dx = \int_\Omega A^{\frac{\alpha}{2}} \phi A^{\frac{\alpha}{2}} (\phi^{q-1}) dx \geq \frac{4(q-1)}{q^2} \int_\Omega [A^{\frac{\alpha}{2}} (\phi^{\frac{q}{2}})]^2 dx. \tag{5.108}$$

Proof. Writing below for simplicity of the notation $\langle \cdot,\cdot \rangle_{L^2(\Omega)}$ for the $L^2(\Omega)$ product, we obtain

$$\begin{aligned}\langle A^\alpha v, v^{q-1} \rangle_{L^2(\Omega)} &= \left\langle \frac{\sin \pi\alpha}{\pi} \int_0^{+\infty} s^{\alpha-1} (sI + A - sI)(sI + A)^{-1} v\, ds, v^{q-1} \right\rangle_{L^2(\Omega)} \\ &= \frac{\sin \pi\alpha}{\pi} \int_0^{+\infty} s^{\alpha-1} (\|v^q\|_{L^1(\Omega)} - \langle s(sI + A)^{-1} v, v^{q-1} \rangle_{L^2(\Omega)}) ds, \quad v \in C_0^+,\end{aligned} \tag{5.109}$$

where

$$C_0^+ = \{\phi \in C^2(\overline{\Omega}) : \phi \geq 0, \phi_{|\partial\Omega} = 0\}.$$

The properties of K_s and an elementary inequality of [44, p. 68]

$$(s-t)(s^{q-1} - t^{q-1}) \geq \frac{4(q-1)}{q^2} |s^{\frac{q}{2}} - t^{\frac{q}{2}}|^2, \quad s \geq 0,\ t \geq 0,\ q \geq 2,$$

ensure next that

$$\begin{aligned}&\|v^q\|_{L^1(\Omega)} - \langle s(sI + A)^{-1} v, v^{q-1} \rangle_{L^2(\Omega)} \\ &\quad = \|v^q\|_{L^1(\Omega)} - s \int_{\Omega\times\Omega} v^{q-1}(x) v(y) K_s(x,y) dx dy \\ &\quad = \frac{s}{2} \int_{\Omega\times\Omega} [v(x) - v(y)][v^{q-1}(x) - v^{q-1}(y)] K_s(x,y) dx dy \\ &\quad\quad + \|v^q\|_{L^1(\Omega)} - s \int_{\Omega\times\Omega} v^q(x) K_s(x,y) dx dy\end{aligned}$$

$$
\begin{aligned}
&= \frac{s}{2}\int_{\Omega\times\Omega} [v(x)-v(y)][v^{q-1}(x)-v^{q-1}(y)]K_s(x,y)dxdy \\
&\quad + \|v^q\|_{L^1(\Omega)} - s\|(sI+A)^{-1}(v^q)\|_{L^1(\Omega)} \\
&\geq \frac{4(q-1)}{q^2}\Bigg(\frac{s}{2}\int_{\Omega\times\Omega} |v^{\frac{q}{2}}(x)-v^{\frac{q}{2}}(y)|^2 K_s(x,y)dxdy \\
&\quad + \|v^q\|_{L^1(\Omega)} - s\|(sI+A)^{-1}(v^q)\|_{L^1(\Omega)}\Bigg), \quad s>0,\ q\geq 2,\ v\in C_0^+, \qquad (5.110)
\end{aligned}
$$

where in the last line above we have used additionally the inequality $1 \geq \frac{4(q-1)}{q^2}$, $q \geq 2$, and the contraction property of A in $L^1(\Omega)$ (see [44, Theorem 1.3.5]), which guarantees that

$$
\|v^q\|_{L^1(\Omega)} - s\|(sI+A)^{-1}(v^q)\|_{L^1(\Omega)} \geq 0, \quad s>0,\ v\in C_0^+.
$$

Similar calculations show that

$$
\begin{aligned}
\langle A^\alpha(v^{\frac{q}{2}}), v^{\frac{q}{2}}\rangle_{L^2(\Omega)} &= \left\langle \frac{\sin\pi\alpha}{\pi}\int_0^{+\infty} s^{\alpha-1}(sI+A-sI)(sI+A)^{-1}(v^{\frac{q}{2}})\,ds, v^{\frac{q}{2}}\right\rangle_{L^2(\Omega)} \\
&= \frac{\sin\pi\alpha}{\pi}\int_0^{+\infty} s^{\alpha-1}(\|v^q\|_{L^1(\Omega)} - \langle s(sI+A)^{-1}(v^{\frac{q}{2}}), v^{\frac{q}{2}}\rangle_{L^2(\Omega)})ds \\
&= \frac{\sin\pi\alpha}{\pi}\int_0^{+\infty} s^{\alpha-1}\Bigg(\|v^q\|_{L^1(\Omega)} - s\int_{\Omega\times\Omega} v^{\frac{q}{2}}(x)v^{\frac{q}{2}}(y)K_s(x,y)dxdy\Bigg)ds, \\
&= \frac{\sin\pi\alpha}{\pi}\int_0^{+\infty} s^{\alpha-1}\Bigg(\|v^q\|_{L^1(\Omega)} + \frac{s}{2}\int_{\Omega\times\Omega} |v^{\frac{q}{2}}(x)-v^{\frac{q}{2}}(y)|^2K_s(x,y)dxdy \\
&\quad - s\int_{\Omega\times\Omega} v^q(x)K_s(x,y)dxdy\Bigg)ds \\
&= \frac{\sin\pi\alpha}{\pi}\int_0^{+\infty} s^{\alpha-1}\Bigg(\frac{s}{2}\int_{\Omega\times\Omega} |v^{\frac{q}{2}}(x)-v^{\frac{q}{2}}(y)|^2K_s(x,y)dxdy \\
&\quad + \|v^q\|_{L^1(\Omega)} - s\|(sI+A)^{-1}(v^q)\|_{L^1(\Omega)}\Bigg)ds, \quad s>0,\ q\geq 2,\ v\in C_0^+. \qquad (5.111)
\end{aligned}
$$

As a consequence of the relation (5.110) and the obvious estimates when $\alpha = 0$ or $\alpha = 1$, the proof is complete. □

It is well known that the resolvent $(\lambda I - \Delta_D)^{-1}$, $\lambda > 0$, preserves positivity. This property extends directly to the resolvent of $(-\Delta_D)^\alpha$, $\alpha \in (0,1)$, because of the formula ([106, p. 319]):

$$(\lambda I + (-\Delta_D)^\alpha)^{-1} = \frac{\sin \pi\alpha}{\pi} \int_0^{+\infty} \frac{\tau^\alpha}{\lambda^2 + 2\lambda\tau^\alpha \cos \pi\alpha + \tau^{2\alpha}} (\tau I - \Delta_D)^{-1} d\tau, \quad \lambda > 0,$$

since the denominator above is positive. Next, we have an integral lower bound (e. g., [34, 51]), a version of the Kato–Beurling–Deny inequality.

Corollary 5.4.23. *For $\alpha \in [0,1]$, $q \in [2,+\infty)$, $\phi \in D((-\Delta_D)^\alpha)$, with $|\phi|^{q-1} \in D((-\Delta_D)^{\frac{\alpha}{2}})$, the following estimate holds:*

$$\int_\Omega (-\Delta_D)^\alpha \phi \,\mathrm{sgn}\,\phi\, |\phi|^{q-1} dx \geqslant \frac{4(q-1)}{q^2} \int_\Omega [(-\Delta_D)^{\frac{\alpha}{2}} (|\phi|^{\frac{q}{2}})]^2 dx. \tag{5.112}$$

Note that taking $q = 4p$, $p \in \mathbb{N}$, in the latter estimate, the absolute value will not be needed.

Remark 5.4.24. A number of estimates similar to the above, but in the case of the whole $\mathbb{R}^N$, were reported in the review article [98]; see also the source references there. In particular the, corresponding to (5.108), (5.112), estimates in $\mathbb{R}^N$ called the *Strook–Varopoulos inequality* can be found in [98] together with their proofs. In fact, a more general form of the operator, *Lévy operator*, appears in these estimates.

Remark 5.4.25. It will be observed that at the right-hand side of the Kato–Beurling–Deny inequality (5.112) absolute value of ϕ is required. However, using the result of [78, p. 152], we will see that

$$\int_\Omega [(-\Delta)^{\frac{1}{2}} v]^2 dx = \int_\Omega [(-\Delta)^{\frac{1}{2}} |v|]^2 dx \tag{5.113}$$

for every $v \in H_0^1(\Omega)$. Indeed, integrating by parts, for $w \in H^2(\Omega) \cap H_0^1(\Omega)$ we get

$$\int_\Omega |\nabla w|^2 dx = \int_\Omega (-\Delta) w w dx = \int_\Omega (-\Delta)^{\frac{1}{2}} w (-\Delta)^{\frac{1}{2}} w dx = \int_\Omega [(-\Delta)^{\frac{1}{2}} w]^2 dx, \tag{5.114}$$

where the middle equality follows by the self-adjointness. By density, the equality above, $\||\nabla w|\|_{L^2(\Omega)} = \|(-\Delta)^{\frac{1}{2}} w\|_{L^2(\Omega)}$, extends to all $w \in H_0^1(\Omega)$. Further, by [78, Lemma 7.6],

$$\nabla |v| = \begin{cases} \nabla v & \text{if } v > 0, \\ 0 & \text{if } v = 0, \\ -\nabla v & \text{if } v < 0. \end{cases}$$

Since for any $v \in H_0^1(\Omega)$ also $|v| \in H_0^1(\Omega)$, then

$$\int_\Omega \left(\frac{\partial}{\partial x_i} |v|\right)^2 dx = \int_{\Omega \cap \{v>0\}} \left(\frac{\partial}{\partial x_i} v\right)^2 dx + \int_{\Omega \cap \{v<0\}} \left(-\frac{\partial}{\partial x_i} v\right)^2 dx = \int_\Omega \left(\frac{\partial}{\partial x_i} v\right)^2 dx, \tag{5.115}$$

and (5.113) is proved (taking $w = v$ and $w = |v|$ in the consequence of (5.114)).

Since also $\||v|-|w|\|_{L^2(\Omega)} \le \|v-w\|_{L^2(\Omega)}$, we will apply [165, Lemma 28.1] on nonlinear interpolation, where we specify: $E_0 = E_1 = L^2(\Omega)$, $F_0 = F_1 = H_0^1(\Omega)$ and $A(u) = |u|$, to conclude that

$$\|(-\Delta)^\mu |v|\|_{L^2(\Omega)} \le C\|(-\Delta)^\mu v\|_{L^2(\Omega)}, \quad \text{for each } \mu \in \left(0, \frac{1}{2}\right)$$

(in case of Hilbert spaces the real and complex interpolation functors coincide: $(F_0, F_1)_{\nu,2} = [F_0, F_1]_\nu = D((-\Delta_2)^{\frac{\nu}{2}})$ where $0 < \nu < 1$). We leave the details for the reader.

Bibliographical notes

There is a growing number of publications devoted to various technical tools needed in the studies of equations with fractional powers of operators. Besides the main reference [130], we quote here only the review articles [45] and [98]. The recent results concerning function spaces, in particular Sobolev and Nirenberg–Gagliardo estimates, are collected in [7]. We sketched also, from [109], a valuable comparison of various existing definitions of fractional powers of negative Laplacian in $\mathbb{R}^N$. A more complete discussion of such kind, in case of the Dirichlet problem, is not available yet.

6 Navier–Stokes equation in 2D and 3D

6.1 Introduction

Following Dan Henry [86], extension of *semilinear parabolic equations*, namely semilinear equations with sectorial positive operator in the main part, will be studied inside the semigroup approach. Such technique was successfully applied to several classical problems, some of them were reported in [32, 86]. The present chapter is devoted to a generalization of that approach to a *critical problem*. We will deal with the celebrated 2-D Navier–Stokes equation (N-S equation, for short), an example of a problem critical with respect to the a priori estimate in L^2. In order, the 3-D N-S equation is *supercritical* in a sense specified in Subsection 6.1.3. Its *weak solution* will be obtained regularizing the Stokes operator by adding to it higher order diffusion term $-\epsilon(-P\Delta)^s$, $\epsilon > 0$, $s > \frac{5}{4}$, solving the regularized equation, then letting $\epsilon \to 0$.

The Dirichlet problem for classical 3-D *Navier–Stokes equation* considered here has the form:

$$\begin{aligned} &u_t = \nu\Delta u - \nabla p - (u \cdot \nabla)u + f, \quad \operatorname{div} u = 0, \quad x \in \Omega,\ t > 0, \\ &u = 0, \quad t > 0, \quad x \in \partial\Omega, \\ &u(0,x) = u_0(x), \end{aligned} \tag{6.1}$$

(see also (6.11)) where $\nu > 0$ is the viscosity coefficient, $u = (u_1(t,x), u_2(t,x), u_3(t,x))$ denotes velocity, $p = p(t,x)$ pressure, $f = (f_1(x), f_2(x), f_3(x))$ external force, and Ω is a bounded domain with C^2 boundary. It is impossible to recall even the most important publications devoted to that problem, since the corresponding literature is too large; see anyway [28, 31, 63–65, 67, 73, 75, 82, 89, 97, 104, 105, 110, 114, 121, 128, 170, 172] together with the references cited there.

In space dimensions 2 and 3, the N-S equation possess local in time regular solutions, as stated in Theorems 6.1.4 and 6.1.5. We analyze further *criticality* of the Navier–Stokes equation (compare [178]), in a sense that available for it $L^2(\Omega)$ a priori estimate (6.8) is not sufficient to control its nonlinearity through the viscosity term $\nu P\Delta u$. Consequently, possible is a balance between the income from nonlinearity and the stabilizing action of the viscosity so that the local solutions cannot be extended globally in time. For small initial data, the decisive role is played by viscosity, while for larger initial data the nonlinear term is strong enough to destroy regularization of the solutions through the main part operator. We will try to see such effect through the estimates obtained further in the text.

We look at the 3-D N-S equation as a *supercritical problem* in a sense that a stronger diffusion term, with $-P\Delta$ operator in the power $s > \frac{5}{4}$, is needed to guarantee the control on the nonlinear term with the use of the standard $L^2(\Omega)$ a priori estimate. The approximation proposed in 3-D is given in (6.37) with $s > \frac{5}{4}$.

https://doi.org/10.1515/9783110599831-006

While in the 2-D case, the known $L^2(\Omega)$ a priori estimate makes the N-S equation critical, better a priori estimates in $H^1(\Omega)$ are known, allowing for global in time extendibility of the local solutions; see Subsection 6.2.3. If we use such better a priori estimate, the theory presented in Section 4 applies and the 2-D Navier–Stokes equation becomes subcritical.

There were several tries to replace the original N-S equation with another equation with better behaving solutions, starting with J. Leray *α-regularization* reported in [121]; see also [64]. Modification of one factor in nonlinearity, using mollifier, was sufficient to significantly improve properties of solutions. Another modification of the N-S equation was proposed by J.-L. Lions in [124, Chapter 1, Remarque 6.11], where the $-\Delta$ operator was replaced with $-\Delta + \kappa(-\Delta)^l$, $l \geq \frac{5}{4}$ (in [124] exponent $l \in \mathbb{N}$). Further modifications can be easily found in the literature (see, e. g., [111]). In fact, we follow here the idea of J.-L. Lions to replace the classical diffusion with a higher order diffusion term, but with general, fractional diffusion.

We should finally mention, that working with fractional extensions of the classical Navier–Stokes equation corresponding to strengthen diffusion $(-P\Delta)^s$ with $s \in (1,2)$, we need formally to assume higher regularity of the boundary; namely that $\partial\Omega \in C^4$, since the estimates in that case are given in connection with the square of the Stokes operator.

Recall that for $r \in \mathbb{R}$, r^- denotes a number strictly less than r but close to it. Similarly, $r^+ > r$ and r^+ close to r. When needed for clarity of the presentation, we mark the dependence of the solution u of (6.37) on $\epsilon > 0$, calling it u^ϵ.

6.1.1 Properties of the Stokes operator

The following spaces are frequently used in the theory of the N-S equation:

$$\begin{aligned} \mathcal{L}^r(\Omega) &= [L^r(\Omega)]^3, \\ \mathcal{W}^{2,r}(\Omega) &= [W^{2,r}(\Omega)]^3, \\ X_r &= \mathrm{cl}_{\mathcal{L}^r(\Omega)}\{\phi \in [C_0^\infty(\Omega)]^3; \operatorname{div}\phi = 0\}, \end{aligned} \tag{6.2}$$

$1 < r < \infty$. We define also the *Stokes operator*

$$A_r = -\nu P_r \begin{bmatrix} \Delta & 0 & 0 \\ 0 & \Delta & 0 \\ 0 & 0 & \Delta \end{bmatrix},$$

where P_r denotes the projection from $\mathcal{L}^r(\Omega)$ to X_r given by the decomposition of $\mathcal{L}^r(\Omega)$ onto the space of divergence-free vector fields and scalar-function gradient (e. g., [171]). The following property is also well known [75, Lemma 1.1].

Proposition 6.1.1. *The operator $-A_r$ considered with the domain*

$$D(A_r) = X_r \cap \{\phi \in \mathcal{W}^{2,r}(\Omega); \phi = 0 \text{ on } \partial\Omega\},$$

generates on X_r an analytic semigroup $\{e^{-tA_r}\}$ for arbitrary $1 < r < \infty$.

A complete description of the domains of fractional powers of the Stokes operator $A_r = -\nu P_r \Delta$, $D(A_r) = D(-\Delta) \cap X_r$, can be found in [75, p. 269], or in [74]. Note further, that the domains of negative powers of the operator A_r are introduced through the relation (e. g., [75, p. 269]):

$$D(A_r^\beta) = D(A_{r'}^{-\beta})^*, \quad \frac{1}{r} + \frac{1}{r'} = 1.$$

Thus, $D(A_r^\beta)$, $\beta < 0$, is the completion of X_r under the norm $\|A_r^\beta \cdot\|_{0,r'}$.

It is also easy to see, that the resolvent of the operator A_2 fulfills an estimate: $\mathbb{R}(\sigma(A_2)) \geq \nu\lambda_1$, where λ_1 is the first positive eigenvalue of $-\Delta$ in $L^2(\Omega)$ considered with the Dirichlet boundary condition. The same estimate remains valid for the operator considered in $L^r(\Omega)$ with any $r \in (1, \infty)$. It follows further from [75, Lemma 3.1], that the resolvent of A_r is compact. Also the embeddings $D(A_r^\beta) \subset D(A_r^\alpha)$ are compact when $0 < \alpha < \beta$ ([86, Theorem 1.4.8]). In fact, the operator $A = A_2$ (we skip the subscript further for simplicity) is self-adjoint in the Hilbert space $\mathcal{L}^2(\Omega)$; see, for example, [73, 75]. For such type operators, the powers of the order $(1 + \alpha)$ have similar properties; in particular they are also sectorial operators.

Consequently, the operators A_r, $1 < r < \infty$, are *sectorial positive. Fractional powers* of the order $1 + \alpha$ for such operators are introduced through the Balakrishnan formula ([106]):

$$A^{1+\alpha}\phi = \frac{\sin(\pi\alpha)}{\alpha\pi} \int_0^\infty \lambda^\alpha [A(\lambda + A)^{-1}]^2 \phi d\lambda, \quad \phi \in D(A^2). \tag{6.3}$$

We recall that in case of the N-S equation the, specific for that problem, *a priori estimate* is obtained multiplying (6.1) in $[L^2(\Omega)]^3$ through u, to obtain

$$\frac{1}{2}\frac{d}{dt}\|u\|^2_{\mathcal{L}^2(\Omega)} = -\nu\|\nabla u\|^2_{\mathcal{L}^2(\Omega)} - \int_\Omega \nabla p \cdot u dx + \int_\Omega f \cdot u dx, \tag{6.4}$$

since the nonlinear component vanishes due to condition $\operatorname{div} u = 0$:

$$\sum_{j=1}^3 \int_\Omega \sum_{i=1}^3 u_i \frac{\partial u_j}{\partial x_i} u_j dx = -\frac{1}{2} \int_\Omega \sum_{i=1}^3 \frac{\partial u_i}{\partial x_i} \sum_{j=1}^3 u_j^2 dx = 0. \tag{6.5}$$

The term $\int_\Omega \nabla p \cdot u dx$, for regular solutions, is transformed as follows:

$$\int_\Omega \nabla p \cdot u dx = \int_\Omega \left(\frac{\partial p}{\partial x_1} u_1 + \frac{\partial p}{\partial x_2} u_2 + \frac{\partial p}{\partial x_3} u_3 \right) dx$$
$$= -\int_\Omega p \operatorname{div} u dx = 0. \tag{6.6}$$

Consequently, an $\mathcal{L}^2(\Omega)$ estimate is obtained:

$$\frac{1}{2} \frac{d}{dt} \|u\|^2_{\mathcal{L}^2(\Omega)} = -\nu \|\nabla u\|^2_{\mathcal{L}^2(\Omega)} + \int_\Omega f \cdot u dx$$
$$\leq -\frac{\nu}{2} \|\nabla u\|^2_{\mathcal{L}^2(\Omega)} + c_\nu \|f\|^2_{\mathcal{L}^2(\Omega)} \leq -\frac{\nu}{2c_P} \|u\|^2_{\mathcal{L}^2(\Omega)} + c_\nu \|f\|^2_{\mathcal{L}^2(\Omega)}, \tag{6.7}$$

thanks to the Poincaré inequality (e. g., (5.63)). We finally obtain a global in time estimate of the solution:

$$\|u(t)\|^2_{\mathcal{L}^2(\Omega)} \leq \max\left\{ \|u_0\|^2_{\mathcal{L}^2(\Omega)}; \frac{2c_\nu c_P \|f\|^2_{\mathcal{L}^2(\Omega)}}{\nu} \right\}, \tag{6.8}$$

where c_P denotes the constant in the Poincaré inequality. Having already the last estimate, one can return to (6.7) to see that

$$\|u\|^2_{L^2(0,T;[H^1_0(\Omega)]^3)} \leq \frac{1}{2\nu} (c_\nu T \|f\|^2_{\mathcal{L}^2(\Omega)} + \|u_0\|^2_{\mathcal{L}^2(\Omega)}), \tag{6.9}$$

for arbitrary $T > 0$. These are the strongest natural a priori estimates that can be obtained for, sufficiently regular, solutions of the N-S equation.

6.1.2 Local in time solvability of the 3-D and 2-D N-S problems

We will rewrite N-S equation in a form of an abstract parabolic equation with sectorial positive operator. Let us comment that using Henry's approach we have fairly large choice of eventual *phase space*. In fact, recalling here an interesting paper [145], as far as we consider the *linear Cauchy problem* with sectorial positive operator:

$$u_t + Au = 0, \quad t > 0,$$
$$u(0) = u_0,$$

we can *pose* it at any level of the *fractional power scale* $X^\beta = D((-A)^\beta)$, $\beta \in \mathbb{R}$, corresponding to $-A$ (see [6, Section V.2] for extension of that idea). When we move to the *semilinear problem* (6.10) below, with nonlinearity F subordinated to A, it is an art to choose the proper level at that scale to be the phase space. For that, we need to analyze a priori estimates available for the specific equation (usually of physical origin, e. g., following from energy decay or conservation of mass valid in the process described through the equation). The full semilinear problem will be next written abstractly as

$$\begin{aligned} &u_t + Au = F(u), \quad t > 0, \\ &u(0) = u_0. \end{aligned} \tag{6.10}$$

The standard way to set the problem (6.1) in the above setting, in $L^2(\Omega)$ (see, e. g., [73, 75, 171]), is to apply to the first equation the projector $P = P_2 : [L^2(\Omega)]^3 \to H$, where H is the closure in $[L^2(\Omega)]^3$ of the set of divergence-free functions $\{u \in [C_0^\infty(\Omega)]^3; \operatorname{div} u = 0\}$. The pressure term disappears then from the equation. The realization A of the diagonal matrix operator $\nu P[-\Delta]_{3\times3}$ acts from $D(A) \to H$. We also introduce the energy space $V = \{u \in [H_0^1(\Omega)]^3; \operatorname{div} u = 0\}$, and the commonly used notation for the nonlinearity; $F(u) = -P(u \cdot \nabla)u$.

Operator $A = A_2$ has an associated scale of *fractional-order spaces* $X^\beta \subset [H^{2\beta}(\Omega)]^3$, $\beta \geq 0$. The realizations of A in X^β act from $D(A^\beta) = X^{\beta+1} \to X^\beta$ and are *sectorial positive* operators (see, e. g., [173, 73, 86, 130]).

We will rewrite the classical N-S equation in an equivalent form, using the property of the divergence-free functions. We have

$$\begin{aligned} &u_{jt} = \nu\Delta u_j - \frac{\partial p}{\partial x_j} - \sum_{i=1}^{3} \frac{\partial(u_i u_j)}{\partial x_i} + f_j, \quad \operatorname{div} u = 0, \quad x \in \Omega,\, t > 0,\, j = 1, 2, 3, \\ &u = 0, \quad t > 0,\, x \in \partial\Omega, \\ &u(0, x) = u_0(x), \end{aligned} \tag{6.11}$$

where $u = (u_1, u_2, u_3)$. Equivalently, using the notation introduced above and applying the projector P, problem (6.11) will be written as

$$\begin{aligned} &u_t + Au = -P(u \cdot \nabla)u + Pf, \quad t > 0, \\ &u(0) = u_0. \end{aligned} \tag{6.12}$$

We will recall next an important estimate borrowed from [75, Lemma 2.1]. A similar observation was given in [86, p. 18] in dimension one.

Corollary 6.1.2. For each j, $1 \leq j \leq N$, the operator $A^{-\frac{1}{2}} P \frac{\partial}{\partial x_j}$ extends uniquely to a bounded linear operator from $[L^r(\Omega)]^N$ to X_r, $1 < r < \infty$. Consequently, the following estimate holds:

$$\|A^{-\frac{1}{2}} P(u \cdot \nabla)v\|_{[L^r(\Omega)]^N} \leq M(r) \| |u||v| \|_{[L^r(\Omega)]^N}. \tag{6.13}$$

We formulate next an important estimate needed further in the text, a consequence of Corollary 6.1.2.

Remark 6.1.3. For all $N \in \mathbb{N}$:

$$\|A^{-\frac{1}{2}}P(u\cdot\nabla)v\|_{[L^2(\Omega)]^N} \le c\||u||v|\|_{[L^2(\Omega)]^N} \le c\|u\|_{[L^4(\Omega)]^N}\|v\|_{[L^4(\Omega)]^N},$$
$$\|P(u\cdot\nabla)v\|_{[L^2(\Omega)]^N} \le c\|u\|_{[L^4(\Omega)]^N}\|\nabla v\|_{[L^4(\Omega)]^N}. \tag{6.14}$$

Now, for any $\delta \in (0, \frac{1}{2})$, using the theory of interpolation:

$$\begin{aligned}\|A^{-\delta}P(u\cdot\nabla)v\|_{[L^2(\Omega)]^N} &\le c\|A^{-\frac{1}{2}}P(u\cdot\nabla)v\|^{2\delta}_{[L^2(\Omega)]^N}\|P(u\cdot\nabla)v\|^{1-2\delta}_{[L^2(\Omega)]^N}\\ &\le c\|u\|^{1+2\delta}_{[L^4(\Omega)]^N}\|\nabla v\|^{1-2\delta}_{[L^4(\Omega)]^N}.\end{aligned} \tag{6.15}$$

In a similar way, starting from the estimates:

$$\|A^{-\frac{1}{2}}P(u\cdot\nabla)v\|_{[L^2(\Omega)]^N} \le c\||u||v|\|_{[L^2(\Omega)]^N} \le c\|u\|_{[L^6(\Omega)]^N}\|v\|_{[L^3(\Omega)]^N},$$
$$\|P(u\cdot\nabla)v\|_{[L^2(\Omega)]^N} \le c\|u\|_{[L^6(\Omega)]^N}\|\nabla v\|_{[L^3(\Omega)]^N}, \tag{6.16}$$

for any $\delta \in (0, \frac{1}{2})$, we get

$$\begin{aligned}\|A^{-\delta}P(u\cdot\nabla)v\|_{[L^2(\Omega)]^N} &\le c\|A^{-\frac{1}{2}}P(u\cdot\nabla)v\|^{2\delta}_{[L^2(\Omega)]^N}\|P(u\cdot\nabla)v\|^{1-2\delta}_{[L^2(\Omega)]^N}\\ &\le c\|u\|_{[L^6(\Omega)]^N}\|u\|^{2\delta}_{[L^3(\Omega)]^N}\|\nabla v\|^{1-2\delta}_{[L^3(\Omega)]^N}.\end{aligned} \tag{6.17}$$

While estimates (6.15), (6.17), are valid for all the space dimensions $N = 2, 3, 4, \dots$, we extend them further, using Sobolev-type estimates, in a way depending on N.

For local in time solvability, we will set the problem (6.11) in the base space $X^{-\frac{1}{4}}$ for the space dimension $N = 2$, and in the base space $X^{-\frac{1}{8}}$ for the space dimension $N = 3$. The corresponding *phase spaces* will be $X^{\frac{1}{2}^+} \subset [H^{1^+}(\Omega)]^2$ in case $N = 2$, and $X^{\frac{3}{4}^+} \subset [H^{\frac{3}{2}^+}(\Omega)]^3$ in case $N = 3$ (e. g., [75, Proposition 1.4]). Note that, in both cases, the phase spaces are contained in the space $[L^\infty(\Omega)]^N$. There is another possible choice of the phase spaces (e. g., [75]), if we decide to work in the scale build on $[L^r(\Omega)]^N$, $N = 2, 3$ with $r > N$.

We will formulate now the corresponding local existence results for $N = 2, 3$.

Case $N = 3$. The main tool is the estimate taken from [75, Lemma 2.2] (with $\delta = \frac{1}{8}$, $\theta = \rho = \frac{3}{4}$):

$$\|A^{-\frac{1}{8}}P(u\cdot\nabla)v\|_{[L^2(\Omega)]^3} \le M\|A^{\frac{3}{4}^+}u\|_{[L^2(\Omega)]^3}\|A^{\frac{3}{4}^+}v\|_{[L^2(\Omega)]^3}. \tag{6.18}$$

Since the form above is bilinear, we have also the following consequences of the last estimate:

$$\|A^{-\frac{1}{8}}P((u-v)\cdot\nabla)v\|_{[L^2(\Omega)]^3} \le M\|A^{\frac{3}{4}^+}(u-v)\|_{[L^2(\Omega)]^3}\|A^{\frac{3}{4}^+}v\|_{[L^2(\Omega)]^3},$$
$$\|A^{-\frac{1}{8}}P(u\cdot\nabla)(u-v)\|_{[L^2(\Omega)]^3} \le M\|A^{\frac{3}{4}^+}u\|_{[L^2(\Omega)]^3}\|A^{\frac{3}{4}^+}(u-v)\|_{[L^2(\Omega)]^3}. \tag{6.19}$$

Consequently, the nonlinear term $F(u) + Pf = -P(u \cdot \nabla)u + Pf$ acts from $D(A^{\frac{3}{4}^+}) \subset [H^{\frac{3}{2}^+}(\Omega)]^3$ into $D(A^{-\frac{1}{8}})$, as a map Lipschitz continuous on bounded subsets of $D(A^{\frac{3}{4}^+})$. This suffices (see Chapter 3) to obtain a *local in time solution* of the 3-D equation (6.1) (also of 3-D equation (6.37)). More precisely, we have the following.

Theorem 6.1.4. *When $Pf \in D(A^{-\frac{1}{8}})$, $u_0 \in D(A^{\frac{3}{4}^+})$, then there exists a unique local in time mild solution $u(t)$ to (6.1) in the phase space $D(A^{\frac{3}{4}^+}) \subset [H^{\frac{3}{2}^+}(\Omega)]^3$. Moreover,*

$$u \in C([0,\tau); D(A^{\frac{3}{4}^+})) \cap C((0,\tau); D(A^{\frac{7}{8}})), \quad u_t \in C((0,\tau); D(A^{\frac{7}{8}^-})). \tag{6.20}$$

Here, $\tau > 0$ is the "lifetime" of that local in time solution. Moreover, the Cauchy formula is satisfied:

$$u(t) = e^{-At}u_0 + \int_0^t e^{-A(t-s)}F(u(s))ds, \quad t \in [0,\tau),$$

where e^{-At} denotes the linear semigroup corresponding to the operator A. We need also to mention that the considered here mild solutions have additional regularity properties, as described in Corollary 3.3.1; here, $u \in C^1((0,\tau); D(A^{\frac{7}{8}^-}))$. This property is used in further calculations.

Case $N = 2$. We will use a version of the estimate of [75, Lemma 2.2] (with $\delta = \frac{1}{4}$, $\theta = \rho = \frac{1}{2}$) which reads

$$\|A^{-\frac{1}{4}}P(u \cdot \nabla)v\|_{[L^2(\Omega)]^2} \le M\|A^{\frac{1}{2}^+}u\|_{[L^2(\Omega)]^3}\|A^{\frac{1}{2}^+}v\|_{[L^2(\Omega)]^2}. \tag{6.21}$$

Since the form above is bilinear, we have also the following consequences of the last estimate:

$$\begin{aligned}\|A^{-\frac{1}{4}}P((u-v)\cdot\nabla)v\|_{[L^2(\Omega)]^2} &\le M\|A^{\frac{1}{2}^+}(u-v)\|_{[L^2(\Omega)]^2}\|A^{\frac{1}{2}^+}v\|_{[L^2(\Omega)]^2},\\ \|A^{-\frac{1}{4}}P(u\cdot\nabla)(u-v)\|_{[L^2(\Omega)]^2} &\le M\|A^{\frac{1}{2}^+}u\|_{[L^2(\Omega)]^2}\|A^{\frac{1}{2}^+}(u-v)\|_{[L^2(\Omega)]^2}.\end{aligned} \tag{6.22}$$

Consequently, the nonlinear term $F(u) + Pf = -P(u \cdot \nabla)u + Pf$ acts from $D(A^{\frac{1}{2}^+}) \subset [H^{1^+}(\Omega)]^2$ into $D(A^{-\frac{1}{4}})$ as a map, Lipschitz continuous on bounded subsets of $D(A^{\frac{1}{2}^+})$. According to Theorem 3.1.2, this suffices to obtain a *local in time solution* of the 2-D equation (6.1) (also of 2-D equation (6.12)). More precisely, we have the following.

Theorem 6.1.5. *When $Pf \in D(A^{-\frac{1}{4}})$, $u_0 \in D(A^{\frac{1}{2}^+}) \subset [H^{1^+}(\Omega)]^2$, then there exists a unique local in time mild solution $u(t)$ to (6.1) in the phase space $D(A^{\frac{1}{2}^+}) \subset [H^{1^+}(\Omega)]^2$. Moreover,*

$$u \in C([0,\tau); D(A^{\frac{1}{2}^+})) \cap C((0,\tau); D(A^{\frac{3}{4}})), \quad u_t \in C((0,\tau); D(A^{\frac{3}{4}^-})). \tag{6.23}$$

Here, $\tau > 0$ is the "lifetime" of that local in time solution. Moreover, the Cauchy formula is satisfied:

$$u(t) = e^{-At}u_0 + \int_0^t e^{-A(t-s)}F(u(s))ds, \quad t \in [0, \tau),$$

where e^{-At} denotes the linear semigroup corresponding to the operator A.

Remark 6.1.6. Another choice of the pairs base/phase spaces is possible. In particular, setting for $N = 2$: $(\delta = 1 - \nu, \theta = \rho = \frac{\nu}{2}, \nu = 0^+)$ in [75, Lemma.2.2], one has that $F : D(A^{\frac{\nu}{2}}) \to D(A^{-1+\nu})$. For $N = 3$, setting there: $(\delta = \frac{1}{2}, \theta = \rho = \frac{3}{8})$, one has that $F : D(A^{\frac{3}{8}}) \to D(A^{-\frac{1}{2}})$, and F is Lipschitz continuous on bounded sets. Corresponding local solvability results follow immediately.

6.1.3 Supercriticality of the N-S equation in 3-D, and criticality in 2-D

We will consider now *criticality* of the N-S equation (6.1) in the sense stated in Definition 1.0.4.

Case $N = 3$. The 3-D N-S equation is *supercritical* in the sense stated in Definition 1.0.4. Recall that a general problem (6.10) will be called *supercritical with respect to the a priori estimate in Y*, if for all its possible local solutions an estimate holds

$$\exists_{\theta_0 > 1} \exists_{\text{nondecreasing } g:[0,\infty)\to[0,\infty)} \|F(u(t))\|_{X^\alpha} \le g(\|u(t)\|_Y)(1 + \|u(t)\|_{X^{\alpha+1}}^{\theta_0}), \tag{6.24}$$

valid for all t in "lifetime" of that solution, and such type estimate *is not true* for exponents $\theta < \theta_0$.

In case of the Navier–Stokes equation (6.1), the mentioned above space $Y = [L^2(\Omega)]^N$; see (6.8) for the corresponding a priori estimate.

We will see that the problem (6.1) is supercritical. Indeed, the estimate (6.17) written for the local solution $u = u(t)$ obtained in Theorem 6.1.4 extends for $N = 3$, with the use of the Nirenberg–Gagliardo-type estimates to:

$$\begin{aligned} \|A^{-\frac{1}{8}}P(u \cdot \nabla)u\|_{[L^2(\Omega)]^3} &\le c\|u\|_{[L^6(\Omega)]^3}\|u\|_{[L^3(\Omega)]^3}^{\frac{1}{4}}\|\nabla u\|_{[L^3(\Omega)]^3}^{\frac{3}{4}} \\ &\le c\|u\|_{[H^{\frac{7}{4}}(\Omega)]^3}^{\frac{9}{7}}\|u\|_{[L^2(\Omega)]^3}^{\frac{5}{7}}, \end{aligned} \tag{6.25}$$

and the estimate is sharp in the sense there are functions fulfilling it with equality. This shows the nonlinearity of the 3-D N-S equation is supercritical as a map from $X^{\frac{7}{8}}$ to $X^{-\frac{1}{8}}$. Exponent θ_0 obtained above equals $\frac{9}{7}$.

Using the estimate of Corollary 6.1.2, we will find now the value θ_0 in case of the 3-D N-S equation. We obtain

$$\begin{aligned}\|P(u\cdot\nabla)u\|_{X^{-\frac{1}{2}}} &= \|A^{-\frac{1}{2}}P(u\cdot\nabla)u\|_{[L^2(\Omega)]^3} \le M\||u|^2\|_{[L^2(\Omega)]^3} \le c\|u\|^2_{[L^4(\Omega)]^3}\\ &\le c\|u\|^{\frac{3}{2}}_{[H^1(\Omega)]^3}\|u\|^{\frac{1}{2}}_{[L^2(\Omega)]^3},\end{aligned}\tag{6.26}$$

and the estimates are sharp, therefore, $\theta_0 = \frac{3}{2}$ in that case.

Finally, for the further use, we will check how large the exponent $s > 1$ should be, for the "strengthen diffusion" of the form A^s replacing the usual operator A in (6.1) to make the $L^2(\Omega)$ estimate critical. Using again Corollary 6.1.2 and the following Nirenberg–Gagliardo inequality,

$$\|\phi\|_{L^4(\Omega)} \le c\|\phi\|^{\frac{3}{4(2s-1)}}_{H^{2s-1}(\Omega)}\|\phi\|^{\frac{8s-7}{8s-4}}_{L^2(\Omega)}$$

we get that

$$\begin{aligned}\|A^{-\frac{1}{2}}P(u\cdot\nabla)u\|_{[L^2(\Omega)]^3} &\le M\||u|^2\|_{[L^2(\Omega)]^3} = M\|u\|^2_{[L^4(\Omega)]^3}\\ &\le c\|u\|^{\frac{3}{2(2s-1)}}_{[H^{2s-1}(\Omega)]^3}\|u\|^{\frac{8s-7}{4s-2}}_{[L^2(\Omega)]^3}\\ &\le c'\|A^{s-\frac{1}{2}}u\|^{\frac{3}{2(2s-1)}}_{[L^2(\Omega)]^3}\|u\|^{\frac{8s-7}{4s-2}}_{[L^2(\Omega)]^3}.\end{aligned}\tag{6.27}$$

We find such critical value of s from the condition; $\frac{3}{2(2s-1)} = 1$. Consequently, $s = \frac{5}{4}$.

Case $N = 2$. Using Corollary 6.1.2, it is easy to see *criticality of the 2-D Navier–Stokes equation.* More precisely, to verify that its nonlinearity is critical with respect to the standard $L^2(\Omega)$ a priori estimates, as a map from $X^{\frac{1}{2}} \subset [H^1(\Omega)]^2$ to $X^{-\frac{1}{2}}$. Indeed,

$$\begin{aligned}\|P(u\cdot\nabla)u\|_{X^{-\frac{1}{2}}} &\le \|A^{-\frac{1}{2}}P(u\cdot\nabla)u\|_{[L^2(\Omega)]^2} \le M\||u|^2\|_{[L^2(\Omega)]^2}\\ &\le c\|u\|^2_{[L^4(\Omega)]^2} \le \|u\|_{[H^1(\Omega)]^2}\|u\|_{[L^2(\Omega)]^2}\\ &= c(\|u\|_{[L^2(\Omega)]^2})\|u\|^1_{[H^1(\Omega)]^2},\end{aligned}\tag{6.28}$$

and *no better estimate (with exponent smaller than 1) is possible.*

Remark 6.1.7. Using the estimate (6.15) with $\delta = \frac{1}{4}$, we will show that the nonlinearity in 2-D N-S equation is *critical as a map from* $X^{\frac{3}{4}} \subset [H^{\frac{3}{2}}(\Omega)]^2$ *to* $X^{-\frac{1}{4}}$. Indeed, from (6.15) with $\delta = \frac{1}{4}$, $N = 2$,

$$\|A^{-\frac{1}{4}}P(u\cdot\nabla)u\|_{[L^2(\Omega)]^2} \le c\|u\|^{\frac{3}{2}}_{[L^4(\Omega)]^2}\|\nabla u\|^{\frac{1}{2}}_{[L^4(\Omega)]^2} \le c\|u\|_{[L^2(\Omega)]^2}\|u\|_{[H^{\frac{3}{2}}(\Omega)]^2},\tag{6.29}$$

where the Nirenberg–Gagliardo-type estimates were used:

$$\begin{aligned}\|\phi\|_{L^4(\Omega)} &\le c\|\phi\|^{\frac{1}{3}}_{H^{\frac{3}{2}}(\Omega)}\|\phi\|^{\frac{2}{3}}_{L^2(\Omega)},\\ \|\phi\|_{W^{1,4}(\Omega)} &\le c\|\phi\|_{H^{\frac{3}{2}}(\Omega)}.\end{aligned}\tag{6.30}$$

No better estimates are possible.

Note further, that in 2-D the nonlinearity is also critical with respect to the $L^2(\Omega)$ a priori estimates as a map between $X^1 = D(A) \subset [H^2(\Omega)]^2$ and X_2. Indeed, the following estimate holds:

$$\|P(u \cdot \nabla)u\|_{X_2} \leq c\|u\|_{[L^4(\Omega)]^2}\|\nabla u\|_{[L^4(\Omega)]^2} \leq c\|u\|_{[H^2(\Omega)]^2}\|u\|_{[L^2(\Omega)]^2}, \tag{6.31}$$

and *no smaller exponent on the* $[H^2(\Omega)]^2$ *norm is possible.*

6.2 Global in time solutions in 3-D with small data

As is well known, global in time extendibility of the local mild solution constructed in Theorem 6.1.4 is possible provided we have sufficiently well a priori estimates that prevents the $D(A^{\frac{3}{4}^+}) \subset [H^{\frac{3}{2}^+}(\Omega)]^3$ norm of the solution to blow up in a finite time. We will propose next such type estimate, in $[H^1(\Omega)]^3$, for solutions of the 3-D N-S equation when the data; u_0 and f are smooth and sufficiently small. Such estimate will be used later to construct global in time solutions of the N-S with small data. Another approach to that problem in arbitrary dimension N, using estimates on integral equation, was presented in [31].

Theorem 6.2.1. *If* $u_0 \in D(A) \subset [H^2(\Omega)]^3$ *and* $f \in [L^2(\Omega)]^3$ *fulfill the "smallness restriction"* (6.35), *then the* $[H^1(\Omega)]^3$ *norms of the solutions* u *are bounded uniformly in time* $t \geq 0$.

Proof. It is known (e. g., [86, 32]), that the local in time solution u is regularized through the equation for $t > 0$ entering the space $D(A) \subset [H^2(\Omega)]^3$. When the initial data $u_0 \in D(A)$, they simply vary in $D(A)$ for $t \geq 0$ small, until possible blow-up time $t(\alpha, u_0)$. We want to show that, to small u_0 and f correspond small solutions, in the $[H^1(\Omega)]^3$ norm (uniformly in time).

To get an estimate of the solution u in $[H^1(\Omega)]^3$ multiply (6.12) by Au to obtain

$$\begin{aligned}\langle u_t, Au\rangle_{[L^2(\Omega)]^3} &= -\langle Au, Au\rangle_{[L^2(\Omega)]^3} - \langle P(u \cdot \nabla)u, Au\rangle_{[L^2(\Omega)]^3} \\ &\quad + \langle Pf, Au\rangle_{[L^2(\Omega)]^3},\end{aligned} \tag{6.32}$$

since the pressure term vanishes. Thanks to (6.14), this gives

$$\begin{aligned}\frac{1}{2}\frac{d}{dt}\|u\|^2_{[H^1(\Omega)]^3} &\leq -cv\|u\|^2_{[H^2(\Omega)]^3} + c\|u\|_{[W^{1,4}(\Omega)]^3}\|u\|_{[L^4(\Omega)]^3}\|u\|_{[H^2(\Omega)]^3} \\ &\quad + \|Pf\|_{[L^2(\Omega)]^3}\|u\|_{[H^2(\Omega)]^3} \\ &\leq -cv\|u\|^2_{[H^2(\Omega)]^3} + c_1\|u\|^{\frac{5}{4}}_{[H^1(\Omega)]^3}\|u\|^{\frac{7}{4}}_{[H^2(\Omega)]^3} \\ &\quad + \|Pf\|_{[L^2(\Omega)]^3}\|u\|_{[H^2(\Omega)]^3} \\ &\leq -\frac{c'v}{2}\|u\|^2_{[H^1(\Omega)]^3} + C_v(\|u\|^{10}_{[H^1(\Omega)]^3} + \|Pf\|^2_{[L^2(\Omega)]^3}),\end{aligned} \tag{6.33}$$

with the standard use of Young's inequality, and the embedding $[H^2(\Omega)]^3 \subset [H^1(\Omega)]^3$ (constant c'). Denoting; $y(t) := \|u(t)\|^2_{[H^1(\Omega)]^3}$, we arrive at the *differential inequality* (e. g., [164, 177]):

$$\frac{1}{2}y'(t) \leq -\frac{c'\nu}{2}y(t) + C_\nu y^5(t) + C_\nu \|Pf\|^2_{[L^2(\Omega)]^3},$$
$$y(0) = \|u(0)\|^2_{[H^1(\Omega)]^3}. \tag{6.34}$$

Analyzing its right-hand side, real function $g(z) = -\frac{c'\nu}{2}z + C_\nu z^5 + C_\nu\|Pf\|^2_{[L^2(\Omega)]^3}$, we see that $g(0) = C_\nu\|Pf\|^2_{[L^2(\Omega)]^3} > 0$, $g'(0) < 0$, and g has a minimum for the argument $z_{\min} = (\frac{c'\nu}{10C_\nu})^{\frac{1}{4}}$, with $g(z_{\min}) < 0$ when the "free term" $C_\nu\|Pf\|^2_{[L^2(\Omega)]^3}$ is small. More precisely, to keep the value of $y(t)$ bounded for all positive times, we need to assume the *smallness hypothesis*: Let the data: $\|u(0)\|^2_{[H^1(\Omega)]^3}$ and $\|Pf\|^2_{[L^2(\Omega)]^3}$ be so small that

$$g(z_{\min}) < 0, \text{ equivalently } \|Pf\|^2_{[L^2(\Omega)]^3} < 4\left(\frac{c'\nu}{10C_\nu}\right)^{\frac{5}{4}}, \quad \text{and}$$
$$\|u_0\|^2_{[H^1(\Omega)]^3} \leq z_{\min} = \left(\frac{c'\nu}{10C_\nu}\right)^{\frac{1}{4}}. \tag{6.35}$$

Note that $(\frac{\nu}{C_\nu})^{\frac{1}{4}}$ is proportional to ν^2. Consequently, we obtain the bound

$$\|u(t)\|^2_{[H^1(\Omega)]^3} \leq z_{\min} \quad \text{valid for all } t \geq 0. \tag{6.36}$$

With the last assumption, the smooth local solutions $u(t)$, as in Theorem 6.2.1 are bounded in $[H^1(\Omega)]^3$ uniformly in $t \geq 0$. Note that a bit more accurate bounds in (6.35) are possible if one compares the data with the two positive zeros of the function g. See also the corresponding restrictions formulated in [171, Theorem 3.7]. □

6.2.1 Regularization of the 3-D N-S equation

It was shown above that in the 3-D case the N-S equation (6.1) behaves with respect to the standard estimate in a supercritical way. Global in time extendibility of the local solutions constructed in Theorem 6.1.4 is in general unknown (unless for small data). The viscosity term in the equation (6.1) together with the standard a priori estimate are not strong enough to control the nonlinearity. In the present subsection, we discuss an approximation/regularization (with stronger diffusion term) of the 3-D N-S equation, for which the $[L^2(\Omega)]^3$ estimates (same as for the original equation) are sufficient to make such problems *subcritical*. The idea of such regularization was proposed first by J.-L. Lions in [124, Chapter 1, Remarque 6.11].

We consider approximation of the original 3-D Navier–Stokes equation having global in time, unique and regular solutions. Consider namely the approximation/regularization of (6.1) having the form:

$$u_t = -(A + \epsilon A^s)u - P(u \cdot \nabla)u + Pf, \quad t > 0,$$
$$u(0, x) = u_0(x), \tag{6.37}$$

with a parameter $s > 1$ (to be chosen), and $\epsilon > 0$. For fixed (for a moment) parameter ϵ, denote the solution to the above problem as u^s. We mean here the solution on the base space $D(A^{-\frac{1}{4}})$, obtained in a similar way as for the original N-S equation (6.1) in Subsection 6.1.2.

It is clear how to determine the proper, sufficiently large, value of the exponent $s > 1$ to guarantee, together with the $[L^2(\Omega)]^3$ estimates valid also for solutions of (6.37), that the nonlinear term is subordinated to the main part operator $-(A + \epsilon A^s)$. We need to compare the bound obtained for the nonlinearity with the income from the improved viscosity term. Estimate of the nonlinearity, obtained from (6.15), reads

$$\|A^{-\frac{1}{4}}F(u^s(t))\|_{[L^2(\Omega)]^3} \le c\|u^s(t)\|^{\frac{3}{2}}_{[L^4(\Omega)]^3}\|u^s(t)\|^{\frac{1}{2}}_{[W^{1,4}(\Omega)]^3}. \tag{6.38}$$

Let $s > \frac{5}{4}$. Estimating as in (6.27), we obtain the *subordination condition*

$$\begin{aligned}\|A^{-\frac{1}{4}}F(u^s(t))\|_{[L^2(\Omega)]^3} &\le c(\|u^s(t)\|_{[L^2(\Omega)]^3})\|u^s(t)\|^{\theta}_{[H^{2s-\frac{1}{2}}(\Omega)]^3}\\ &\le c'(\|u^s(t)\|_{[L^2(\Omega)]^3})\|A^{s-\frac{1}{4}}u^s(t)\|^{\theta}_{[L^2(\Omega)]^3},\end{aligned} \tag{6.39}$$

where $\theta = \frac{2}{2s-\frac{1}{2}} < 1$. Consequently, the standard $[L^2(\Omega)]^3$ a priori estimate is sufficient to assure the global in time extendibility of such local solutions u^s when $s > \frac{5}{4}$. We will study such approximation next.

Remark 6.2.2. The exponent $\frac{5}{4}$ proposed for regularization of (6.1) in [124, Chapter 1, Remarque 6.11] is the same (also, as in (6.27)).

6.2.2 3-D N-S equation

Let us study the introduced above regularization (6.37) of the original 3-D N-S equation *with exponent* $s > \frac{5}{4}$. We formulate first the corresponding existence result.

Theorem 6.2.3. *Consider* (6.37) *with* $s > \frac{5}{4}$ *as equation in* $D(A^{-\frac{1}{8}})$. *When* $Pf \in D(A^{-\frac{1}{8}})$, $u_0 \in D(A^{\frac{3}{4}^+}) \subset [H^{\frac{3}{2}^+}(\Omega)]^3$, *then there exists a unique local in time mild solution* $u^\epsilon(t)$ *to* (6.37), $s > \frac{5}{4}$, *in the phase space* $D(A^{\frac{3}{4}^+})$. *Moreover,*

$$u^\epsilon \in C([0,\tau); D(A^{\frac{3}{4}^+})) \cap C((0,\tau); D(A^{s-\frac{1}{4}})), \quad u^\epsilon_t \in C((0,\tau); D(A^{(s-\frac{1}{4})^-})). \tag{6.40}$$

Here, $\tau > 0$ is the "lifetime" of that local in time solution. Also, the corresponding Cauchy formula is satisfied.

Furthermore, thanks to the standard $[L^2(\Omega)]^3$ a priori estimate valid for solutions of (6.37) *uniformly in $\epsilon > 0$, the obtained above local solution will be extended globally in time in the class* (6.40).

The proof, similar as for Theorem 6.1.4, is omitted.

Passing to the limit $\epsilon \to 0^+$ in (6.41)

We will describe next shortly the process of passing to the limit, as $\epsilon \to 0^+$, in such approximations. A similar idea, called *parabolic regularization technique*, was used earlier in [48, 50, 55, 124].

We write the approximating equation:

$$u_t = -Au - \epsilon A^s u + F(u) + Pf, \tag{6.41}$$

with $s > \frac{5}{4}$ (but close to $\frac{5}{4}$), and denote its solution by u^ϵ. Applying to it the operator $A^{-s+\frac{1}{2}}$, we obtain

$$A^{-s+\frac{1}{2}} u_t^\epsilon = -A^{-s+\frac{3}{2}} u^\epsilon - \epsilon A^{\frac{1}{2}} u^\epsilon + A^{-s+\frac{1}{2}} F(u^\epsilon) + A^{-s+\frac{1}{2}} Pf. \tag{6.42}$$

Using to nonlinear term the estimate of [75, Lemma 2.2] with $\delta = s - \frac{1}{2}$, $\epsilon = s - 1$, we get

$$\begin{aligned} \|A^{-\delta} F(u^\epsilon)\|_{[L^2(\Omega)]^3} &\le c \| |u^\epsilon|^2 \|_{[L^z(\Omega)]^3} \le c' \|u^\epsilon\|^2_{[L^{2z}(\Omega)]^3} \\ &\le c'' \|u^\epsilon\|^{\frac{7}{2}-2s}_{[H^1(\Omega)]^3} \|u^\epsilon\|^{2s-\frac{3}{2}}_{[L^2(\Omega)]^3} \\ &= c''' \|A^{\frac{1}{2}} u^\epsilon\|^{\frac{7}{2}-2s}_{[L^2(\Omega)]^3} \|u^\epsilon\|^{2s-\frac{3}{2}}_{[L^2(\Omega)]^3}, \end{aligned} \tag{6.43}$$

where $\frac{1}{z} = \frac{1}{2} + \frac{2\epsilon}{3}$, consequently $z = \frac{6}{3+4(s-1)} < \frac{3}{2}$ (but close). Due to the standard $[L^2(\Omega)]^3$ estimate, all of the right-hand side components in (6.42) belong to $L^2(0,T;[L^2(\Omega)]^3)$ (in particular $-\epsilon A^{\frac{1}{2}} u$; note that $\frac{7}{2} - 2s < 1$ in (6.43)). Consequently,

$$A^{-s+\frac{1}{2}} u_t^\epsilon \in L^2(0,T;[L^2(\Omega)]^3), \quad \text{or} \quad u_t^\epsilon \in L^2(0,T;D(A^{-s+\frac{1}{2}})).$$

Thus the Lions–Aubin compactness lemma (see section 5.4.2) will be used to conclude that any sequence $\{u^{\epsilon_n}\}$ has a subsequence convergent almost everywhere in $(0,T) \times \Omega$, as $\epsilon_n \to 0^+$. Also, any sequence $\{u^{\epsilon_n}\}$ has a subsequence convergent in $L^2(0,T;[H_0^{1^-}(\Omega)]^3)$.

Denote by U the limit of u^ϵ in $L^2(0,T;[H_0^{1^-}(\Omega)]^3)$, over a sequence $\epsilon_n \to 0^+$. Remember that from the standard a priori estimate written for the problem (6.37) it follows that the $D(A^{\frac{s}{2}})$ norms of u^ϵ fulfill

$$\sqrt{\epsilon} \|A^{\frac{s}{2}} u^\epsilon\|_{L^2(0,T;[L^2(\Omega)]^3)} \le \text{const}, \tag{6.44}$$

with const independent on $\epsilon > 0$. We recall next a *weak formulation* of the approximating equations (6.41):

$$\begin{aligned}\langle u_t^\epsilon, v\rangle_{[L^2(\Omega)]^3} &= -\langle A^{\frac{1}{2}}u^\epsilon, A^{\frac{1}{2}}v\rangle_{[L^2(\Omega)]^3} - \epsilon\langle A^{\frac{s}{2}}u^\epsilon, A^{\frac{s}{2}}v\rangle_{[L^2(\Omega)]^3} \\ &\quad + \langle F(u^\epsilon), v\rangle_{[L^2(\Omega)]^3} + \langle Pf, v\rangle_{[L^2(\Omega)]^3},\end{aligned} \tag{6.45}$$

where $v \in D(A^{\frac{1}{2}})$ arbitrary. Note that, due to (6.44), the second component on the right-hand side vanishes when $\epsilon \to 0^+$. For the nonlinear component, we have

$$\begin{aligned}\int_0^T \langle F(u^{\epsilon_n}), v\rangle_{[L^2(\Omega)]^3} dt &= \int_0^T \langle A^{-\frac{1}{2}}F(u^{\epsilon_n}), A^{\frac{1}{2}}v\rangle_{[L^2(\Omega)]^3} dt \\ &\to \int_0^T \langle A^{-\frac{1}{2}}F(U), A^{\frac{1}{2}}v\rangle_{[L^2(\Omega)]^3} dt,\end{aligned} \tag{6.46}$$

due to the estimate

$$\begin{aligned}&\int_0^T \|A^{-\frac{1}{2}}[F(u^{\epsilon_n}) - F(U)]\|_{[L^2(\Omega)]^3} dt \le \int_0^T \||u^{\epsilon_n} - U|(|u^{\epsilon_n}| + |U|)\|_{[L^2(\Omega)]^3} dt \\ &\le \int_0^T \|u^{\epsilon_n} - U\|_{[L^4(\Omega)]^3}(\|u^{\epsilon_n}\|_{[L^4(\Omega)]^3} + \|U\|_{[L^4(\Omega)]^3}) dt \\ &\le \int_0^T \|u^{\epsilon_n} - U\|_{[H^{1^-}(\Omega)]^3}(\|u^{\epsilon_n}\|_{[H^{1^-}(\Omega)]^3} + \|U\|_{[H^{1^-}(\Omega)]^3}) dt \\ &\le \|u^{\epsilon_n} - U\|_{L^2(0,T;[H^{1^-}(\Omega)]^3)}(\|u^{\epsilon_n}\|_{L^2(0,T;[H^{1^-}(\Omega)]^3)} + \|U\|_{L^2(0,T;[H^{1^-}(\Omega)]^3)}) \\ &\to 0, \quad \text{as } \epsilon_n \to 0^+.\end{aligned} \tag{6.47}$$

Finally, thanks to [171, Lemma 1.1] and the convergence $u^{\epsilon_n} \to U$ in $L^2(0, T; [H_0^{1^-}(\Omega)]^3)$ (or weakly in $L^2(0, T; [H_0^1(\Omega)]^3)$), we have

$$\langle u_t^{\epsilon_n}, v\rangle_{[L^2(\Omega)]^3} = \frac{d}{dt}\langle u^{\epsilon_n}, v\rangle_{[L^2(\Omega)]^3} \to \frac{d}{dt}\langle U, v\rangle_{[L^2(\Omega)]^3}, \tag{6.48}$$

in the sense of "scalar distributions" (e. g., [124, 171]); the derivative $\frac{d}{dt}$ in the sense of distributions.

Consequently, we were able to pass to the limit in the weakly formulated regularization (6.41) of the 3-D N-S equation. We obtain

$$\begin{aligned}\frac{d}{dt}\langle U, v\rangle_{[L^2(\Omega)]^3} &= -\langle A^{\frac{1}{2}}U, A^{\frac{1}{2}}v\rangle_{[L^2(\Omega)]^3} + \langle A^{-\frac{1}{2}}F(U), A^{\frac{1}{2}}v\rangle_{[L^2(\Omega)]^3} \\ &\quad + \langle Pf, v\rangle_{[L^2(\Omega)]^3},\end{aligned} \tag{6.49}$$

where $v \in D(A^{\frac{1}{2}})$ is arbitrary, and the convergence of the first component on the right-hand side is in the weak sense. We obtain a weak solution to the 3-D N-S equation in the spirit of the J. Leray original definition (e. g., [178, p. 139]). Such solution is global in time as was first observed in [89].

Note that the weak solution obtained above is global in time, while eventually not unique, since it depends on the subsequence chosen in the approximating process. For regular initial data $u_0 \in D(A_2^{\frac{3}{4}^+})$, it must coincide (for small times) with the unique local strong solution described in Theorem 6.1.4, since the last exists on certain time interval $t \in [0, \tau)$ and fulfills (6.49) for $T < \tau$.

6.2.3 2-D Navier–Stokes equation

As was observed in Subsection 6.1.3, the 2-D Navier–Stokes equation is critical with respect to the standard a priori estimate (6.8) in $[L^2(\Omega)]^2$. However, we have a better norm a priori estimate for such problem in 2-D, which allows to treat it as subcritical problem. The mentioned estimate, while not very elegant, uses a standard Gronwall lemma providing a priori estimate of solutions in $[H_0^1(\Omega)]^2$. Unfortunately, a similar estimate is not true in higher than two space dimension.

The following estimate holds for local solutions constructed in 2D case in Theorem 6.1.5.

Proposition 6.2.4. *Let $Pf \in D(A^{-\frac{1}{4}})$, $u_0 \in D(A^{\frac{1}{2}^+}) \subset [H^{1^+}(\Omega)]^2$ and $u(t)$ be the unique local in time solution to (6.1) in the phase space $D(A^{\frac{1}{2}^+}) \subset [H^{1^+}(\Omega)]^2$. The following a priori estimate then holds:*

$$\forall_{T>0} \exists_{M(T)>0} \; \|u\|_{L^\infty(0,T;[H_0^1(\Omega)]^2) \cap L^2(0,T;D(A))} \le M(T). \tag{6.50}$$

Proof. Multiplying (6.12) ($N = 2$) by Au, we get

$$\begin{aligned}\frac{1}{2}\frac{d}{dt}\|A^{\frac{1}{2}}u\|^2_{[L^2(\Omega)]^2} &= -\|Au\|^2_{[L^2(\Omega)]^2} - \langle P(u\cdot\nabla)u, Au\rangle_{[L^2(\Omega)]^2} \\ &\quad + \langle Pf, Au\rangle_{[L^2(\Omega)]^2}.\end{aligned} \tag{6.51}$$

Further, due to (6.15), Nirenberg–Gagliardo and Young inequalities,

$$\begin{aligned}&|\langle P(u\cdot\nabla)u, Au\rangle_{[L^2(\Omega)]^2}| \le \|P(u\cdot\nabla)u\|_{[L^2(\Omega)]^2}\|Au\|_{[L^2(\Omega)]^2} \\ &\le c\|u\|_{[L^4(\Omega)]^2}\|\nabla u\|_{[L^4(\Omega)]^2}\|Au\|_{[L^2(\Omega)]^2} \le \epsilon\|Au\|^2_{[L^2(\Omega)]^2} + C_\epsilon\|u\|^4_{[H^1(\Omega)]^2}\|u\|^6_{[L^2(\Omega)]^2}.\end{aligned}$$

For small $\epsilon > 0$, we thus have

$$\begin{aligned}\frac{1}{2}\frac{d}{dt}\|A^{\frac{1}{2}}u\|^2_{[L^2(\Omega)]^2} &\le -\frac{1}{4}\|Au\|^2_{[L^2(\Omega)]^2} + C'\|u\|^4_{[H^1(\Omega)]^2}\|u\|^6_{[L^2(\Omega)]^2} \\ &\quad + \|Pf\|^2_{[L^2(\Omega)]^2}.\end{aligned} \tag{6.52}$$

It follows from the natural a priori estimate (6.8), (6.9) of u in $L^\infty(0,T;[L^2(\Omega)]^2) \cap L^2(0,T;[H_0^1(\Omega)]^2)$ and the standard Gronwall inequality (Lemma 2.1.7) where we specify:

$$y(t) = \|u\|^2_{[H^1(\Omega)]^2}, \quad v(t) = \|Pf\|^2_{[L^2(\Omega)]^2}, \quad w(t) = C'\|u\|^2_{[H^1(\Omega)]^2}\|u\|^6_{[L^2(\Omega)]^2},$$

that u is bounded in the space

$$L^\infty(0,T;[H_0^1(\Omega)]^2) \cap L^2(0,T;D(A)). \tag{6.53}$$

□

As a result of Proposition 6.2.4, setting for clarity of calculations $\frac{1}{2}^+ = \frac{1}{2} + \delta$ with $0 < \delta < \frac{1}{8}$, for the local solution as in the above proposition, a *subordination condition* is satisfied

$$\begin{aligned}\|P(u\cdot\nabla)u\|_{X^{-\frac{1}{4}}} &= \|A^{-\frac{1}{4}}P(u\cdot\nabla)u\|_{[L^2(\Omega)]^2}\\ &\le M\|A^{\frac{1}{2}^+}u\|^2_{[L^2(\Omega)]^2} \le cM\|A^{\frac{1}{2}}u\|^{2-16\delta}_{[L^2(\Omega)]^2}\|A^{\frac{5}{8}}u\|^{16\delta}_{[L^2(\Omega)]^2}.\end{aligned} \tag{6.54}$$

Consequently (see Theorem 4.1.3), such local solutions will be extended globally in time. We are thus able to formulate the following.

Theorem 6.2.5. *The local in time solution $u(t)$ to* (6.1) *constructed as in Theorem* 6.1.5 *will be extended globally in time in the specified class. Moreover, the a priori estimate* (6.50) *is valid for each $T > 0$.*

Remark 6.2.6. As is well known (e. g., [124, Theorem 6.2]), the $L^\infty(0,T;[L^2(\Omega)]^2) \cap L^2(0,T;D(A^{\frac{1}{2}}))$ solution of the 2-D N-S equation with $u_t \in L^2(0,T;D(A^{-\frac{1}{2}}))$ is unique. Indeed, if u_1, u_2 are two such solutions, applying projector P, taking the difference of the equations and multiplying the result by $w = u_1 - u_2$, we get

$$\begin{aligned}\frac{1}{2}\frac{d}{dt}\|w(t)\|^2_{[L^2(\Omega)]^2} \le -\|A^{\frac{1}{2}}w\|^2_{[L^2(\Omega)]^2} - \langle P(w\cdot\nabla)u_1, w\rangle_{[L^2(\Omega)]^2}\\ - \langle P(u_2\cdot\nabla)w, w\rangle_{[L^2(\Omega)]^2}.\end{aligned} \tag{6.55}$$

The last term vanishes for divergence-free functions. The earlier term, using an equivalent form of the nonlinearity in (6.11) and the Nirenberg–Gagliardo inequality, is estimated as follows:

$$\begin{aligned}|\langle P(w\cdot\nabla)u_1, w\rangle_{[L^2(\Omega)]^2}| &\le c\|w\|^2_{[L^4(\Omega)]^2}\|A^{\frac{1}{2}}u_1\|_{[L^2(\Omega)]^2}\\ &\le c\|w\|_{[L^2(\Omega)]^2}\|A^{\frac{1}{2}}w\|_{[L^2(\Omega)]^2}\|A^{\frac{1}{2}}u_1\|_{[L^2(\Omega)]^2}.\end{aligned} \tag{6.56}$$

Inserting the last estimate into (6.55), using Cauchy's inequality, we obtain

$$\begin{aligned}\frac{1}{2}\frac{d}{dt}\|w(t)\|^2_{[L^2(\Omega)]^2} &\le -\|A^{\frac{1}{2}}w\|^2_{[L^2(\Omega)]^2} + c\|w\|_{[L^2(\Omega)]^2}\|A^{\frac{1}{2}}w\|_{[L^2(\Omega)]^2}\|A^{\frac{1}{2}}u_1\|_{[L^2(\Omega)]^2}\\ &\le C\|w\|^2_{[L^2(\Omega)]^2}\|A^{\frac{1}{2}}u_1\|^2_{[L^2(\Omega)]^2}.\end{aligned}$$

Since $\|w(0)\|^2_{[L^2(\Omega)]^2} = 0$, then $\|w(t)\|^2_{[L^2(\Omega)]^2} = 0$ for all $t \in [0,T]$, due to the Gronwall lemma.

Remark 6.2.7. To understand better the utility of the method used in the study of the global extendibility of solutions, we will discuss shortly an example of the *Burgers-type system in 3-D* (with nonlinear term as in 6.11), obtained by neglecting the viscosity in homogeneous 3-D N-S equation:

$$
\begin{aligned}
&U_t = \nu\Delta U - (U\cdot\nabla)U, \quad x\in\Omega,\ t>0,\\
&U = 0, \quad t>0, \quad x\in\partial\Omega,\\
&U(0,x) = U_0(x).
\end{aligned} \tag{6.57}
$$

It is easy to see that each component of the sufficiently regular (i. e., varying in $[L^\infty(\Omega)]^3$) solution of (6.57) fulfills maximum principle:

$$
\|U_i(t,\cdot)\|_{L^\infty(\Omega)} \le \|U_{0i}\|_{L^\infty(\Omega)}, \quad i = 1,2,3. \tag{6.58}
$$

Indeed, classical proofs of the maximum principle (e. g., [154, p. 84], [115, Chapter I, Section 2]) are applicable to any component equation; $i = 1,\dots,N$:

$$
\frac{\partial U_i}{\partial t} = \nu\Delta U_i - \sum_{j=1}^N U_j\frac{\partial U_i}{\partial x_j},
$$

separately. We are thus given a natural a priori estimate in $Y = [L^\infty(\Omega)]^3$ for such system. The nonlinear term is as in the N-S equation, and we have an estimate (for U satisfying (6.58)):

$$
\|(U\cdot\nabla)V\|_{[L^2(\Omega)]^3} \le \|U\|_{[L^\infty(\Omega)]^3}\|V\|_{[H^1_0(\Omega)]^3} \le \|U_0\|_{[L^\infty(\Omega)]^3}\|V\|_{[H^1_0(\Omega)]^3}. \tag{6.59}
$$

It is easy to construct a local in time solution varying in $[H^{\frac{3}{2}^+}(\Omega)]^3 \cap [H^1_0(\Omega)]^3$ (as in Theorem 6.1.4, but in the base space $[L^2(\Omega)]^3$). Since $H^{\frac{3}{2}^+}(\Omega) \subset L^\infty(\Omega)$, $N=3$, the maximum principle works for such solutions. Moreover (6.59) and the Nirenberg–Gagliardo inequality give

$$
\begin{aligned}
&\|(U\cdot\nabla)U\|_{[L^2(\Omega)]^3} \le c\|U_0\|_{[L^\infty(\Omega)]^3}\|U\|_{[H^1_0(\Omega)]^3}\\
&\le c\|U_0\|_{[L^\infty(\Omega)]^3}\|U\|^{\frac{1}{3}}_{[L^2(\Omega)]^3}\|U\|^{\frac{2}{3}}_{[H^{\frac{3}{2}}(\Omega)]^3} \le c(\|U_0\|_{[L^\infty(\Omega)]^3})\|U\|^{\frac{2}{3}}_{[H^{\frac{3}{2}}(\Omega)]^3},
\end{aligned} \tag{6.60}
$$

which shows the nonlinearity is subcritical in that case. Consequently, the local solution in $[H^{\frac{3}{2}^+}(\Omega)]^3 \cap [H^1_0(\Omega)]^3$ will be extended globally in time.

The above example shows that our approach is sensitive not only on the form of nonlinearity, but also on another specific properties. It also indicates *the role of the pressure* in the classical N-S equation, the term which seems responsible for the delicate properties of solutions of that equation.

Remark 6.2.8. It is evident from the considerations above that the phenomenon of *loosing regularity by local smooth solutions of 3-D N-S* is possible only if they enter the supercritical range of the nonlinearity (compare (6.25)). It would be therefore interesting to consider local solutions corresponding to initial data u_0 fulfilling

$$\|A^{-\frac{1}{8}}F(u_0)\|_{[L^2(\Omega)]^3} \geq g(\|u_0\|_{[L^2(\Omega)]^3})\|A^{\frac{7}{8}}u_0\|^{1+\epsilon}_{[L^2(\Omega)]^3}, \tag{6.61}$$

where $0 < \epsilon < \frac{2}{7}$, as eventual candidates for such phenomenon.

Remark 6.2.9. The viscosity term in the 3-D classical N-S equation is too weak to prevent better norms of its solutions from blowing-up in a finite time. The 2-D equation is still *critical with respect to the standard* $[L^2(\Omega)]^2$ *estimates* (while, of course, better norm estimates are available in that case).

For small initial data, we enjoy the standard property of problems with *quadratic nonlinearity* (compare Corollary 6.1.2) that since the estimates of solutions are proportional to the square of the norm, there is a ball centered at zero such that the solutions originated in it will never leave that ball (like in Theorem 6.2.1). For such solutions, the nonlinear term in N-S is subordinated to the main part operator.

To solve the problem connected with the 3-D N-S, one needs to find a better than the standard $[L^2(\Omega)]^3$ a priori estimate, following from the “symmetry” of that equation. Then the nonlinear term would be subordinated to the main part operator preventing possible blow-up of better norms of solutions, and the global in time existence-uniqueness theory will apply. In particular, if the estimate of solutions in $[L^N(\Omega)]^N$ ($[L^{N^+}(\Omega)]^N$) is known, it will make the N-D Navier–Stokes equation critical (subcritical) due to the following simple calculation:

$$\begin{aligned}\|P(u\cdot\nabla)u\|_{X^{-\frac{1}{2}}} &\leq c\||u|^2\|_{[L^2(\Omega)]^N} \leq c'\|u\|_{[L^{\frac{2N}{N-2}}(\Omega)]^N}\|u\|_{[L^N(\Omega)]^N}\\ &\leq c''\|u\|_{X^{\frac{1}{2}}}\|u\|_{[L^N(\Omega)]^N}.\end{aligned} \tag{6.62}$$

The example discussed in Remark 6.2.7 teaches us that the problem with the N-S equation is connected with the *pressure term* appearing in it. A similar form of the nonlinearity present in that example together with better than in the N-S a priori estimate (in L^∞) is sufficient to keep control on solutions, allowing to extend them globally in time.

Bibliographical notes

The literature devoted to the Navier–Stokes equations is huge. It is impossible to list even the most important positions. We call therefore only the classical reference [121], the monographs [111, 124, 158, 171, 175], and a nice review article [74]. The regularization of the original Navier–Stokes equation discussed here was given first in [124, Chapter 1, Remarque 6.11]. Recently, similar regularization was used by J. Wu in [185], in case of the MHD equation, and by J. Avrin in [10].

7 N-D Navier–Stokes equation, an extended discussion

7.1 Introduction

The present chapter, reported earlier in [33], extends the studies of the Navier–Stokes equation presented in Chapter 6. We consider Dirichlet problem for fractional generalization of the Navier–Stokes equation in a bounded smooth domain $\Omega \subset \mathbb{R}^N$, $N \geq 2$. We provide detail analysis of the local and global solvability, uniqueness, regularization, and continuation properties of the solutions. Global in time *viscosity solutions* of the original Navier–Stokes problem are obtained as limits of solutions of such fractional generalizations. The Navier–Stokes equation in $\mathbb{R}^N$ has been investigated in the last few years, for example, in [24, 125, 179].

Recall the original N-D Navier–Stokes problem:

$$\begin{cases} u_t = \nu\Delta u - (u \cdot \nabla)u + f - \nabla p, & x \in \Omega,\ t \geq 0, \\ \operatorname{div} u = 0, & x \in \Omega,\ t \geq 0, \\ u = 0, & x \in \partial\Omega,\ t \geq 0, \\ u(0,x) = u_0(x), & x \in \Omega, \end{cases} \tag{7.1}$$

where Ω is a bounded domain in $\mathbb{R}^N$, $N \geq 2$, $\partial\Omega \in C^\infty$, $u = (u^1(t,x), \dots, u^N(t,x))$ denotes velocity, ν viscosity constant, $p(t,x)$ pressure, and $f = (f^1(x), \dots, f^N(x))$ is external force.

Following the approach of J. Leray [120, 121], introducing the operator

$$A_r = -\nu P_r \Delta \quad \text{and} \quad F_r(u) = -P_r(u \cdot \nabla)u,$$

where P_r is a projector (see [68, 70]) onto the space

$$X_r = \operatorname{cl}_{[L^r(\Omega)]^N}\{\phi \in [C_0^\infty(\Omega)]^N : \operatorname{div}\phi = 0\}, \quad 1 < r < \infty,$$

the original problem (7.1) will be rewritten in an abstract form

$$u_t + A_r u = F_r(u) + P_r f, \quad t > 0, \qquad u(0) = u_0 \tag{7.2}$$

of equation with sectorial operator.

Simultaneously with (7.2), we consider the whole family of its fractal generalizations with power σ,

$$u_t + A_r^\sigma u = F_r(u) + P_r f, \quad t > 0, \qquad u(0) = u_0, \tag{7.3}$$

where $N \geq 2$ and $r \in (1,\infty)$ are given constants and the parameter σ is restricted in the following way:

$$\max\left\{\frac{1}{2}, \frac{N}{4r} + \frac{1}{4}\right\} < \sigma; \tag{7.4}$$

https://doi.org/10.1515/9783110599831-007

in particular, for $r = 2$ and $N = 2$ we consider $\frac{1}{2} < \sigma$.

We are using the notation of spaces associated with the Stokes operator, as in [75]:

$$X_r^\xi = D(A_r^\xi) \quad \text{for } \xi > 0 \qquad \text{and} \qquad X_r^\xi = (X_{r'}^{-\xi})^* \quad \text{for } \xi < 0.$$

Recall that for $\sigma = 1$, in a distinguished work [95], these ideas were used to solve (7.2) locally in time for $u_0 \in X_r^0$ if $N = 2$ and for $u_0 \in X_2^{\frac{1}{4}}$ if $N = 3$. The latter result was then significantly improved in [75] (see also [182]), where solutions were obtained for $u_0 \in X_r^\gamma$ with $\gamma \in [\frac{N}{2r} - \frac{1}{2}, 1)$ for any $r \in (1, \infty)$.

We follow here the results of [75], which allow to include the local existence-uniqueness considerations for (7.3) inside the abstract approach originated in [86] and used in this book. Further, in Subsection 7.3, we formulate a global existence result for the regularization (7.25) of the original Navier–Stokes problem. The obtained there global solutions to (7.25) are used next, in Subsections 7.3.1.1 and 7.3.1.2, to obtain a weak global solution to the original Navier–Stokes equation. In the first subsection, we describe in Theorem 7.3.4 passing to the limit in weakly formulated regularized equation, and passing to the limit in the Cauchy's integral formula is discussed in Theorem 7.3.6 of the second subsection.

We also obtain additional information about the solution, which includes, for example, its continuous dependence on initial data. On the other hand, several conditions sufficient for global in time continuation of the solution follow from our analysis of (7.3).

7.2 Local solutions and their properties

Following the approach of Chapter 3 we are searching for mild solution of (7.3) fulfilling Cauchy's integral formula

$$u(t) = e^{-A_r^\sigma t} u_0 + \int_0^t e^{-A_r^\sigma (t-s)} (F_r(u(s)) + P_r f) ds. \tag{7.5}$$

Here, $\{e^{-A_r^\sigma t} : t \geq 0\}$ is a linear C^0 analytic semigroup generated by $-A_r^\sigma$ in X_r such that[1]

$$\|e^{-A_r^\sigma t}\|_{\mathcal{L}(X_r^\beta, X_r^\alpha)} \leq Ct^{-\frac{\alpha-\beta}{\sigma}}, \quad t > 0, \, \sigma \geq \alpha \geq \beta \geq -\sigma, \tag{7.6}$$

where, as in [75, p. 269], $X_r^\xi = D(A_r^\xi)$ for $\xi > 0$ and $X_r^\xi = (X_{r'}^{-\xi})^*$ for $\xi < 0$. According to the concept of extrapolation spaces in [6, Chapter V], we also treat X_r^ξ for $\xi < 0$ as a completion of $(X_r, \|A_r^\xi \cdot \|_{X^r})$.

1 Note that due to Proposition 2.2.29 fractional powers $(A_r^\sigma)^\xi$ are equal to $A_r^{\sigma\xi}$ (for $\sigma > 0, \xi > 0$). Hence the semigroup associated with $-A_r^\sigma$ satisfies (7.6).

Let

$$P_r f \in X_r^0,$$

let $\alpha, \beta \in \mathbb{R}$ satisfy

$$0 \le \alpha - \beta < \sigma, \tag{7.7}$$

and $F_r : X_r^\alpha \to X_r^\beta$ fulfills the Lipschitz condition

$$\|F_r(u) - F_r(v)\|_{X_r^\beta} \le L\|u - v\|_{X_r^\alpha}(\|u\|_{X_r^\alpha} + \|v\|_{X_r^\alpha}), \qquad u, v \in X_r^\alpha. \tag{7.8}$$

With assumptions (7.4), (7.6), (7.7)–(7.8), we study the following notion of *mild* γ*-solution* to (7.3).

Definition 7.2.1. For $r \in (1, \infty)$, $P_r f \in X_r^\beta$, γ satisfying

$$\alpha - \frac{\sigma}{2} < \gamma \in [2\alpha - \beta - \sigma, \alpha], \tag{7.9}$$

and $u_0 \in X_r^\gamma$,a mild γ-solution of (7.3) in $[0, T]$ is a function u such that

A) $u \in L^\infty_{\text{loc}}(0, \tau; X_r^\alpha)$,

B) $t^{\frac{\alpha-\gamma}{\sigma}} \|u(t)\|_{X_r^\alpha}$ is bounded as $t \to 0^+$ and if $\gamma = 2\alpha - \beta - \sigma > \alpha - \frac{1}{2}$ then, in addition,

$$\text{either } t^{\frac{\alpha-\gamma}{\sigma}} \|u(t)\|_{X_r^\alpha} \to 0 \text{ as } t \to 0^+ \quad \text{or} \quad u(t) \text{ is right-continuous in } X_r^\gamma \text{ at } t = 0,$$

C) $u(0) = u_0$ and (7.5) holds in $(0, \tau]$ as the equality in X_r^α.

Concerning the above notion of solution observe that for positive times, due to (7.8), such solution should take values in X_r^α. From this and (7.6), observe also that if we let in (7.5) $u_0 \in X_r^\gamma$ then, applying X_r^α-norm and multiplying both sides by $t^{\frac{\alpha-\gamma}{\sigma}}$, we have to ensure that, on the one hand, the integral $\int_0^t (t-s)^{\frac{\beta-\alpha}{\sigma}} s^{\frac{2(\gamma-\alpha)}{\sigma}} ds$ is convergent and also the right-hand side, containing the power $t^{\frac{-2\alpha+\beta+\gamma+\sigma}{\sigma}}$, is not singular at $t = 0$. Hence we get the restriction on γ as in Definition 7.2.1 and on α, β as in (7.7) (for more detailed analysis, see [35]).

Our main technical tool in the proof of local solvability is the estimate (7.10) following from [75, Lemma 2.1, 2.2].

Proposition 7.2.2. *The following estimate holds:*

$$\|P_r(u \cdot \nabla)v\|_{X_r^\beta} \le M\|u\|_{X_r^\alpha}\|v\|_{X_r^\alpha}, \quad u, v \in X_r^\alpha, \tag{7.10}$$

with a constant $M = M(\alpha, \beta, r)$ *provided that*

$$0 \le -\beta < \frac{1}{2} + \frac{1}{2}\left(N - \frac{N}{r}\right), \quad 2\alpha - \beta \ge \frac{1}{2}\left(\frac{N}{r} + 1\right), \quad \alpha > 0, \quad \alpha - \beta > \frac{1}{2}.$$

Actually, each $A_r^{-\frac{1}{2}} P_r \frac{\partial}{\partial x_j}, j = 1, \dots, N$, *extends uniquely to a bounded linear operator from* $[L^r(\Omega)]^N$ *to* X_r, *so that there exists a constant* $M_r > 0$ *such that*

$$\left\| P_r \frac{\partial u}{\partial x_j} \right\|_{X_r^{-\frac{1}{2}}} \leq M_r \|u\|_{[L^r(\Omega)]^N}, \quad u \in [L^r(\Omega)]^N.$$

7.2.1 Existence-uniqueness of mild γ-solution

We formulate now the existence result in X_r^γ.

Theorem 7.2.3. *Assume that:*

(i) $r \in (1, \infty), \quad \sigma > \max\left\{\frac{N}{4r} + \frac{1}{4}, \frac{1}{2}\right\}, \quad P_r f \in X_r^0, \quad N \geq 2,$

(ii) $\gamma \in \left[\frac{N}{2r} + \frac{1}{2} - \sigma, \sigma\right)$ *if* $r \leq N$, *alternatively if* $r > N$ *and* $\sigma < \frac{N}{r} + 1$,

(iii) $\gamma \in \left(-\frac{\sigma}{2}, \sigma\right)$ *if* $r > N$ *and* $\sigma \geq \frac{N}{r} + 1$. (7.11)

Then there exist $\alpha, \beta \in (-\sigma, \sigma)$ *such that (7.7), (7.8), (7.9) hold and for each* $u_0 \in X_r^\gamma$ *the problem (7.3) has a unique mild* γ*-solution satisfying (A), (B), and (C) in Definition 7.2.1.*

Proof. Due to Proposition 7.2.2, F_r satisfies (7.8) whenever

$$0 \leq -\beta < \frac{1}{2} + \frac{N}{2r}, \quad 2\alpha - \beta \geq \frac{1}{2} + \frac{N}{2r}, \quad \alpha > 0, \quad \alpha - \beta > \frac{1}{2} \tag{7.12}$$

and, as mentioned in the context of Definition 7.2.1, we also impose the restriction

$$0 \leq \alpha - \beta < \sigma. \tag{7.13}$$

We transform the scale of spaces in such a way that we can directly apply the argument in [35]. Hence we let

$$\tilde{\alpha} = \frac{\alpha}{\sigma}, \qquad \tilde{\beta} = \frac{\beta}{\sigma} \tag{7.14}$$

and, drawing on the $\tilde{\alpha}\tilde{\beta}$-plane the region described by (7.12), (7.13), and (7.14), we denote it $\tilde{S}$. We recall $\tilde{S}$ below as *admissible region.*

Observe that $\tilde{S}$ varies depending on the parameters, namely,

- if $N \geq r$, then $\tilde{S}$ is a triangle without the right edge; also without the upper left vertex when $N = r$,
- if $N < r$, then $\tilde{S}$ is a trapezoid, a pentagon, or a hexagon without the parallel edges, also without the edge lying on $\tilde{\beta}$-axis and without the bottom edge whenever such edges appear.

The role of $\tilde{\mathcal{S}}$ becomes clear if one notice that, due to the definition of $\tilde{\mathcal{S}}$ and Proposition 7.2.2, for any $(\tilde{\alpha}, \tilde{\beta}) \in \tilde{\mathcal{S}}$, the nonlinear term satisfies

$$\|F_r(w) - F_r(v)\|_{X_r^{\sigma\tilde{\beta}}} \leq M\|w - v\|_{X_r^{\sigma\tilde{\alpha}}}(\|w\|_{X_r^{\sigma\tilde{\alpha}}} + \|v\|_{X_r^{\sigma\tilde{\alpha}}}), \quad w, v \in X_r^{\sigma\tilde{\alpha}}$$

whereas the linear semigroup is such that

$$\|e^{-A_r^\sigma t}\|_{\mathcal{L}(X_r^{\sigma\tilde{\beta}}, X_r^{\sigma\tilde{\alpha}})} \leq Ct^{-(\tilde{\alpha}-\tilde{\beta})}, \quad t > 0,\ 1 \geq \tilde{\alpha} \geq \tilde{\beta} \geq -1.$$

Hence the approach of [35] can be directly applied to (7.5) in the transformed scale of spaces

$$\tilde{X}^{\tilde{\gamma}} = X^{\sigma\tilde{\gamma}}, \quad \tilde{\gamma} \in [-1, 1],$$

which we recall further as *tilde-scale*. In particular, if

$$\tilde{\alpha} - \frac{1}{2} < \tilde{\gamma} \in [2\tilde{\alpha} - \tilde{\beta} - 1, \tilde{\alpha}], \tag{7.15}$$

the operator

$$\mathcal{L}(u)(t) = e^{-A_r^\sigma t}u_0 + \int_0^t e^{-A_r^\sigma (t-s)}(F_r(u(s)) + P_r f)ds$$

has a unique fixed point in a complete metric space

$$\mathcal{X} = \left\{\varphi \in L^\infty_{\text{loc}}(0, \tau; X_r^{\tilde{\alpha}\sigma}) : \sup_{t\in(0,\tau]} t^{\tilde{\alpha}-\tilde{\gamma}}\|u(t)\|_{X_r^{\tilde{\alpha}\sigma}} \leq K\right\}$$

for suitably chosen $\tau > 0$ and $K > 0$ (see [35] for details).

The "bottom" $\tilde{\gamma}$ in (7.15), denoted here $\tilde{\gamma}_c$, is now obtained due to the geometry of $\tilde{\mathcal{S}}$. Namely,
- if $r \leq N$, or if $r > N$ and $\sigma < \frac{N}{r} + 1$, then $\tilde{\gamma}_c$ is achieved for admissible pairs $(\tilde{\alpha}, \tilde{\beta})$ lying on the portion of the edge of $\tilde{\mathcal{S}}$ which is contained in the line $2\tilde{\alpha} - \tilde{\beta} = \frac{N}{2r\sigma} + \frac{1}{2\sigma}$ below the line $\tilde{\alpha} - \tilde{\beta} = \frac{1}{2}$, that is, $\tilde{\gamma}_c = \frac{N}{2r\sigma} + \frac{1}{2\sigma} - 1$,
- if $\sigma \geq \frac{N}{r} + 1$ and $r > N$, then "bottom $\tilde{\gamma}$" in (7.15) equals $-\frac{\sigma}{2}$, and is not achieved.

On the other hand, looking for the largest possible $\tilde{\gamma}$ in (7.15), we would like to choose $(\tilde{\alpha}, \tilde{\beta})$ from the edge of $\tilde{\mathcal{S}}$ which is contained in the line $\tilde{\beta} = 0$, that is, such $\tilde{\gamma}$ can be arbitrarily close to 1, but is not achieved.

Returning finally to the nontilde scale, thus letting $\gamma = \sigma\tilde{\gamma}$, we get the result for γ as in the statement. □

Remark 7.2.4. Observe that if (7.4) is violated then the admissible region in the proof of Theorem 7.2.3 is empty.

Remark 7.2.5. Assumption $P_r f \in X_r^0$ in Theorem 7.2.3 can be replaced by

$$P_r f \in X_r^\beta \tag{7.16}$$

with $\beta \leq 0$ such that a certain $\alpha > 0$ can be found for which

$$(\tilde{\alpha}, \tilde{\beta}) := \left(\frac{\alpha}{\sigma}, \frac{\beta}{\sigma}\right) \in \tilde{\mathcal{S}} \quad \text{and} \quad \tilde{\alpha} - \frac{1}{2} < \tilde{\gamma} = \frac{\gamma}{\sigma} \in [2\tilde{\alpha} - \tilde{\beta} - 1, \tilde{\alpha}].$$

For example, if $\gamma = \frac{N}{2r} + \frac{1}{2} - \sigma$, $N \geq r$ and $\sigma \leq 1$, it suffices to have (7.16) for $\beta > \frac{N}{2r} + \frac{1}{2} - 2\sigma$ being arbitrarily close to $\frac{N}{2r} + \frac{1}{2} - 2\sigma$.

7.2.2 Regularization of mild γ-solution

We now present additional properties of the mild γ-solution and show that it regularizes for positive times.

Theorem 7.2.6. *Assume* (7.11) *holds. Let $u_0 \in X_r^\gamma$ and u be the mild γ-solution of* (7.3) *on* $[0,\tau]$ *which was obtained in Theorem* 7.2.3. *Then $u \in C([0,\tau]; X_r^\gamma)$, and there exists $\xi^* \leq \sigma$ as in* (7.18) *such that for any $\xi \in (\gamma, \xi^*)$,*

$$u \in C((0,\tau]; X_r^\xi)$$

and

$$t^{\frac{\xi-\gamma}{\sigma}} \|u(t)\|_{X_r^\xi} \to 0 \quad \text{as } t \to 0^+.$$

Also, if $\xi \in [\gamma, \xi^*)$ and $u_{0n} \to u_0 \in X_r^\gamma$, then

$$\sup_{t \in (0,\tau]} t^{\frac{\xi-\gamma}{\sigma}} \|u(t, u_{0n}) - u(t, u_0)\|_{X_r^\xi} \to 0.$$

Proof. For γ as in (7.11), we let

$$\tilde{\gamma} = \frac{\gamma}{\sigma} \tag{7.17}$$

and recall that the solution in Theorem 7.2.3 is obtained as a fixed point of $\mathcal{L}$ in $\mathcal{X}$, whenever $(\tilde{\alpha}, \tilde{\beta}) \in \tilde{\mathcal{S}}$ is chosen such that $\tilde{\alpha} - \frac{1}{2} < \tilde{\gamma} \in [2\tilde{\alpha} - \tilde{\beta} - 1, \tilde{\alpha}]$. Recall also from the proof of Theorem 7.2.3 that the approach of [35] can be directly applied to (7.5) in the tilde scale $\tilde{X}^{\tilde{\gamma}} = X^{\sigma\tilde{\gamma}}$, $\tilde{\gamma} \in [-1, 1]$. After returning to nontilde scale we obtain that $u \in C([0,\tau]; X_r^\gamma)$ and that all other properties written in the statement hold for $\xi \in (\gamma, \tilde{\beta}\sigma + \sigma)$.

Note that there are several pairs $(\tilde{\alpha}, \tilde{\beta})$ in $\tilde{\mathcal{S}}$ for which the above argument is applicable. Also, to find the optimal range of ξ, the bootstrapping procedure as in [35, Section 4] can be applied. Hence we get the result for $\xi \in (\gamma, \xi^*)$, where due to the geometry of $\tilde{\mathcal{S}}$,

$$\xi^* = \sigma \quad \text{if } N \geq r \quad \text{and} \quad \gamma < \xi^* \in \left[\frac{N}{2r} - \frac{1}{2} + \sigma, \sigma\right) \quad \text{if } N < r, \tag{7.18}$$

which fulfills the statement. □

We now show that a mild γ-solution of (7.3) is in fact a strong γ-solution.

Definition 7.2.7. Let (7.11) hold and $u_0 \in X_r^\gamma$. A function

$$u \in C([0,\tau); X_r^\gamma) \cap C^1((0,\tau); X_r^0) \cap C((0,\tau); X_r^\sigma),$$

such that $u(0) = u_0$ and the differential equation in (7.3) holds in X_r^0 on $(0,\tau)$ is called a strong γ-solution of (7.3) until time τ.

Theorem 7.2.8. *Let (7.11) be satisfied. In addition, suppose that*

$$\sigma > 1 - \frac{N}{2r} \quad \textit{whenever } r > N. \tag{7.19}$$

If $u_0 \in X_r^\gamma$ and u is a mild γ-solution of (7.3) on $[0,\tau]$ obtained in Theorem 7.2.3, then u is a strong γ-solution of (7.3). In addition, $u \in C^1((0,\tau); X_r^\xi)$ for each $\xi < \sigma$.

Proof. Due to Theorem 7.2.6, we have that $u(\epsilon) \in X_r^\alpha$ and that $u(t+\epsilon)$, as a function of variable $t \in [0, \tau - \epsilon]$, is of the class $C([0,\tau-\epsilon]; X_r^\alpha)$ for any $\epsilon > 0$ small enough and any $\alpha < \xi^*$ (ξ^* given in (7.18)).

Using this, we can now choose α strictly less than ξ^* (but close enough to it), such that $F_r + P_r f$ is Lipschitz continuous on bounded sets as a map from X_r^α into X_r^0 (see Proposition 7.2.2). Hence the result follows from [32, Lemma 2.2.1 and Corollary 2.3.1]. □

Specifying the results of Theorems 7.2.3 and 7.2.8 to $r = 2$, we get the existence-uniqueness result in the Hilbert scale.

Theorem 7.2.9. *If $P_2 f \in X_2^0$ and*

$$\frac{N}{8} + \frac{1}{4} < \sigma, \qquad \gamma \in \left[\frac{N}{4} + \frac{1}{2} - \sigma, \sigma\right),$$

then for each $u_0 \in X_2^\gamma$ there exists a unique local strong γ-solution of (7.3), that is, a unique function $u \in C([0,\tau); X_2^\gamma) \cap C^1((0,\tau); X_2^0) \cap C((0,\tau); X_2^\sigma)$ with some $\tau > 0$, such that $u(0) = u_0$ and the differential equation in (7.3) holds in X_2^0 on $(0,\tau)$.

7.2.3 Continuation of mild γ-solution

In this subsection, we discuss continuation in time of a mild γ-solution.

Theorem 7.2.10. *Assume (7.11) holds. If $u_0 \in X_r^\gamma$ and u is a mild γ-solution of (7.3) on $[0,\tau]$ obtained in Theorem 7.2.3, then there is a maximal time interval $[0,\tau_{u_0})$, such that for each subinterval $[0,T] \subset [0,\tau_{u_0})$ function u extends to a mild γ-solution on $[0,T]$.*

Furthermore, under the assumptions of Theorem 7.2.8, such extension of u is actually a strong γ-solution of (7.3) until time τ_{u_0} and, in addition, it belongs to $C^1((0,\tau); X_r^\xi)$ for each $\xi < \sigma$.

Proof. Due to Theorem 7.2.6, u is continuous in X_r^ξ for $0 < t \le \tau$ whenever $\xi < \xi^*$ (ξ^* is the constant appearing in (7.18)). Solving (7.5) with $u(\tau)$ as the "new initial condition," we observe that, after concatenation, u has an extension to a certain time interval larger than $[0,\tau]$. Hence we define a set, denoted by $I(u_0)$, consisting of all positive times T such that u extends onto $[0,T] \supset [0,\tau]$. Letting

$$\tau_{u_0} := \sup I(u_0),$$

we observe that $0 < \tau_{u_0} \le \infty$ and $\tau_{u_0} \notin I(u_0)$. Otherwise, we can extend the solution to some larger interval.

Due to the above argument, we have thus defined an extension of u to $[0,\tau_{u_0})$ as in the statement. This extension preserves continuity properties of u from Theorem 7.2.6. Using [32, Lemma 2.2.1 and Corollary 2.3.1], we obtain, in a similar manner as in the proof of Theorem 7.2.8, that this extension is actually a strong γ-solution of (7.3) until time τ_{u_0}. □

Definition 7.2.11.
(i) The extension of γ-solution u onto $[0,\tau_{u_0})$ which is described in Theorem 7.2.10 (and which below is still denoted by u) is called a maximally defined γ-solution.
(ii) If $\tau_{u_0} = \infty$, then a maximally defined γ-solution is called a global γ-solution.

In what follows, we determine conditions on u which lead to a conclusion that $\tau_{u_0} = \infty$. We start from the estimate near a blow-up point which occurs provided that $\tau_{u_0} < \infty$.

Theorem 7.2.12. *Assume (i) and (ii) of (7.11) hold. Let $u_0 \in X^\gamma$ and u be a maximally defined γ-solution of (7.3).*

Then either $\tau_{u_0} = \infty$ or, otherwise, given any ζ satisfying

$$\gamma_c := \frac{N}{2r} + \frac{1}{2} - \sigma < \zeta \begin{cases} \le \frac{N}{4r} + \frac{1}{4}, & N > r,\ \sigma < \frac{N}{2r} + \frac{1}{2}, \\ < \frac{N}{2r} + \frac{1}{2} - \frac{\sigma}{2}, & N \ge r,\ \frac{N}{2r} + \frac{1}{2} \le \sigma, \\ < \frac{N}{4r} + \frac{1}{4}, & N = r,\ \sigma < \frac{N}{2r} + \frac{1}{2}, \\ < \frac{N}{2r}, & N < r,\ \sigma \le 1, \\ < \frac{N}{2r} + \frac{1}{2} - \frac{\sigma}{2}, & N < r,\ \sigma > 1, \end{cases} \tag{7.20}$$

for all times close enough to τ_{u_0} we have the estimate

$$\|u(t)\|_{X_r^\zeta} \geq \frac{\text{const.}}{(\tau_{u_0} - t)^{\frac{\zeta}{\sigma} - \frac{N}{2r\sigma} - \frac{1}{2\sigma} + 1}}. \tag{7.21}$$

Proof. If $r \leq N$, alternatively if $r > N$ and $\sigma < \frac{N}{r} + 1$, since $X_r^\gamma \subset X_r^{\gamma_c}$, we infer from Theorem 7.2.6 that u can be regarded, in particular, as a mild γ_c-solution.[2] Using (7.14) and (7.17), we consider the tilde scale and the region $\tilde{S}$ as in the proof of Theorem 7.2.3. For ζ as in (7.20), we let $\tilde{\zeta} = \frac{\zeta}{\sigma}$ and choose from a portion of the edge of $\tilde{S}$ lying on the line $2\tilde{\alpha} - \tilde{\beta} = \frac{N}{2r\sigma} + \frac{1}{2\sigma}$ and below the line $\tilde{\alpha} - \tilde{\beta} = \frac{1}{2}$ some admissible pair $(\tilde{\alpha}, \tilde{\beta}) \in \tilde{S}$ such that $\tilde{\zeta} \in [\frac{N}{2\sigma r} + \frac{1}{2\sigma} - 1, \tilde{\alpha}]$. After this is done, [35, Proposition 3.12] can be directly applied to (7.3) in the tilde scale, and hence for $\tilde{X}_r^{\tilde{\zeta}} = X_r^{\sigma\tilde{\zeta}}$ we get

$$\|u(t)\|_{\tilde{X}_r^{\tilde{\zeta}}} \geq \frac{\text{const.}}{(\tau_{u_0} - t)^{\tilde{\zeta} - (2\tilde{\alpha} - \tilde{\beta} - 1)}} = \frac{\text{const.}}{(\tau_{u_0} - t)^{\tilde{\zeta} - \frac{N}{2r\sigma} - \frac{1}{2\sigma} + 1}}.$$

Inserting $\sigma\tilde{\zeta} = \zeta$, we obtain (7.21). On the other hand, admissible pairs $(\tilde{\alpha}, \tilde{\beta})$ considered above lead to the characterization in (7.20). □

Remark 7.2.13. A version of Theorem 7.2.12 holds true assuming (7.11) (iii) instead of (7.11) (ii). But the exponent in (7.21) will be then different, which we do not pursue here.

In what follows, we give some consequences of (7.21). We start from the corollary which is in the vein of the geometric theory in [9, 32, 35, 86, 176].

Corollary 7.2.14. *Assuming (7.11) (i) and (7.11) (ii), and given ζ as in (7.20), we have that either $\|u(t)\|_{X_r^\zeta}$ norm of a mild γ-solution to (7.3) blows-up in some finite time or, otherwise, $\tau_{u_0} = \infty$.*

The next two results remain in the vein of [112, 139, 143, 150, 186]. Note that for $q \in [1, \infty)$ these results come out here by integrating q-power of both sides of (7.21) over time interval $(\tau_{u_0} - \epsilon, \tau_{u_0})$ for arbitrarily small $\epsilon > 0$ whenever τ_{u_0} is finite. On the other hand, note that for $q = \infty$ we get a version of Corollary 7.2.14.

Corollary 7.2.15. *Let (i) and (ii) of (7.11) be satisfied, and ζ be as in (7.20).*

If it is a priori known that a maximally defined γ-solution u to (7.3) satisfies: for arbitrary $T > 0$ there exists (small) $\epsilon > 0$ such that

$$u \in L^q(T - \epsilon, T; X_r^\zeta),$$

2 Following the proof of Theorem 7.2.6, we choose a pair (α, β) such that (7.12)–(7.13) hold and $\alpha - \frac{\sigma}{2} < \gamma_c = 2\alpha - \beta - \sigma$. If $\gamma \in (\gamma_c, \alpha]$, then $t^{\frac{\alpha - \gamma_c}{\sigma}} \|u\|_{X_r^\alpha} = t^{\frac{\alpha - \gamma}{\sigma}} \|u\|_{X_r^\alpha} t^{\frac{\gamma - \gamma_c}{\sigma}}$, which in turn implies (see Definition 7.2.1) that a mild γ-solution is a mild γ_c-solution. On the other hand, if $\gamma > \alpha$, due to Theorem 7.2.6, a mild γ-solution is in the class $C([0, \tau]; X^\gamma)$ which for $\gamma > \alpha$ is a subset of $C([0, \tau]; X^\alpha)$. Hence, due to the previous argument, u is a mild γ_c-solution.

where

$$q \in [1, \infty] \quad and \quad \frac{2\sigma}{q} + \frac{N}{r} \le 2\zeta + 2\sigma - 1,$$

then $\tau_{u_0} = \infty$.

We remark that if $\frac{N}{r} < 2\sigma - 1$ then (7.20) allows us to choose $\zeta = 0$. We state the result in this case separately as it involves $L^q(0, T; L^r(\Omega))$-norm.

Corollary 7.2.16. *Let (i) and (ii) of (7.11) be satisfied. If it is a priori known that a maximally defined γ-solution u to (7.3) satisfies: for arbitrary $T > 0$, there exists (small) $\epsilon > 0$ such that*

$$u \in L^q(T - \varepsilon, T; X_r^0),$$

where

$$q \in [1, \infty], \quad \frac{2\sigma}{q} + \frac{N}{r} \le (2\sigma - 1) \quad and \quad \frac{N}{r} < 2\sigma - 1,$$

then $\tau_{u_0} = \infty$.

Note that (7.20) above and Corollary 7.2.15 exclude $\zeta = \frac{N}{2r} + \frac{1}{2} - \sigma$. Hence we give another result for the continuation of a mild γ-solution (see [26, 176]).

Theorem 7.2.17. *Assume (7.11) holds. If a maximally defined γ-solution u to (7.3) is uniformly continuous in X_r^γ-norm on each bounded time interval $[0, T) \subset [0, \tau_{u_0})$, then $\tau_{u_0} = \infty$.*

Proof. Note that the time of existence in Theorem 7.2.3 can be made uniform on bounded sets of a positive Hausdorff's measure of noncompactness (see [35, Theorem 3.8]).

If τ_{u_0} is finite then, by assumption, for each $\epsilon > 0$ the set $\{u(t) \in X_r^\gamma : t \in [0, \tau_{u_0})\}$ can be covered in X_r^γ by a finite number of balls of radius ϵ. Since ϵ can be as small as we wish, given $t_n \to \tau_{u_0}^-$ and solving (7.5) with u_0 replaced by $u(t_n)$, we can obtain a continuation of the solution beyond τ_{u_0}, which is a contradiction with definition of τ_{u_0}. □

Corollary 7.2.18. *Assume (7.11) holds for $r = 2$ and $\sigma = \frac{N}{4} + \frac{1}{2}$. If we know about a maximally defined γ-solution u to (7.3) that it satisfies an a priori bound: for every $T > 0$,*

$$\|u\|_{L^\infty(0,T;X_2^0)} + \|u\|_{L^2(0,T;X_2^{\frac{\sigma}{2}})} \le C(T), \tag{7.22}$$

then $\tau_{u_0} = \infty$.

Proof. Using Proposition 7.2.2 (with $\beta = -\frac{\sigma}{2}$, $\alpha = \frac{\sigma}{4}$) and an interpolation inequality for fractional power spaces (see [74, Theorem 1] and [32, Proposition 1.3.9]), we obtain

$$\|F_2(u)\|_{X_2^{-\frac{\sigma}{2}}} \le M\|u\|^2_{X_2^{\frac{\sigma}{4}}} \le c\|u\|_{X_2^{\frac{\sigma}{2}}}\|u\|_{X_2^0}. \tag{7.23}$$

From (7.3), (7.22), and (7.23), we now have

$$\|u_t\|_{L^2(0,T;X_2^{-\frac{\sigma}{2}})} \le \hat{C}(T), \quad T > 0. \tag{7.24}$$

Due to [171, p. 260], we can combine (7.22) with (7.24) to conclude that u is a uniformly continuous function from each bounded time interval $[0,T) \subset [0,\tau_{u_0})$ into X_2^0. Thus the result follows from Theorem 7.2.17. □

7.3 Global solutions of *N*-D Navier–Stokes equations

In this section, we focus on proving the existence of global solutions of (7.2).

7.3.1 Dimension $N \ge 3$

We follow here the main idea of this monograph to approximate the original Navier–Stokes problem by its fractal regularizations. Hence we consider

$$u_t^\epsilon = -\epsilon A_2^\sigma u^\epsilon - A_2 u^\epsilon + F_2(u^\epsilon) + P_2 f, \qquad u(0) = u_0, \tag{7.25}$$

with $\sigma \ge \frac{N}{4} + \frac{1}{2}$ fixed and with arbitrary $\epsilon > 0$.

Remark 7.3.1. Let us make a few observations concerning (7.25) in dimension $N \ge 3$.

(i) Note that there is an operator $-\epsilon A_2^\sigma - A_2$ in the linear main part of (7.25) instead of A_2^σ. However, this operator will still generate an analytic C^0 semigroup satisfying the same smoothing estimate as the semigroup generated by A_2^σ (this is because the scale will be built on the powers of A_2 with equivalent norms; see [86, Theorem 1.4.8]). Hence the existence-uniqueness, regularization, and continuation theory of Section 7.2 will extend to the present situation with minor changes in the proofs. In particular, whenever $\sigma \ge \frac{N}{4} + \frac{1}{2}$ and $\gamma \in [\frac{N}{4} + \frac{1}{2} - \sigma, \sigma)$, there exists a unique mild γ-solution of (7.25).

(ii) A mild γ-solution of (7.25) regularizes to a strong γ-solution. In addition, given $u_0 \in X_2^0$ and $T > 0$, we obtain a uniform with respect to ϵ a priori estimate (see section 6.1.1)

$$\|u^\epsilon\|_{L^\infty(0,T;X_2^0)} + \|u^\epsilon\|_{L^2(0,T;X_2^{\frac{1}{2}})} + \sqrt{\epsilon}\|u^\epsilon\|_{L^2(0,T;X_2^{\frac{\sigma}{2}})} \le C(T). \tag{7.26}$$

(iii) The natural estimate (7.26) implies that a maximally defined γ-solutions of (7.25) actually exists for all $t \ge 0$ (if $\sigma > \frac{N}{4} + \frac{1}{2}$, see Theorem 7.2.12 with $\zeta = 0$ in (7.20); if $\sigma = \frac{N}{4} + \frac{1}{2}$, see Corollary 7.2.18).

The following two lemmas will be useful in the limiting procedure.

Lemma 7.3.2. *Assume that $N \geq 3$, $1 \leq \sigma \in (\frac{N}{2} - 1, \frac{N}{2} + 1)$ and let*

$$\eta_0 = \frac{N}{4} + \frac{1}{2} - \frac{\sigma}{2}. \tag{7.27}$$

Then the following estimates hold true:

(i) *for each $\eta \in [\eta_0, 1)$ there is a constant $M > 0$ such that*

$$\|F_2(v) - F_2(w)\|_{X_2^{-\frac{\sigma}{2}}} \leq M(\|v\|_{X_2^{\frac{\eta}{2}}} + \|w\|_{X_2^{\frac{\eta}{2}}})\|v - w\|_{X_2^{\frac{\eta}{2}}}, \quad \textit{for all } v, w \in X_2^{\frac{\eta}{2}},$$

(ii) *in particular, there is a constant $C > 0$ such that*

$$\|F_2(z)\|^{\frac{1}{\eta}}_{X_2^{-\frac{\sigma}{2}}} \leq C\|z\|^{2(\frac{1}{\eta}-1)}_{X_2^0}\|z\|^2_{X_2^{\frac{1}{2}}}, \quad \textit{whenever } z \in X_2^{\frac{1}{2}}.$$

Proof. Part (i) follows from Proposition 7.2.2 (used here with $\alpha = \frac{\eta}{2}$ and $\beta = -\frac{\sigma}{2}$).

Part (ii) follows from part (i) and interpolation inequality for fractional power spaces (see [74, Theorem 1], or section 5.4.3). □

Lemma 7.3.3. *Let $N \geq 3$, $P_2 f \in X_2^0$, $\frac{N}{4} + \frac{1}{2} \leq \sigma \in (\frac{N}{2} - 1, \frac{N}{2} + 1)$ and $0 \leq \gamma \in [\frac{N}{4} + \frac{1}{2} - \sigma, \sigma)$.*

Then, for each $u_0 \in X_2^\gamma$ there is a global strong γ-solution of (7.25) satisfying (7.26) and

$$\|u_t^\epsilon\|_{L^{\theta_0}(0,T;X_2^{-\frac{\sigma}{2}})} \leq C(T), \quad T > 0, \tag{7.28}$$

where $\theta_0 = \min\{\frac{1}{\eta}, 2\}$ and η is as in Lemma 7.3.2.

There is also a sequence $\epsilon_n \to 0$, and a function u such that, for any $T > 0$ and $\eta < 1$,

(i) $u^{\epsilon_n} \longrightarrow u$ *weakly in* $L^2(0, T; X_2^{\frac{1}{2}})$ *and* $*$*-weakly in* $L^\infty(0, T; X_2^0)$,

(ii) $u^{\epsilon_n} \longrightarrow u$ *in* $L^2(0, T; X_2^{\frac{\eta}{2}})$; *in particular* $u^{\epsilon_n} \to u$ *in* $L^2((0, T) \times \Omega)$,

(iii) $u^{\epsilon_n} \longrightarrow u$ *in* $X_2^{\frac{\eta}{2}}$ *e. a. in* $(0, T)$,

(iv) $u^{\epsilon_n} \longrightarrow u$ *e. a. in* $(0, T) \times \Omega$,

(v) $F_2(u^{\epsilon_n}) \longrightarrow F_2(u)$ *in* $X_2^{-\frac{\sigma}{2}}$ *e. a. in* $(0, T)$,

(vi) $F_2(u^{\epsilon_n}) \longrightarrow F_2(u)$ *in* $L^1(0, T; X_2^{-\frac{\sigma}{2}})$ *and there exists* $h \in L^1(0, T; \mathbb{R})$ *such that for each* $n \in \mathbb{N}$ $\|F_2(u^{\epsilon_n})\|_{X_2^{-\frac{\sigma}{2}}} \leq h$ *a. e. in* $(0, T)$,

(vii) $F_2(u^{\epsilon_n}) \longrightarrow F_2(u)$ *weakly in* $L^{\theta_0}(0, T; X_2^{-\frac{\sigma}{2}})$,

(viii) $u_t^{\epsilon_n} \longrightarrow u_t$ *weakly in* $L^{\theta_0}(0, T; X_2^{-\frac{\sigma}{2}})$.

In particular, for arbitrary $T > 0$ we have that

(a) $u \in L^\infty(0, T; X_2^0) \cap L^2(0, T; X_2^{\frac{1}{2}})$,

(b) $u_t \in L^{\theta_0}(0, T; X_2^{-\frac{\sigma}{2}})$,

(c) $u : [0, T] \to X_2^{-\frac{\sigma}{2}}$ *is absolutely continuous,*
(d) u *with values in* X_2^0 *is weakly continuous on* $(0, T)$.

Proof. For the existence of a strong y-solution u^{ϵ_n} and the uniform with respect to parameter estimate (7.26), see Remark 7.3.1.

From (7.25), we get $A_2^{-\frac{\sigma}{2}} u_t^\epsilon = -\epsilon A_2^{\frac{\sigma}{2}} u^\epsilon - A_2^{-\frac{\sigma}{2}+1} u^\epsilon + A_2^{-\frac{\sigma}{2}} F_2(u^\epsilon) + A_2^{-\frac{\sigma}{2}} P_2 f$, which together with part (ii) of Lemma 7.3.2 and (7.26) gives (7.28).

Parts (i)–(ii) and (viii) are due to the well known compactness criteria (see, e. g., [124]). Then (iii)–(iv) follow from (ii). Since, due to Lemma 7.3.2(ii) and (7.26), $F(u^{\epsilon_n})$ is bounded in $L^{\frac{1}{\eta}}(0, T; X_2^{-\frac{\sigma}{2}})$, part (vii) follows in a similar manner.

Parts (v)–(vi) are consequences of (iii), part (i) of Lemma 7.3.2 and (7.26) (see [18, Theorem IV.9]).

Using properties of the weak limit, we obtain from above that $u \in W^{1,\theta_0}(0, T; X_2^{-\frac{\sigma}{2}}) \cap L^\infty(0, T; X_2^0) \cap L^2(0, T; X_2^{\frac{1}{2}})$. In particular, (a), (b), and (c) hold. Finally, due to parts (a) and (c), function u with values in X_2^0 is weakly continuous (see [162, Theorem 2.1]). □

Note that the Cauchy integral formula associated with (7.25):

$$u^\epsilon(t) = e^{-(\epsilon A_2^\sigma + A_2)t} u_0 + \int_0^t e^{-(\epsilon A_2^\sigma + A_2)(t-s)} (F_2(u^\epsilon(s)) + P_2 f) ds. \tag{7.29}$$

Hence the passage to the limit can be carried out both in (7.25) and in (7.29), which actually can be merged via [12, Theorem, p. 371]. In either case, the resulting solution of (7.2), although global in time, is potentially not unique, since it may depend on the chosen subsequence.

In the following two subsections, we will describe two possible ways of passing to the limit, $\epsilon \to 0$, in the approximating equations (7.25); passing to the limit in a weakly formulated equation, or passing to the limit in the Cauchy integral formula.

7.3.1.1 Passing to the limit in (7.25)

We consider a sequence of approximating problems

$$u_t^{\epsilon_n} = -\epsilon_n A_2^\sigma u^{\epsilon_n} - A_2 u^{\epsilon_n} + F_2(u^{\epsilon_n}) + P_2 f, \qquad u(0) = u_0 \in X_2^0, \tag{7.30}$$

where $\sigma \geq \frac{N}{4} + \frac{1}{2}$ is fixed and $\epsilon_n \to 0$ is as in Lemma 7.3.3.

Theorem 7.3.4. *Under the assumptions and notation of Lemma 7.3.3, for the limit function* u *defined therein and any* $v \in X_2^{\frac{\sigma}{2}}$, *the equality*

$$\frac{d}{dt}\langle u, v\rangle = -\langle A_2^{\frac{1}{2}} u, A_2^{\frac{1}{2}} v\rangle + \langle A_2^{-\frac{\sigma}{2}} F_2(u), A_2^{\frac{\sigma}{2}} v\rangle + \langle P_2 f, v\rangle$$

holds in the sense of scalar distributions, where $\langle\cdot,\cdot\rangle$ *is the scalar product in* X_2^0 *and* $\frac{d}{dt}$ *above stands for the distributional derivative.*

Proof. Due to (7.30), for arbitrary $v \in X_2^{\frac{\sigma}{2}}$ we have

$$\langle u_t^{\epsilon_n}, v\rangle = -\langle A_2^{\frac{1}{2}} u^{\epsilon_n}, A_2^{\frac{1}{2}} v\rangle - \epsilon_n \langle A_2^{\frac{\sigma}{2}} u^{\epsilon_n}, A_2^{\frac{\sigma}{2}} v\rangle + \langle A_2^{-\frac{\sigma}{2}} F_2(u^{\epsilon_n}), A_2^{\frac{\sigma}{2}} v\rangle + \langle P_2 f, v\rangle. \tag{7.31}$$

On the left-hand side in (7.31), using first regularity of u^{ϵ_n} and then Lemma 7.3.3(ii), for each smooth scalar test function $\chi \in D((0,T))$ we get

$$\int_0^T \langle u_t^{\epsilon_n}, v\rangle \chi dt = \int_0^T \frac{d}{dt}\langle u^{\epsilon_n}, v\rangle \chi dt = -\int_0^T \langle u^{\epsilon_n}, v\rangle \chi' dt \longrightarrow -\int_0^T \langle u, v\rangle \chi' dt.$$

On the right-hand side in (7.31), using first part (i) and then part (vi) of Lemma 7.3.3, we have

$$\int_0^T \langle A_2^{\frac{1}{2}} u^{\epsilon_n}, A_2^{\frac{1}{2}} v\rangle \chi dt \longrightarrow \int_0^T \langle A_2^{\frac{1}{2}} u, A_2^{\frac{1}{2}} v\rangle \chi dt$$

and

$$\int_0^T \langle A_2^{-\frac{\sigma}{2}} F_2(u^{\epsilon_n}), A_2^{\frac{\sigma}{2}} v\rangle \chi dt \longrightarrow \int_0^T \langle A_2^{-\frac{\sigma}{2}} F_2(u), A_2^{\frac{\sigma}{2}} v\rangle \chi dt.$$

Thanks to (7.26), the third term on the right-hand side in (7.31) vanishes in the limit:

$$\epsilon_n \int_0^T \langle A_2^{\frac{\sigma}{2}} u^{\epsilon_n}, A_2^{\frac{\sigma}{2}} v\rangle \chi dt \le \sqrt{\epsilon_n} \|\chi\|_{L^\infty(0,T)} \|\sqrt{\epsilon_n} u^{\epsilon_n}\|_{L^2((0,T),X_2^{\frac{\sigma}{2}})} \|v\|_{L^2((0,T),X_2^{\frac{\sigma}{2}})} \to 0.$$

Hence the equality

$$-\int_0^T \langle u, v\rangle \chi' dt = \int_0^T (\langle A_2^{\frac{1}{2}} u, A_2^{\frac{1}{2}} v\rangle + \langle A_2^{-\frac{\sigma}{2}} F_2(u), A_2^{\frac{\sigma}{2}} v\rangle + \langle P_2 f, v\rangle)\chi dt,$$

holds for each scalar test function χ, which gives the result. □

7.3.1.2 Passing to the limit in Cauchy integral formula

We consider now Cauchy's formula

$$u^{\epsilon_n}(t) = e^{-(\epsilon_n A_2^{\sigma} + A_2)t} u_0 + \int_0^t e^{-(\epsilon_n A_2^{\sigma} + A_2)(t-s)} (F_2(u^{\epsilon_n}(s)) + P_2 f) ds, \tag{7.32}$$

where $u_0 \in X_2^0$ and $\epsilon_n \to 0$ is as in Lemma 7.3.3.

Remark 7.3.5.

(i) As in [138, p. 85], we will use the notation $A \in G(M, \omega)$ for an operator A which is the infinitesimal generator of a C_0 semigroup $T(t)$ satisfying $\|T(t)\| \le Me^{\omega t}$. Notifying that $\epsilon A_2^\sigma + A_2$, $\epsilon \ge 0$, are positive definite self-adjoint operators with compact resolvent in a Hilbert space X_2^0, and bottom of their spectra is the first positive eigenvalue of A_2, then the operator $\epsilon A_2^\sigma + A_2 \in G(M, \omega)$. In particular, the $\|e^{-(\epsilon A_2^\sigma + A_2)t}\|_{\mathcal{L}(X_2^0)}$-norms are bounded by the same constant, independent of $\epsilon > 0$ and $t \ge 0$. Applying [138, Chapter 3, Theorem 4.5], we get

$$e^{-(\epsilon A_2^\sigma + A_2)t} u_0 \xrightarrow{X_2^0} e^{-A_2 t} u_0 \quad \text{as } \epsilon \to 0^+ \text{ for all } t \ge 0,\ u_0 \in X_2^0.$$

(ii) Due to [6, Theorem 2.1.3 and Corollary 2.1.4], the semigroups $\{e^{-(\epsilon A_2^\sigma + A_2)t}\}$ (not changing the notation) will be extended to analytic C^0 semigroups on spaces belonging to a negative part of the scale, thus the operators $\epsilon A_2^\sigma + A_2$ are still of the class $G(M, \omega)$. In particular, the $\|e^{-(\epsilon A_2^\sigma + A_2)t}\|_{\mathcal{L}(X_2^{-\frac{\sigma}{2}})}$-norms are bounded by the same constant, independent of $\epsilon > 0$ and $t \ge 0$, and applying [138, Chapter 3, Theorem 4.5], we get

$$e^{-(\epsilon A_2^\sigma + A_2)t} u_0 \xrightarrow{X_2^{-\frac{\sigma}{2}}} e^{-A_2 t} u_0 \quad \text{as } \epsilon \to 0^+ \text{ for all } t \ge 0,\ u_0 \in X_2^{-\frac{\sigma}{2}}. \tag{7.33}$$

Now we show that the limit function u satisfies Cauchy's formula associated with (7.2).

Theorem 7.3.6. *With the assumptions and notation of Lemma 7.3.3, the limit function u defined therein satisfies the equality*

$$u(t) = e^{-A_2 t} u_0 + \int_0^t e^{-A_2(t-s)} (F_2(u(s)) + P_2 f) ds, \tag{7.34}$$

in $X_2^{-\frac{\sigma}{2}}$ and in each time interval $[0, T]$.

In addition, $u(t) \to u_0$ in $X_2^{-\frac{\sigma}{2}}$ as $t \to 0^+$ and $u(0) = u_0$.

Proof. Given $t > 0$, we consider

$$\begin{aligned} J^{\epsilon_n}(t) &= \int_0^t e^{-(\epsilon_n A_2^\sigma + A_2)(t-s)} (F_2(u^{\epsilon_n}(s)) + P_2 f) ds - \int_0^t e^{-A_2(t-s)} (F_2(u(s)) + P_2 f) ds \\ &= \int_0^t (e^{-(\epsilon_n A_2^\sigma + A_2)(t-s)} - e^{-A_2(t-s)}) (F_2(u^{\epsilon_n}(s)) + P_2 f) ds \\ &\quad + \int_0^t e^{-A_2(t-s)} (F_2(u^{\epsilon_n}(s)) - F_2(u(s))) ds =: J_1^{\epsilon_n}(t) + J_2^{\epsilon_n}(t). \end{aligned}$$

Using Remark 7.3.5 and parts (v)–(vi) of Lemma 7.3.3, we apply Lebesgue's dominated convergence theorem to get

$$\|J_1^{\epsilon_n}(t)\|_{X^{-\frac{\sigma}{2}}} \le \int_0^t \|(e^{-(\epsilon_n A_2^\sigma + A_2)(t-s)} - e^{-A_2(t-s)})(F_2(u^{\epsilon_n}(s)) + P_2 f)\|_{X_2^{-\frac{\sigma}{2}}} ds \to 0$$

and

$$\|J_2^{\epsilon_n}(t)\|_{X^{-\frac{\sigma}{2}}} \le \int_0^t \|e^{-A_2(t-s)}(F_2(u^{\epsilon_n}(s)) - F_2(u(s)))\|_{X_2^{-\frac{\sigma}{2}}} ds \to 0.$$

Consequently, $\|J^{\epsilon_n}(t)\|_{X_2^{-\frac{\sigma}{2}}} \to 0$. Using this and (7.33), we observe that for every $t > 0$, the right-hand side of (7.32) converges to the right-hand side of (7.34) in $X_2^{-\frac{\sigma}{2}}$.

Lemma 7.3.3(iii) ensures that, a.e in $[0, T]$, the left-hand side of (7.29) converges to $u(t)$ in $X_2^{-\frac{\sigma}{2}}$. Since u is continuous from $[0, T]$ to $X_2^{-\frac{\sigma}{2}}$ (see Lemma 7.3.3) and the right-hand side of (7.34) is also continuous from $[0, T]$ to $X_2^{-\frac{\sigma}{2}}$, the equality (7.34) holds on $[0, T]$.

We remark that, due to Lemma 7.3.3, $F_2(u) \in L^1(0, T; X_2^{-\frac{\sigma}{2}})$. Hence

$$\left\| \int_0^t e^{-A_2(t-s)}(F_2(u(s)) + P_2 f) ds \right\|_{X_2^{-\frac{\sigma}{2}}} \to 0 \quad \text{as } t \to 0^+$$

and from (7.34) we conclude that $u(t) \to u_0$ in $X_2^{-\frac{\sigma}{2}}$ as $t \to 0^+$. In particular, recalling Lemma 7.3.3 c), $u(0) = u_0$. □

7.4 Regularization of global solutions of the 3-D Navier–Stokes equations

Adapting to the case of bounded domain formulation of an open problem in [60, (A), p. 2], we will prove the following regularization result. Its proof will be given in a number of lemmas.

Theorem 7.4.1. *If $N = 3$, $f \equiv 0$, and $u_0 \in X_2^0$, then there exists a function u such that $u(0) = u_0$, for each $\tau > 0$,*

$$u \in L^\infty(0, \tau; X_2^0) \cap L^2(0, \tau; X_2^{\frac{1}{2}})$$

and u solves

$$u(t) = e^{-A_2 t} u_0 + \int_0^t e^{-A_2(t-s)} F_2(u(s)) ds \quad \text{in } [0, \tau].$$

In addition, u is weakly continuous on $(0,\infty)$ with values in X_2^0, and satisfies the estimate:

$$\|u(t)\|_{X_2^0} \le e^{-\lambda_1 t}\|u_0\|_{X_2^0}, \quad t \ge 0.$$

Furthermore, after a certain time $\mathcal{T} = \mathcal{T}(\|u_0\|_{X_2^0})$, the norms $\|u(t)\|_{X_2^0}$ and $\|u(t)\|_{X_2^{\frac{1}{2}}}$ are nonincreasing,

$$u \in C^1((\mathcal{T},\infty); X_2^0) \cap C((\mathcal{T},\infty); X_2^1)$$

and the equality

$$u_t + A_2 u = F_2(u)$$

holds in X_2^0 for every $t > \mathcal{T}$.

Actually, $u \in C^\infty((\mathcal{T},\infty) \times \bar{\Omega})$.

We formulate the lemmas forming the proof of Theorem 7.4.1.

Lemma 7.4.2. *Assume that $N = 3$ and $f \equiv 0$. Then, given $\frac{5}{2} \ge \sigma \ge \frac{5}{4}$, $0 \le \gamma \in [\frac{5}{4} - \sigma, \sigma)$ and $u_0 \in X_2^\gamma$, there exist a function u and a sequence $\{u^{\epsilon_n}\}$ of the global strong γ-solutions u^{ϵ_n} to*

$$u_t^{\epsilon_n} = -\epsilon_n A_2^\sigma u^{\epsilon_n} - A_2 u^{\epsilon_n} + F_2(u^{\epsilon_n}), \qquad u(0) = u_0, \tag{7.35}$$

such that for each $T > 0$, parts (i)–(viii) and (a)–(d) of Lemma 7.3.3 hold true.

Furthermore, $u(0) = u_0$ and

(I) *for any $v \in X_2^{-\frac{\sigma}{2}}$, u satisfies $\frac{d}{dt}\langle u, v\rangle = -\langle A_2^{\frac{1}{2}} u, A_2^{\frac{1}{2}} v\rangle + \langle A_2^{-\frac{\sigma}{2}} F_2(u), A_2^{\frac{\sigma}{2}} v\rangle$ in the sense of scalar distributions, where $\langle\cdot,\cdot\rangle$ denotes the scalar product in X_2^0;*

(II) *u satisfies the equation $u(t) = e^{-A_2 t}u_0 + \int_0^t e^{-A_2(t-s)} F_2(u(s))ds$ in $X_2^{-\frac{\sigma}{2}}$ and in $[0,T]$.*

Proof. The result is a consequence of Lemma 7.3.3 and Theorems 7.3.4, 7.3.6. □

Lemma 7.4.3. *Under the assumptions and notation of Lemma 7.4.2, for each $n \in \mathbb{N}$ and $t \ge 0$, the norm $\|u^{\epsilon_n}(t)\|_{X_2^0}$ is nonincreasing and*

$$\|u^{\epsilon_n}(t)\|_{X_2^0} \le e^{-\lambda_1 t}\|u_0\|_{X_2^0}, \tag{7.36}$$

where $\lambda_1 > 0$ is the first eigenvalue of A_2.

Proof. Multiplying (7.35) by u^{ϵ_n} in X_2^0, we obtain

$$\frac{1}{2}\frac{d}{dt}\|u^{\epsilon_n}\|_{X_2^0}^2 + \|u^{\epsilon_n}\|_{X_2^{\frac{1}{2}}}^2 \le 0. \tag{7.37}$$

Hence $\|u^{\epsilon_n}\|_{X_2^0}^2$ is nonincreasing for $t \ge 0$ and after taking square roots we get (7.36). □

Lemma 7.4.4. *Under the assumptions and notation of Lemma 7.4.2, there exists a certain time $\mathcal{T} \geq 1$, independent of $n \in \mathbb{N}$, such that for $t \geq \mathcal{T}$, each norm $\|u^{\epsilon_n}(t)\|_{X_2^{\frac{1}{2}}}$ is nonincreasing and*

$$\|u^{\epsilon_n}(t)\|_{X_2^{\frac{1}{2}}} \leq \|u_0\|_{X_2^0} e^{-\lambda_1(t-1)}, \quad \text{for } t \geq \mathcal{T}, \tag{7.38}$$

where $\mathcal{T}$ is independent of $n \in \mathbb{N}$, and depends only on the constant M in Proposition 7.2.2, constants in the interpolation and embedding inequalities, and the $\|u_0\|_{X_2^0}$-norm. In particular,

$$\mathcal{T} \to \infty \quad as\ \|u_0\|_{X_2^0} \to \infty. \tag{7.39}$$

Proof. Coming back to (7.37), we infer that

$$\left(\int_\tau^{\tau+1} \|u^{\epsilon_n}(s)\|^2_{X_2^{\frac{1}{2}}} ds\right)^{\frac{1}{2}} \leq \frac{1}{\sqrt{2}} \|u^{\epsilon_n}(\tau)\|_{X_2^0} \leq \|u_0\|_{X_2^0} e^{-\lambda_1 \tau} \quad \text{for each } \tau \geq 0. \tag{7.40}$$

We now multiply (7.35) in X_2^0 by $A_2 u^{\epsilon_n}$ to get

$$\frac{1}{2}\frac{d}{dt}\|u^{\epsilon_n}\|^2_{X_2^{\frac{1}{2}}} \leq -\|u^{\epsilon_n}\|^2_{X_2^1} + \langle A_2 u^{\epsilon_n}, F_2(u^{\epsilon_n})\rangle. \tag{7.41}$$

With the aid of Proposition 7.2.2 (used here with $\beta = 0$ and $\alpha = \frac{5}{8}$) and then with the interpolation inequality (see [32, Proposition 1.3.9] and [74, Theorem 1]), the last term in (7.41) is estimated as follows:

$$\begin{aligned}\langle A_2 u^{\epsilon_n}, F_2(u^{\epsilon_n})\rangle &\leq \|u^{\epsilon_n}\|_{X_2^1} \|F_2(u^{\epsilon_n})\|_{X_2^0} \leq M \|u^{\epsilon_n}\|_{X_2^1} \|u^{\epsilon_n}\|^2_{X_2^{\frac{5}{8}}} \\ &\leq M \|u^{\epsilon_n}\|_{X_2^1} \Big(c \|u^{\epsilon_n}\|^{\frac{1}{4}}_{X_2^1} \|u^{\epsilon_n}\|^{\frac{3}{4}}_{X_2^{\frac{1}{2}}}\Big)^2 = c^2 M \|u^{\epsilon_n}\|^{\frac{3}{2}}_{X_2^1} \|u^{\epsilon_n}\|^{\frac{3}{2}}_{X_2^{\frac{1}{2}}}.\end{aligned} \tag{7.42}$$

Inserting (7.42) into (7.41), using Young's inequality and the embedding $X_2^1 \subset X_2^{\frac{1}{2}}$, we have

$$\frac{d}{dt}\|u^{\epsilon_n}\|^2_{X_2^{\frac{1}{2}}} \leq -\|u^{\epsilon_n}\|^2_{X_2^1} + C\|u^{\epsilon_n}\|^6_{X_2^{\frac{1}{2}}} \leq -\lambda \|u^{\epsilon_n}\|^2_{X_2^{\frac{1}{2}}} + C\|u^{\epsilon_n}\|^6_{X_2^{\frac{1}{2}}} \tag{7.43}$$

for some constants $C > 0$, $\lambda > 0$, independent of $n \in \mathbb{N}$. Therefore, $y(t) := \|u^{\epsilon_n}(t)\|^2_{X_2^{\frac{1}{2}}}$ satisfies

$$y' \leq Cy\left(y^2 - \frac{\lambda}{C}\right). \tag{7.44}$$

If $\|u_0\|_{X_2^0} > \sqrt{\frac{\lambda}{C}}$, let τ_0 be the positive time such that

$$\|u_0\|_{X_2^0} e^{-\lambda_1 \tau_0} = \sqrt{\frac{\lambda}{C}}. \tag{7.45}$$

If $\|u_0\|_{X_2^0} \le \sqrt{\frac{\lambda}{C}}$, we take $\tau_0 = 0$.

Observe from (7.40) that, for each $\tau > \tau_0$ and any $n \in \mathbb{N}$, there exists time $\tau_n \in [\tau, \tau+1]$ (obtained using the mean value theorem for integral) such that

$$\|u^{\epsilon_n}(\tau_n)\|_{X_2^{\frac{1}{2}}} = \left(\int_\tau^{\tau+1} \|u^{\epsilon_n}(s)\|^2_{X_2^{\frac{1}{2}}} ds \right)^{\frac{1}{2}} \le \|u_0\|_{X_2^0} e^{-\lambda_1 \tau} < \sqrt{\frac{\lambda}{C}}. \tag{7.46}$$

Due to (7.44), $y(t) = \|u^{\epsilon_n}(t)\|^2_{X_2^{\frac{1}{2}}}$ is nonincreasing for $t \ge \tau_0 + 1$ and then it follows from (7.46) that

$$\|u^{\epsilon_n}(\tau+1)\|_{X_2^{\frac{1}{2}}} \le \|u^{\epsilon_n}(\tau_n)\|_{X_2^{\frac{1}{2}}} \le \|u_0\|_{X_2^0} e^{-\lambda_1 \tau}.$$

Hence, letting $\mathcal{T} := \tau_0 + 1$ and $t := \tau + 1$, we get (7.38). Finally, (7.39) follows from (7.45). □

Lemma 7.4.5. *For u introduced in Lemma 7.4.2, there is a time $\mathcal{T} = \mathcal{T}(\|u_0\|_{X_2^0})$, as in Lemma 7.4.4, such that*

$$u \in C^1((\mathcal{T}, \infty); X_2^0) \cap C((\mathcal{T}, \infty); X_2^1), \tag{7.47}$$

and the equation

$$u_t + A_2 u = F_2(u) \quad \text{for each } t > \mathcal{T}, \tag{7.48}$$

holds in X_2^0.

Actually, we have

$$u \in C^\infty((\mathcal{T}, \infty) \times \bar{\Omega}). \tag{7.49}$$

Proof. Let $\mathcal{T}$ be the time in Lemma 7.4.4. Then the limit u of $\{u^{\epsilon_n}\}$ in Lemma 7.4.2 can also be viewed as $*$-weak limit in $L^\infty(\mathcal{T}, \infty; X_2^{\frac{1}{2}})$. In particular, we have

$$\|u\|_{L^\infty(\mathcal{T}, \infty; X_2^{\frac{1}{2}})} \le \|u_0\|_{X_2^0} e^{-\lambda_1(\mathcal{T}-1)} \tag{7.50}$$

and we can choose $t_0 \ge \mathcal{T}$, arbitrarily close to $\mathcal{T}$, such that $u(t_0) \in X_2^{\frac{1}{2}}$.

Since u satisfies (*II*) of Lemma 7.4.2, $v(t) := u(t + t_0)$ satisfies

$$v(t) = e^{-A_2 t} v_0 + \int_0^t e^{-A_2(t-s)} F_2(v(s)) ds, \quad t > 0, \tag{7.51}$$

with $v(0) = u(t_0) := v_0 \in X_2^{\frac{1}{2}}$. By Definition 7.2.1 (letting therein $r = 2$, $f = 0$, $\gamma = \alpha = \frac{1}{2}$, $\beta = -\frac{1}{4}$ and $\sigma = 1$), we observe that v is a mild $\frac{1}{2}$-solution of

$$v_t + A_2 v = F_2(v), \quad t > 0, \qquad v(0) = v_0. \tag{7.52}$$

Using Theorem 7.2.8, we conclude that v is a strong $\frac{1}{2}$-solution of (7.52) (see Definition 7.2.7 with $r = 2, f = 0, \gamma = \frac{1}{2}$ and $\sigma = 1$). Hence we get (7.47) and (7.48).

Applying finally [75, Theorem 3.4], we obtain that $v \in C^{\infty}((0,T] \times \bar{\Omega}), T > 0$. □

Lemma 7.4.6. *Under the assumptions and notation of Lemma 7.4.2, for the limit function u therein we have:*

(i) $\|u(t)\|_{X_2^0} \le e^{-\lambda_1 t}\|u_0\|_{X_2^0}$ *for each* $t \ge 0$,

(ii) $\|u(t)\|_{X_2^0}$ *and* $\|u(t)\|_{X_2^{\frac{1}{2}}}$ *are nonincreasing after time* $\mathcal{T}$ *as in Lemma 7.4.4.*

Proof. Combining (7.36) with part (iii) of Lemma 7.3.3, we obtain (i) for $t \in [0,\infty) \setminus E$, where E is a certain set of measure zero. Now, if $t_0 \in E$ then there is a sequence $\{t_n\} \subset [0,\infty) \setminus E$ such that $t_n \to t_0$ and $\|u(t_n)\|_{X_2^0} \le e^{-\lambda_1 t_n}\|u_0\|_{X_2^0}$. Due to part (c) of Lemma 7.3.3, $u(t_n) \to u(t_0)$ in $X_2^{-\frac{\sigma}{2}}$. On the other hand, since $\{u(t_n)\}$ is bounded in X_2^0, there is a subsequence $\{u(t_{n_k})\}$ convergent weakly in X_2^0 to a certain $z \in X_2^0$. Hence $z = u(t_0)$ and $\|u(t_0)\|_{X_2^0} \le \liminf \|u(t_{n_k})\|_{X_2^0} \le e^{-\lambda_1 t_0}\|u_0\|_{X_2^0}$, which completes the proof of part (i).

Given $t_0 > \mathcal{T}$, function $v(t) = u(t + t_0)$ satisfies (7.52) and the proof of part (ii) follows as the proofs of Lemmas 7.4.3 and 7.4.4. □

Due to Lemmas 7.4.2–7.4.6, Theorem 7.4.1 is proved.

7.5 Closing remarks

Theorems 7.2.9, 7.2.10, and 7.4.1 together imply that, given $\gamma \in [\frac{1}{4}, 1)$ and $u_0 \in X_2^{\gamma}$, the problem (7.2) has a unique strong γ-solution u existing until τ_{u_0} and either $\tau_{u_0} = \infty$ or, otherwise, u has (a potentially not unique) extension onto $[0,\infty)$, which becomes again a strong solution after time $\mathcal{T} = \mathcal{T}(\|u\|_{X_2^0})$. Hence, if $N = 3$ and $f \equiv 0$, then for sufficiently smooth initial data the lack of regularity of the solution to (7.2) can actually occur only for $t \in [\tau_{u_0}, \mathcal{T}(\|u\|_{X_2^0})]$. Note that for small regular data this interval is empty.

Bibliographical notes

The literature devoted to the Navier–Stokes equation is huge indeed. Therefore, we limit ourself to the results important in the preparation of this book. Evidently, such a role was played by the monograph [124] by J.-L. Lions in which the 3-D Navier–Stokes problem was solved in weak formulation, and also the critical exponents (1.11) were found in Remarque 6.11. We further recall the results of W. von Wahl [175, 176] in which the role of *subordination condition* was explained.

The semigroup approach to the original Navier–Stokes problem using fractional powers of operators was studied initially by P. E. Sobolevskii (see [108, 155, 156]), and by Japan mathematicians (see [190] for older references) and extended mostly by them to the present form. We are using two of the important results of Y. Giga and T. Miyakawa to treat the N-D Navier–Stokes problem inside the approach of this monograph. The first one is the description of the action of the nonlinearity in (7.3) on the fractional scale of spaces associated with the Stokes operator. Such information was obtained in [75]; see also [67, 92, 95, 155] for the earlier contributions. We are using further essentially the properties of linear analytic semigroup generated by the Stokes operator and description of the associated fractional power spaces as reported in [73–75].

8 Cauchy's problem for 2-D quasi-geostrophic equation

8.1 Introduction

The dissipative quasi-geostrophic equation considered here has the form:

$$\begin{aligned} &\theta_t + u \cdot \nabla\theta + \kappa(-\Delta)^\alpha \theta = f, \quad x \in \mathbb{R}^2, t > 0, \\ &\theta(0,x) = \theta_0(x), \end{aligned} \tag{8.1}$$

where θ represents the potential temperature, $\kappa > 0$ is a diffusivity coefficient, $\alpha \in [\frac{1}{2}, 1]$ is a fractional exponent, and $u = (u_1, u_2)$ is the *velocity field* determined by θ through the relation:

$$u = \left(-\frac{\partial\psi}{\partial x_2}, \frac{\partial\psi}{\partial x_1}\right), \quad \text{where } (-\Delta)^{\frac{1}{2}}\psi = -\theta, \tag{8.2}$$

or, in a more explicit way,

$$u = (-R_2\theta, R_1\theta), \tag{8.3}$$

where R_i, $i = 1, 2$ are the *Riesz transforms.*

This chapter is devoted to the global in time solvability and properties of solutions to the Cauchy problem (8.1). A large literature was devoted to that problem through the last 20 years; compare [39, 40, 59, 21, 71, 101, 119, 174, 183, 184] for more references. The basic approach was to obtain a *weak solution* to (8.1) using the *viscosity technique* (e. g., [124]), which means adding the viscosity term $\epsilon\Delta\theta$ to the right-hand side of the equation, solving the regularized problem and letting $\epsilon \to 0^+$. Our approach is different. We consider first a *family of subcritical problems* (8.1) with $\alpha \in (\frac{1}{2}, 1]$, which can be treated in the framework of [32, 86] as semilinear equations with sectorial operator. Thanks to a maximum principle valid for (8.1) (Lemma 8.2.4), we have a *uniform in* $\alpha \in (\frac{1}{2}, 1]$ estimates of solutions to that subcritical problems in $L^p(\mathbb{R}^2)$, $1 \le p \le +\infty$. Letting $\alpha \to \frac{1}{2}^+$ over a sequence of regular $H^{2\alpha^- + s}(\mathbb{R}^2)$, $s > 1$, solutions θ^α to (8.1), this property allows us to introduce in Theorem 8.6.2 a "*weak L^p solution*" of the limiting *critical problem* (8.1). Our considerations relate most closely to the J. Wu papers [183, 184], using however another, semigroup approach of the monographs [32, 86].

8.1.1 Description of the results

The quasi-geostrophic equation (8.1) is a challenging problem to study. Many papers devoted to it were published very recently; some of them are listed in the references; see anyway [101, 102, 117, 188].

https://doi.org/10.1515/9783110599831-008

There are different possible choices of the phase space for that problem. Following the considerations of J. Wu [183, 184], we choose $L^p(\mathbb{R}^2)$ with p sufficiently large, or $H^s(\mathbb{R}^2)$ with $s > 1$, as the *base spaces* (in which the equation is fulfilled) for (8.1); see also [101]. Our aim is to include, in a subcritical case of exponents $\alpha \in (\frac{1}{2}, 1]$, the problem (8.1) into the frame of semilinear parabolic equations with sectorial operator. This offers a simple but formalized proof of the local solvability and regularity in subcritical case. The presented approach in Section 8.6 to critical nonlinearity is new here. However, using weak compactness of bounded sets in the reflexive Banach space as a tool for getting convergence to a weak solution of the critical problem, we will not be able to show that the limit of the nonlinearities of subcritical problems equals to nonlinearity of the limiting critical problem. The existing uniform in $\alpha \in (\frac{1}{2}, 1]$ a priori estimates are too weak to guarantee such property. However, they work well in the case of all the linear components in the equation. The main result obtained in that direction is formulated in Theorem 8.6.2, where we introduce the notion of the *weak L^p solution* to the critical equation (8.1). It is followed by four technical observations used in the proof of that theorem; the main technical result there seems to be Lemma 5.3.3, vital when letting $\alpha \to \frac{1}{2}^+$ in (8.1).

Another result reported in this chapter is the existence of a global attractor for a variant of the subcritical problem (with added linear damping term $\lambda\theta$, required in the case of $\mathbb{R}^2$) in the phase space $W^{2\alpha^-,p}(\mathbb{R}^2)$, where p is given by (8.7). In the proof of asymptotic compactness, the reasoning is based on the *tail estimates technique*; [178] is a source reference for that technique.

When needed for clarity of the presentation, we mark the dependence of the solution θ of (8.1) on $\alpha \in (\frac{1}{2}, 1]$, calling it θ^α. Sometimes we do not mark explicitly the dependence of u on α, since u always stays next to θ^α. In the sequel, C denotes an arbitrary positive constant, which may be different from line to line and even in the same line.

8.2 Solvability of subcritical (8.1), $\alpha \in (\frac{1}{2}, 1]$, in $W^{2\alpha^-,p}(\mathbb{R}^2)$

8.2.1 Formulation of the problem and its local solvability

Our first task is the local in time solvability of the subcritical problem (8.1) when the equation is treated in the *base space* $X := L^p(\mathbb{R}^2)$. We will use a standard approach of this monograph. To work with a *sectorial positive* operator (see [86, 32]), we will rewrite (8.1) in an equivalent form:

$$\begin{aligned} &\theta_t + u \cdot \nabla\theta + \kappa(-\Delta)^\alpha\theta + \kappa\theta = f + \kappa\theta, \quad x \in \mathbb{R}^2, t > 0, \\ &\theta(0, x) = \theta_0(x), \end{aligned} \tag{8.4}$$

adding to both sides the term $\kappa\theta$.

Define $A_\alpha := \kappa[(-\Delta)^\alpha + I]$, $\alpha \in (\frac{1}{2}, 1]$, where $(-\Delta)^\alpha$ is the fractional Laplacian. Also, setting

$$F(\theta) = R_2\theta\frac{\partial\theta}{\partial x_1} - R_1\theta\frac{\partial\theta}{\partial x_2} + f + \kappa\theta, \tag{8.5}$$

the problem (8.4) will be written formally as

$$\begin{aligned} &\theta_t + A_\alpha\theta = F(\theta), \quad t > 0, \\ &\theta(0) = \theta_0, \end{aligned} \tag{8.6}$$

which is an abstract "parabolic" equation with sectorial positive operator. To assure that the nonlinearity F is bounded and Lipschitz continuous on bounded sets as a map from the *phase space* $X^{2\alpha^-} := W^{2\alpha^-,p}(\mathbb{R}^2)$ to X (X^β-domain of the β fractional power of the sectorial operator A_1), we need to take a sufficiently large value of p. More precisely, we assume the following condition known, for example, from the references [174, 183]:

$$2\alpha - \frac{2}{p} > 1. \tag{8.7}$$

Remark 8.2.1. Recall that following the standard notation of H. Triebel [173, Section 2.3.1] we have

$$W^{s,p}(\mathbb{R}^N) = \begin{cases} H^s_p(\mathbb{R}^N) & \text{when } s = 0, 1, 2, \dots, \\ B^s_{p,p}(\mathbb{R}^N) & \text{when } 0 < s \neq \text{natural number}, \end{cases}$$

whenever $1 < p < \infty$. In the latter formula, $H^s_p(\mathbb{R}^N)$, $s > 0$, stands for the space of *Bessel potentials* (or *Lebesgue space*), and $B^s_{p,p}(\mathbb{R}^N)$, $s > 0$, for the *inhomogeneous Besov spaces*. We also remark that the *homogeneous Besov spaces* $\dot{B}^s_{p,p}(\mathbb{R}^N)$ fulfill $B^s_{p,p}(\mathbb{R}^N) = \dot{B}^s_{p,p}(\mathbb{R}^N) \cap L^p(\mathbb{R}^N)$ when $s > 0$, with equivalent norms $\|\phi\|_{B^s_{p,p}(\mathbb{R}^N)} = \|\phi\|_{\dot{B}^s_{p,p}(\mathbb{R}^N)} + \|\phi\|_{L^p(\mathbb{R}^N)}$.

It follows next, from [130, Remark 12.3.2], that the spaces $D((-\Delta_p)^{\frac{\alpha}{2}})$ and $D((I - \Delta_p)^{\frac{\alpha}{2}})$ coincide whenever $1 \le p < \infty$ (the first one considered with the graph norm $\|(-\Delta_p)^{\frac{\alpha}{2}}(\cdot)\|_{L^p(\mathbb{R}^N)} + \|\cdot\|_{L^p(\mathbb{R}^N)}$) and are equivalent to the fractional Sobolev spaces $[L^p(\mathbb{R}^N), W^{m,p}(\mathbb{R}^N)]_{\frac{\alpha}{m}} = H^\alpha_p(\mathbb{R}^N)$, $\alpha > 0$, $\mathbb{N} \ni m > \alpha$ (see also [173, Section 2.5.3, (13)]). In what follows, we denote $H^\gamma(\mathbb{R}^N) := W^{\gamma,2}(\mathbb{R}^N)$ and use the notation:

$$\|\phi\|_{\dot{H}^\gamma(\mathbb{R}^N)} = \|(-\Delta)^{\frac{\gamma}{2}}\phi\|_{L^2(\mathbb{R}^N)}, \quad \phi \in H^\gamma(\mathbb{R}^N),$$

for the norm of the *homogeneous space* $\dot{H}^\gamma(\mathbb{R}^N)$ (note that $H^\gamma(\mathbb{R}^N)$ is a subspace of $\dot{H}^\gamma(\mathbb{R}^N)$).

Theorem 8.2.2. *Assume that $f \in L^p(\mathbb{R}^2)$, $\theta_0 \in W^{2\alpha^-,p}(\mathbb{R}^2)$ and the condition* (8.7) *holds. Then there exists a unique local in time mild solution $\theta(t)$ to the subcritical problem* (8.1) *considered on the phase space $W^{2\alpha^-,p}(\mathbb{R}^2)$. Moreover,*

$$\theta \in C((0,\tau); W^{2\alpha,p}(\mathbb{R}^2)) \cap C([0,\tau); W^{2\alpha^-,p}(\mathbb{R}^2)), \quad \theta_t \in C((0,\tau); W^{2\gamma,p}(\mathbb{R}^2)),$$

with arbitrary $\gamma < \alpha$. Here, $\tau > 0$ is the "life time" of that local in time solution. Moreover, the Cauchy formula is satisfied:

$$\theta(t) = e^{-A_\alpha t}\theta_0 + \int_0^t e^{-A_\alpha(t-s)}F(\theta(s))ds, \quad t \in [0,\tau),$$

where $e^{-A_\alpha t}$ denotes the linear semigroup corresponding to the operator $A_\alpha := \kappa[(-\Delta)^\alpha + I]$ in $L^p(\mathbb{R}^2)$.

Proof. To guarantee local solvability, we need the following Lipschitz condition (valid on bounded sets):

$$\forall_{r>0}\,\exists_{L(r)>0}\,\forall_{\theta_1,\theta_2\in B_{X^\beta}(r)}\ \|F(\theta_1) - F(\theta_2)\|_X \le L(r)\|\theta_1 - \theta_2\|_{X^\beta}, \tag{8.8}$$

where $B_{X^\beta}(r)$ denotes an open ball in X^β centered at zero of radius r. For $\theta_1, \theta_2 \in B_{X^\beta}(r)$, using (8.3) and setting $u_1 = (-R_2\theta_1, R_1\theta_1)$, $u_2 = (-R_2\theta_2, R_1\theta_2)$, we obtain

$$\begin{aligned}\|F(\theta_1) - F(\theta_2)\|_{L^p(\mathbb{R}^2)} &\le \kappa\|\theta_1 - \theta_2\|_{L^p(\mathbb{R}^2)} + \Big\|R_2(\theta_1 - \theta_2)\frac{\partial\theta_1}{\partial x_1} \\ &\quad + R_2\theta_2\frac{\partial(\theta_1 - \theta_2)}{\partial x_1}\Big\|_{L^p(\mathbb{R}^2)} + \Big\|R_1(\theta_1 - \theta_2)\frac{\partial\theta_1}{\partial x_2} + R_1\theta_2\frac{\partial(\theta_1 - \theta_2)}{\partial x_2}\Big\|_{L^p(\mathbb{R}^2)}.\end{aligned} \tag{8.9}$$

Next, using the Hölder inequality and (5.50), we estimate the second term above as follows:

$$\begin{aligned}\Big\|R_2(\theta_1 - \theta_2)\frac{\partial\theta_1}{\partial x_1}\Big\|_{L^p(\mathbb{R}^2)} &\le \|R_2(\theta_1 - \theta_2)\|_{L^{2p}(\mathbb{R}^2)}\Big\|\frac{\partial\theta_1}{\partial x_1}\Big\|_{L^{2p}(\mathbb{R}^2)} \\ &\le C\|\theta_1 - \theta_2\|_{L^{2p}(\mathbb{R}^2)}\|\theta_1\|_{W^{1,2p}(\mathbb{R}^2)}.\end{aligned} \tag{8.10}$$

Note that the embeddings $W^{2\alpha^-,p}(\mathbb{R}^2) \subset W^{1,2p}(\mathbb{R}^2) \subset L^{2p}(\mathbb{R}^2)$ are valid since the condition $2\alpha^- - \frac{1}{p} > 1$ is satisfied (compare (8.7)). Estimating the other components in (8.9) analogously we conclude

$$\|F(\theta_1) - F(\theta_2)\|_{L^p(\mathbb{R}^2)} \le \mathrm{const}(\|\theta_1\|_{W^{2\alpha^-,p}(\mathbb{R}^2)}, \|\theta_2\|_{W^{2\alpha^-,p}(\mathbb{R}^2)})\|\theta_1 - \theta_2\|_{W^{2\alpha^-,p}(\mathbb{R}^2)}, \tag{8.11}$$

which is the required local Lipschitz condition. □

Remark 8.2.3. It is seen from the estimate (8.10) that the quasi-geostrophic equation (8.1) falls into the class of equations with *quadratic nonlinearity*. This is typical also for other equations originated in fluid dynamics, such as the Burgers equation (see, e. g., [16, 32]) or the celebrated Navier–Stokes equation where a significant part of the considerations is devoted (see, e. g., [171, Chapter III]) to the studies of the *bilinear* or *trilinear forms* connected with the nonlinearity there. In our problem, at the first view, the a priori estimate following from the maximum principle seems to be sufficient to assure the global in time solvability. But usually the detail considerations leading to such conclusion require more attention.

8.2.2 Global solvability

Having already the local in time solution of (8.1), $\alpha \in (\frac{1}{2}, 1]$, to guarantee its global extendibility we need to have suitable *a priori estimates*. Such role in case of the viscous quasi-geostrophic equation (8.1) is played by the maximum principle (see [40]).

Lemma 8.2.4. *Let $q \in [2, \infty)$ and $f \in L^q(\mathbb{R}^2)$. Then, for a sufficiently regular solution of (8.1), the following estimate holds:*

$$\|\theta(t,\cdot)\|^q_{L^q(\mathbb{R}^2)} \le e^{(q-1)t}\|\theta_0\|^q_{L^q(\mathbb{R}^2)} + te^{(q-1)t}\|f\|^q_{L^q(\mathbb{R}^2)}. \tag{8.12}$$

When $f = 0$, the corresponding estimate takes the form

$$\|\theta(t,\cdot)\|_{L^q(\mathbb{R}^2)} \le \|\theta_0\|_{L^q(\mathbb{R}^2)}. \tag{8.13}$$

Proof. Note that for smooth solutions after multiplying the nonlinear term $u \cdot \nabla\theta$ by $|\theta|^{q-1}\operatorname{sgn}(\theta)$ and integrating the result over $\mathbb{R}^2$ the resulting term will vanish:

$$\begin{aligned}
&\int_{\mathbb{R}^2} \left(\frac{\partial\theta}{\partial x_1}\frac{\partial}{\partial x_2}[(-\Delta)^{-\frac{1}{2}}\theta] - \frac{\partial\theta}{\partial x_2}\frac{\partial}{\partial x_1}[(-\Delta)^{-\frac{1}{2}}\theta] \right)|\theta|^{q-1}\operatorname{sgn}(\theta)dx \\
&\quad = \frac{1}{q}\int_{\mathbb{R}^2} \frac{\partial(|\theta|^q)}{\partial x_1}\frac{\partial}{\partial x_2}[(-\Delta)^{-\frac{1}{2}}\theta] - \frac{\partial(|\theta|^q)}{\partial x_2}\frac{\partial}{\partial x_1}[(-\Delta)^{-\frac{1}{2}}\theta]dx = 0,
\end{aligned}$$

thanks to integration by parts. Consequently, multiplying (8.1) by $|\theta|^{q-1}\operatorname{sgn}(\theta)$, we obtain

$$\int_{\mathbb{R}^2} \theta_t|\theta|^{q-1}\operatorname{sgn}(\theta)dx + \kappa\int_{\mathbb{R}^2}(-\Delta)^\alpha\theta|\theta|^{q-1}\operatorname{sgn}(\theta)dx = \int_{\mathbb{R}^2} f|\theta|^{q-1}\operatorname{sgn}(\theta)dx.$$

Since for all the values $1 \le q < \infty$ and $0 \le \alpha \le 1$, as was shown in [39] and [40, Lemma 2.5],

$$\int_{\mathbb{R}^2}(-\Delta)^\alpha\theta|\theta|^{q-1}\operatorname{sgn}(\theta)dx \ge 0,$$

using the Young inequality, we get

$$\frac{1}{q}\frac{d}{dt}\int_{\mathbb{R}^2}|\theta|^q dx \le \int_{\mathbb{R}^2} f|\theta|^{q-1}\operatorname{sgn}(\theta)dx \le \frac{1}{q}\|f\|^q_{L^q(\mathbb{R}^2)} + \frac{q-1}{q}\|\theta\|^q_{L^q(\mathbb{R}^2)}.$$

Solving the above differential inequality, we get (8.12) when $f \ne 0$ and (8.13) otherwise. □

Recall (see Chapter 4 or [32, pp. 72–73]) that to be able to extend globally in time the local $W^{2\alpha^-,p}(\mathbb{R}^2)$-solution constructed above, with bounded orbits of bounded sets, we need a subordination condition

$$\|F(\theta(t,\theta_0))\|_{L^p(\mathbb{R}^2)} \leq \text{const}(\|\theta(t,\theta_0)\|_Y)(1+\|\theta(t,\theta_0)\|^{\eta}_{W^{2\alpha^-,p}(\mathbb{R}^2)}), \quad t \in (0,\tau_{\theta_0}),$$

with certain auxiliary Banach space $W^{2\alpha,p}(\mathbb{R}^2) \subset Y$ and certain $\eta \in [0,1)$. The dependence of the solution on θ_0 is marked in the notation here.

We will choose $Y := L^p(\mathbb{R}^2) \cap L^{2p}(\mathbb{R}^2)$ for the problem (8.1). By the Hölder inequality and the Nirenberg–Gagliardo estimate (e. g., [32, p. 25]), we obtain

$$\begin{aligned}\|F(\theta)\|_{L^p(\mathbb{R}^2)} &\leq C\|\theta\|_{L^{2p}(\mathbb{R}^2)}\|\theta\|_{W^{1,2p}(\mathbb{R}^2)} + \|f\|_{L^p(\mathbb{R}^2)} + \kappa\|\theta\|_{L^p(\mathbb{R}^2)}\\ &\leq \text{const}(\|\theta\|_{L^p(\mathbb{R}^2)}, \|\theta\|_{L^{2p}(\mathbb{R}^2)})(1+\|\theta\|^{\eta}_{W^{2\alpha^-,p}(\mathbb{R}^2)}) + \|f\|_{L^p(\mathbb{R}^2)},\end{aligned} \tag{8.14}$$

where $\frac{1}{2\alpha^- - \frac{1}{p}} < \eta < 1$. Due to Lemma 8.2.4, we have the required bound sufficient for the global extendibility of the local solution. We claim the following result.

Theorem 8.2.5. *The local solution constructed in Theorem 8.2.2 can be extended globally in time. Moreover, the orbits of bounded subsets of $W^{2\alpha^-,p}(\mathbb{R}^2)$ are bounded. The global solutions $\theta(t)$ of problem (8.1) corresponding to initial values $\theta_0 \in W^{2\alpha^-,p}(\mathbb{R}^2)$ define a semigroup on the phase space $W^{2\alpha^-,p}(\mathbb{R}^2)$ through the formula:*

$$S(t)(\theta_0) = \theta(t), \quad t \geq 0.$$

8.3 Asymptotic behavior of solutions to (8.15) in $W^{2\alpha^-,p}(\mathbb{R}^2)$

In this section, we investigate the long time behavior of solutions of the following *quasi-geostrophic equation with linear damping term*:

$$\begin{aligned}&\theta_t + \lambda\theta + u \cdot \nabla\theta + \kappa(-\Delta)^{\alpha}\theta = f, \quad x \in \mathbb{R}^2,\ t > 0,\\ &\theta(0,x) = \theta_0(x),\end{aligned} \tag{8.15}$$

where κ, α, and u are like in (8.1) and (8.2) and $\lambda > 0$. We change equation (8.1) by adding to its left-hand side the linear damping term $\lambda\theta$; compare [72] for discussion of the role of such a damping. This modification allows for existence of a global attractor, the proof of this is based on the *tail estimate technique* (e. g., [54, 178]). As for problem (8.1) in Section 8.2, we choose $L^p(\mathbb{R}^2)$ as the base space, with p satisfying the condition (8.7). It is easy to check that the local solvability of the subcritical problem (8.1) obtained in Subsection 8.2.1 is also true for the problem (8.15).

Theorem 8.3.1. *Let the condition (8.7) hold, $f \in L^p(\mathbb{R}^2)$ and $\theta_0 \in W^{2\alpha^-,p}(\mathbb{R}^2)$. Then there exists, in the phase space $W^{2\alpha^-,p}(\mathbb{R}^2)$, a unique local in time mild solution θ to the subcritical problem (8.15), $\alpha \in (\frac{1}{2},1]$. Moreover,*

$$\theta \in C((0,\tau); W^{2\alpha,p}(\mathbb{R}^2)) \cap C([0,\tau); W^{2\alpha^-,p}(\mathbb{R}^2)), \quad \theta_t \in C((0,\tau); W^{2\gamma,p}(\mathbb{R}^2)),$$

with arbitrary $\gamma < \alpha$. Here, $\tau > 0$ is the “lifetime” of that local solution. Furthermore, the Cauchy formula is satisfied:

$$\theta(t) = e^{-A_{\alpha,\lambda}t}\theta_0 + \int_0^t e^{-A_{\alpha,\lambda}(t-s)}F_1(\theta(s))ds, \quad t \in [0,\tau), \tag{8.16}$$

where $e^{-A_{\alpha,\lambda}t}$ denotes the linear semigroup corresponding to the positive operator $A_{\alpha,\lambda} := \kappa(-\Delta)^\alpha + \lambda I$ in $L^p(\mathbb{R}^2)$, and

$$F_1(\theta) = -u\cdot\nabla\theta + f = R_2\theta\frac{\partial\theta}{\partial x_1} - R_1\theta\frac{\partial\theta}{\partial x_2} + f. \tag{8.17}$$

Remark 8.3.2. Arguing as in (8.9)–(8.11), by the Hölder inequality and the embedding $W^{2\alpha^-,p}(\mathbb{R}^2) \subset W^{1,2p}(\mathbb{R}^2)$, we see that

$$\|u\cdot\nabla\theta\|_{L^p(\mathbb{R}^2)} \le \left\|R_2\theta\frac{\partial\theta}{\partial x_1}\right\|_{L^p(\mathbb{R}^2)} + \left\|R_1\theta\frac{\partial\theta}{\partial x_2}\right\|_{L^p(\mathbb{R}^2)} \le C\|\theta\|_{L^{2p}(\mathbb{R}^2)}\|\theta\|_{W^{2\alpha^-,p}(\mathbb{R}^2)}. \tag{8.18}$$

Further, following [86, Theorem 3.4.1], one can show that the map $\theta_0 \to \theta(t,\theta_0)$ is continuous in $W^{2\alpha^-,p}(\mathbb{R}^2)$. Also, thanks to Lemma 8.3.5, every solution of the problem (8.15) corresponding to $\theta_0 \in W^{2\alpha^-,p}(\mathbb{R}^2)$ is globally defined. Hence we will introduce a semigroup $S : \mathbb{R}^+ \times W^{2\alpha^-,p}(\mathbb{R}^2) \to W^{2\alpha^-,p}(\mathbb{R}^2)$ through the formula

$$S(t)\theta_0 = \theta(t,\theta_0) \quad \text{for all } (t,\theta_0) \in \mathbb{R}^+ \times W^{2\alpha^-,p}(\mathbb{R}^2).$$

8.3.1 Uniform estimates of solutions

Consider a family of operators $A_{\gamma,\hat\lambda} = \kappa(-\Delta)^\gamma + \hat\lambda I$ where $\gamma > 0$ and $\hat\lambda > 0$, and let

$$X^{\gamma,\hat\lambda} = \{u \in L^p(\mathbb{R}^2) : A_{\gamma,\hat\lambda}u \in L^p(\mathbb{R}^2)\}$$

be normed by $\|u\|_{X^{\gamma,\hat\lambda}} = \|A_{\gamma,\hat\lambda}u\|_{L^p(\mathbb{R}^2)}$.

We recall the following well-known results for the semigroup generated by the positive operator $A_{\gamma,\hat\lambda}$ (see [6, Chapter V, Theorem 2.1.3]).

Proposition 8.3.3. *Let $1 < p < \infty$, $0 < \beta_1 \le \beta_2$ and $u \in X^{\beta_1,\hat\lambda}$. Then there exists a constant $C_1 = C_1(\beta_1,\beta_2,\gamma,\kappa)$ such that*

$$\|e^{-A_{\gamma,\hat\lambda}t}u\|_{X^{\beta_2,\hat\lambda}} \le C_1 e^{-\hat\lambda t}t^{-\frac{\beta_2-\beta_1}{\gamma}}\|u\|_{X^{\beta_1,\hat\lambda}}, \quad t > 0.$$

First, we notify an a priori estimate in $L^q(\mathbb{R}^2)$. Arguing as in the proof of Lemma 8.2.4, we obtain an estimate.

Lemma 8.3.4. *Let $q \in [2, \infty)$ and $f \in L^q(\mathbb{R}^2)$. Then, for a sufficiently regular solution of (8.15), the following estimate holds:*

$$\|\theta(t,\theta_0)\|^q_{L^q(\mathbb{R}^2)} \le \|\theta_0\|^q_{L^q(\mathbb{R}^2)} e^{-\frac{\lambda q}{2}t} + (C_{\lambda,q})^q \|f\|^q_{L^q(\mathbb{R}^2)}, \tag{8.19}$$

where we denote $C_{\lambda,q} := ((\frac{\lambda q}{2(q-1)})^{-(q-1)}\frac{2}{\lambda q})^{\frac{1}{q}}$.

Next, we show the existence of a bounded absorbing set in $W^{2\alpha^-,p}(\mathbb{R}^2)$.

Lemma 8.3.5. *Fix $\alpha \in (\frac{1}{2}, 1]$. Assume that (8.7) holds and $f \in L^p(\mathbb{R}^2) \cap L^{2p}(\mathbb{R}^2)$. Then there exists a bounded set $D \subset W^{2\alpha^-,p}(\mathbb{R}^2)$, which is an absorbing set for $S(t)$ in $W^{2\alpha^-,p}(\mathbb{R}^2)$. That means, for any bounded set $B \subset W^{2\alpha^-,p}(\mathbb{R}^2)$ there exists $T_1' = T_1'(B) > 0$ such that*

$$S(t)B \subseteq D \quad \text{for all } t \ge T_1'. \tag{8.20}$$

Proof. We rewrite the first equation of (8.15) as

$$\theta_t + \kappa(-\Delta)^\alpha \theta + \tilde{\lambda}\theta = -u \cdot \nabla\theta + (\tilde{\lambda} - \lambda)\theta + f,$$

where $\tilde{\lambda}$ will be fixed later. Thanks to (8.18) and Proposition 8.3.3, we have

$$\begin{aligned}
\|A_{\alpha^-,\tilde{\lambda}}\theta(t)\|_{L^p(\mathbb{R}^2)} &\le Ce^{-\tilde{\lambda}t}\|\theta_0\|_{W^{2\alpha^-,p}(\mathbb{R}^2)} \\
&\quad + C\int_0^t e^{-\tilde{\lambda}(t-s)}(t-s)^{-\frac{\alpha^-}{\alpha}}\|\theta(s)\|_{L^{2p}(\mathbb{R}^2)}\|\theta(s)\|_{W^{2\alpha^-,p}(\mathbb{R}^2)}ds \\
&\quad + C(\tilde{\lambda}-\lambda)\int_0^t e^{-\tilde{\lambda}(t-s)}(t-s)^{-\frac{\alpha^-}{\alpha}}\|\theta(s)\|_{L^p(\mathbb{R}^2)}ds \\
&\quad + C\int_0^t e^{-\tilde{\lambda}(t-s)}(t-s)^{-\frac{\alpha^-}{\alpha}}\|f\|_{L^p(\mathbb{R}^2)}ds \\
&:= I_1 + I_2 + I_3 + I_4.
\end{aligned} \tag{8.21}$$

We proceed to estimate the last three terms in (8.21). First, by (8.19) and the Hölder inequality, we obtain

$$\begin{aligned}
I_2 &\le Ce^{-\frac{\lambda}{2}t}\int_0^t (t-s)^{-\frac{\alpha^-}{\alpha}} e^{-(\tilde{\lambda}-\frac{\lambda}{2})}\Big(\tilde{\lambda}-\frac{\lambda}{2}(t-s)\Big)\|\theta_0\|_{L^{2p}(\mathbb{R}^2)}\|\theta(s)\|_{W^{2\alpha^-,p}(\mathbb{R}^2)}ds \\
&\quad + CC_{\lambda,2p}\int_0^t (t-s)^{-\frac{\alpha^-}{\alpha}} e^{-\tilde{\lambda}(t-s)}\|f\|_{L^{2p}(\mathbb{R}^2)}\|\theta(s)\|_{W^{2\alpha^-,p}(\mathbb{R}^2)}ds
\end{aligned}$$

$$\le M_1 e^{-\frac{\lambda}{2}t}\|\theta_0\|_{L^{2p}(\mathbb{R}^2)}\left(\int_0^t e^{-\frac{\nu}{2}(\tilde{\lambda}-\frac{\lambda}{2})(t-s)}\|\theta(s)\|^{\nu}_{W^{2\alpha^-,p}(\mathbb{R}^2)}ds\right)^{\frac{1}{\nu}}$$

$$+ M_2\left(\int_0^t e^{-\frac{\tilde{\lambda}\nu}{2}(t-s)}\|\theta(s)\|^{\nu}_{W^{2\alpha^-,p}(\mathbb{R}^2)}ds\right)^{\frac{1}{\nu}}, \tag{8.22}$$

where $\nu > \frac{\alpha}{\alpha-\alpha^-}$, and we have used the notation

$$M_1 := C\left(\left(\frac{1}{2}\left(\tilde{\lambda}-\frac{\lambda}{2}\right)\frac{\nu}{\nu-1}\right)^{\frac{\alpha^-\nu}{\alpha(\nu-1)}-1}\Gamma\left(1-\frac{\alpha^-\nu}{\alpha(\nu-1)}\right)\right)^{\frac{\nu-1}{\nu}},$$

$$M_2 := CC_{\lambda,2p}\|f\|_{L^{2p}(\mathbb{R}^2)}\left(\left(\frac{\tilde{\lambda}\nu}{2(\nu-1)}\right)^{\frac{\alpha^-\nu}{\alpha(\nu-1)}-1}\Gamma\left(1-\frac{\alpha^-\nu}{\alpha(\nu-1)}\right)\right)^{\frac{\nu-1}{\nu}}, \tag{8.23}$$

where Γ denotes the Gamma function. In a similar way as in (8.23), we have

$$I_3 \le Ce^{-\frac{\lambda}{2}t}\|\theta_0\|_{L^p(\mathbb{R}^2)} + C\|f\|_{L^p(\mathbb{R}^2)}, \tag{8.24}$$

and

$$I_4 \le C\|f\|_{L^p(\mathbb{R}^2)}. \tag{8.25}$$

Taking all these estimates and (8.19) together, we obtain

$$e^{\frac{\nu}{2}(\tilde{\lambda}-\frac{\lambda}{2})t}\|\theta(t)\|^{\nu}_{W^{2\alpha^-,p}(\mathbb{R}^2)} \le Ce^{\frac{\nu}{2}(\tilde{\lambda}-\frac{3\lambda}{2})t}\|\theta_0\|^{\nu}_{W^{2\alpha^-,p}(\mathbb{R}^2)} + Ce^{\frac{\nu}{2}(\tilde{\lambda}-\frac{\lambda}{2})t}\|f\|^{\nu}_{L^p(\mathbb{R}^2)}$$

$$+ (4^{\nu-1}M_1^{\nu}e^{-\frac{\lambda\nu}{2}t}\|\theta_0\|^{\nu}_{L^{2p}(\mathbb{R}^2)} + 4^{\nu-1}M_2^{\nu})\int_0^t e^{\frac{\nu}{2}(\tilde{\lambda}-\frac{\lambda}{2})s}\|\theta(s)\|^{\nu}_{W^{2\alpha^-,p}(\mathbb{R}^2)}ds. \tag{8.26}$$

Let $B \subset W^{2\alpha^-,p}(\mathbb{R}^2)$ be bounded. Then there exists $T_0' = T_0'(B) > 0$ such that for all $\theta_0 \in B$,

$$4^{\nu-1}M_1^{\nu}e^{-\frac{\lambda\nu}{2}t}\|\theta_0\|^{\nu}_{L^{2p}(\mathbb{R}^2)} < 1, \quad \forall t \ge T_0'.$$

We choose $\tilde{\lambda}$ such that $\frac{\nu}{2}(\tilde{\lambda}-\frac{3\lambda}{2}) - 1 - 4^{\nu-1}M_2^{\nu} > 0$. By Gronwall's inequality, we deduce that for all $t \ge T_0$,

$$\|\theta(t)\|^{\nu}_{W^{2\alpha^-,p}(\mathbb{R}^2)} \le Ce^{-\frac{\lambda\nu}{2}t}\|\theta_0\|^{\nu}_{W^{2\alpha^-,p}(\mathbb{R}^2)} + C\|f\|^{\nu}_{L^p(\mathbb{R}^2)}$$

$$+ C\frac{L}{\frac{\nu}{2}(\tilde{\lambda}-\frac{3\lambda}{2})-L}e^{-\frac{\lambda\nu}{2}t}\|\theta_0\|^{\nu}_{W^{2\alpha^-,p}(\mathbb{R}^2)} + C\frac{L}{\frac{\nu}{2}(\tilde{\lambda}-\frac{\lambda}{2})-L}\|f\|^{\nu}_{L^p(\mathbb{R}^2)}, \tag{8.27}$$

where we have used the notation

$$L := 1 + 4^{\nu-1}M_2^{\nu}.$$

The conclusion follows immediately from (8.27), and thus the proof is complete. □

Remark 8.3.6. Arguing as in the proof of Lemma 8.3.5, we find that if $\theta_0 \in D$ and $\tilde{\lambda}$ is such that

$$\frac{\nu}{2}\left(\tilde{\lambda} - \frac{3\lambda}{2}\right) - 4^{\nu-1}M_2^\nu > \max\{1, 4^{\nu-1}M_1^\nu d_0^\nu\},$$

for $t \geq 0$ we have an estimate

$$\|\theta(t)\|_{W^{2\alpha^-,p}(\mathbb{R}^2)} \leq C,$$

where $d_0 := \sup_{\theta \in D} \|\theta\|_{W^{2\alpha^-,p}(\mathbb{R}^2)}$.

8.3.2 Tail estimates

Now, we derive an estimate in $L^p(\mathbb{R}^2)$ of the tails of solutions (compare [54], and [178] for origin of that technique). Such estimate is used for equations in $\mathbb{R}^N$ eventually to get compactness of the trajectories, since the compact embeddings property described in the Rellich–Kondrachov theorem is not true for functions defined on $\mathbb{R}^N$. Instead we will use the *Kolmogorov–Riesz compactness criterion* in $L^p(\mathbb{R}^N)$ that we recall next.

Proposition 8.3.7. *Let $1 \leq p < \infty$. A subset $\mathcal{F}$ of $L^p(\mathbb{R}^N)$ is precompact (totally bounded) if, and only if:*
1. *$\mathcal{F}$ is bounded,*
2. *for every $\epsilon > 0$, there is r such that, for every $f \in \mathcal{F}$,*

$$\int_{|x|>r} |f(x)|^p dx \leq \epsilon^p,$$

3. *for every $\epsilon > 0$, there is $\rho > 0$ such that, for every $f \in \mathcal{F}$ and $y \in \mathbb{R}^N$ with $|y| < \rho$,*

$$\int_{\mathbb{R}^n} |f(x+y) - f(x)|^p dx \leq \epsilon^p.$$

The above necessary and sufficient condition works perfectly for solutions of Cauchy's problem. We refer to [85] for more information concerning the criterion.

To that end, we choose a smooth function $\eta(z)$ defined for $0 \leq z < \infty$, such that $0 \leq \eta(z) \leq 1$ for $z \in \mathbb{R}^+$ and

$$\eta(z) = \begin{cases} 0, & 0 \leq z \leq \frac{1}{2}, \\ 1, & z \geq 1. \end{cases} \tag{8.28}$$

Note that there exists a positive constant C_0 such that $|\eta''(z)| \leq C_0$ for all $z \geq 0$. Moreover, the cut-off function η has the following properties, (e. g., [53]).

Lemma 8.3.8. *Let η be the smooth function defined by (8.28) and $\alpha \in (0,1)$. Then there exists a positive constant C_0' such that for every $x \in \mathbb{R}^2$ and $K \in \mathbb{N}$,*

$$|(-\Delta)^\alpha \eta_K(x)| \le \frac{C_0'}{K^{2\alpha}}, \tag{8.29}$$

where $\eta_K(\cdot) = \eta(\frac{|\cdot|}{K})$, and C_0' is independent of $K \in \mathbb{N}$ and $x \in \mathbb{R}^2$.

Lemma 8.3.9. *Assume that the hypotheses in Lemma 8.3.5 hold. Then for any $\varepsilon > 0$, there exist $T_1 = T_1(\varepsilon) > 0$ and $K_1 = K_1(\varepsilon) > 0$ such that for all $t \ge T_1$ and $K \ge K_1$, the solution θ of (8.15) satisfies*

$$\int_{\mathcal{O}_K} |\theta(t,\theta_0)|^p dx \le \varepsilon,$$

where $\mathcal{O}_K = \{x \in \mathbb{R}^2 : |x| \ge K\}$ and $\theta_0 \in D$.

Proof. Multiplying the first equation of (8.15) by $|\theta|^{p-1}\operatorname{sgn}(\theta)\eta_K$ and then integrating over $\mathbb{R}^2$, we obtain

$$\begin{aligned}
&\int_{\mathbb{R}^2} \theta_t|\theta|^{p-1}\operatorname{sgn}\theta\eta_K dx + \lambda\int_{\mathbb{R}^2} \theta|\theta|^{p-1}\operatorname{sgn}\theta\eta_K dx \\
&\quad = -\int_{\mathbb{R}^2} u\cdot\nabla\theta|\theta|^{p-1}\operatorname{sgn}\theta\eta_K dx - \kappa\int_{\mathbb{R}^2}(-\Delta)^\alpha\theta|\theta|^{p-1}\operatorname{sgn}\theta\eta_K dx \\
&\qquad + \int_{\mathbb{R}^2} f|\theta|^{p-1}\operatorname{sgn}\theta\eta_K dx.
\end{aligned} \tag{8.30}$$

For the first term on the right-hand side of (8.30), since $\nabla\cdot u = 0$, using integration by parts, Lemma 8.3.8, the Young inequality and (5.51), we have

$$\begin{aligned}
-\int_{\mathbb{R}^2} u\cdot\nabla\theta|\theta|^{p-1}\operatorname{sgn}\theta\eta_K dx &= -\frac{1}{p}\int_{\mathbb{R}^2} u\cdot\nabla|\theta|^p\eta_K dx \\
&= \frac{1}{p}\int_{\mathbb{R}^2}\left(\frac{\partial\psi}{\partial x_2}\frac{\partial|\theta|^p}{\partial x_1} - \frac{\partial\psi}{\partial x_1}\frac{\partial|\theta|^p}{\partial x_2}\right)\eta_K dx \\
&= \frac{1}{p}\int_{\mathbb{R}^2} -|\theta|^p\frac{\partial\psi}{\partial x_2}\frac{\partial\eta_K}{\partial x_1} + |\theta|^p\frac{\partial\psi}{\partial x_1}\frac{\partial\eta_K}{\partial x_2}dx \\
&\le \frac{1}{p}\int_{\mathbb{R}^2}|\theta|^p|u||\nabla\eta_K|dx \le \frac{C}{K}\left(\int_{\mathbb{R}^2}|\theta|^{\frac{p^2}{p-1}}dx + \int_{\mathbb{R}^2}|u|^p dx\right) \\
&\le \frac{C}{K}\left(\int_{\mathbb{R}^2}|\theta|^{\frac{p^2}{p-1}}dx + \int_{\mathbb{R}^2}|\theta|^p dx\right).
\end{aligned} \tag{8.31}$$

Thanks to the pointwise estimate in [94, Proposition 3.3] and Lemma 8.3.8, we find that

$$
\begin{aligned}
-\kappa \int_{\mathbb{R}^2} (-\Delta)^\alpha \theta |\theta|^{p-1} \operatorname{sgn} \theta \eta_K dx &\le -\frac{\kappa}{p} \int_{\mathbb{R}^2} (-\Delta)^\alpha |\theta|^p \eta_K dx \\
&= -\frac{\kappa}{p} \int_{\mathbb{R}^2} |\theta|^p (-\Delta)^\alpha \eta_K dx \le \frac{C}{K^{2\alpha}} \int_{\mathbb{R}^2} |\theta|^p dx.
\end{aligned} \tag{8.32}
$$

For the last term in (8.30), by the Young inequality, we have

$$
\int_{\mathbb{R}^2} f |\theta|^{p-1} \operatorname{sgn} \theta \eta_K dx \le C \int_{\mathbb{R}^2} |f|^p \eta_K dx + \frac{\lambda}{2} \int_{\mathbb{R}^2} |\theta|^p \eta_K dx. \tag{8.33}
$$

Substituting (8.31)–(8.33) into (8.30), we deduce that

$$
\begin{aligned}
&\frac{1}{p}\frac{d}{dt} \int_{\mathbb{R}^2} |\theta|^p \eta_K dx + \frac{\lambda}{2} \int_{\mathbb{R}^2} |\theta|^p \eta_K dx \\
&\le \frac{C}{K}\left(\int_{\mathbb{R}^2} |\theta|^{\frac{p^2}{p-1}} dx + \int_{\mathbb{R}^2} |\theta|^p dx \right) + \frac{C}{K^{2\alpha}} \int_{\mathbb{R}^2} |\theta|^p dx + C \int_{\mathbb{R}^2} |f|^p \eta_K dx.
\end{aligned}
$$

Applying Gronwall's inequality, we obtain

$$
\begin{aligned}
&\int_{\mathbb{R}^2} |\theta(t)|^p \eta_K dx \\
&\quad \le e^{-\frac{\lambda p}{2} t} \int_{\mathbb{R}^2} |\theta_0|^p dx + \left(\frac{C}{K^{2\alpha}} + \frac{C}{K} \right) \int_0^t e^{-\frac{\lambda p}{2}(t-s)} (\|\theta_0\|^p_{L^p(\mathbb{R}^2)} + \|f\|^p_{L^p(\mathbb{R}^2)}) ds \\
&\qquad + \frac{C}{K} \int_0^t e^{-\frac{\lambda p}{2}(t-s)} \|\theta(s)\|^{\frac{p^2}{p-1}}_{W^{2\alpha^-,p}(\mathbb{R}^2)} ds + C \int_0^t e^{-\frac{\lambda p}{2}(t-s)} \int_{\mathbb{R}^2} |f|^p \eta_K dx ds,
\end{aligned}
$$

thanks to Lemma 8.3.4 and the embedding $W^{2\alpha^-,p}(\mathbb{R}^2) \subset L^{\frac{p^2}{p-1}}(\mathbb{R}^2)$. Note that $f \in L^p(\mathbb{R}^2)$ and $\theta_0 \in D$. By Remark 8.3.6, we find that given $\varepsilon > 0$, there exist $T_1 = T_1(\varepsilon)$ and $K_1 = K_1(\varepsilon) > 0$ such that

$$
\int_{\mathbb{R}^2} |\theta(t)|^p \eta_K dx \le \varepsilon, \tag{8.34}
$$

if $t \ge T_1$ and $K \ge K_1$. This completes the proof. □

8.3.3 Existence of the global attractor

First, we formulate the following lemma.

Lemma 8.3.10. *Assume that the hypotheses of Lemma 8.3.5 hold. Then for any ε, $0 < \varepsilon < \alpha - \alpha^-$, there exists a bounded set $D_1 \subset W^{2\alpha-2\varepsilon,p}(\mathbb{R}^2)$ such that D_1 is an absorbing set for $S(t)$. More precisely, for any bounded set $B \subset W^{2\alpha^-,p}(\mathbb{R}^2)$, there exists $T_1^* = T_1^*(B) > 0$ such that*

$$S(t)B \subset D_1 \quad \text{for all } t \geq T_1^*.$$

Proof. Fix arbitrary $\varepsilon \in (0, \alpha - \alpha^-)$. Then it follows from Proposition 8.3.3, (8.16), and (8.18) that

$$\begin{aligned} t^{\frac{\alpha-\varepsilon-\alpha^-}{\alpha}} \|A_{\alpha-\varepsilon,\lambda}\theta(t)\|_{L^p(\mathbb{R}^2)} &\leq t^{\frac{\alpha-\varepsilon-\alpha^-}{\alpha}} \|A_{\alpha-\varepsilon,\lambda} e^{-A_{\alpha,\lambda}t}\theta_0\|_{L^p(\mathbb{R}^2)} \\ &\quad + t^{\frac{\alpha-\varepsilon-\alpha^-}{\alpha}} \int_0^t \|A_{\alpha-\varepsilon,\lambda} e^{-A_{\alpha,\lambda}(t-s)} u(s) \cdot \nabla\theta(s)\|_{L^p(\mathbb{R}^2)} ds \\ &\quad + t^{\frac{\alpha-\varepsilon-\alpha^-}{\alpha}} \int_0^t \|A_{\alpha-\varepsilon,\lambda} e^{-A_{\alpha,\lambda}(t-s)} f\|_{L^p(\mathbb{R}^2)} ds \\ &\leq Ce^{-\lambda t}\|\theta_0\|_{W^{2\alpha^-,p}(\mathbb{R}^2)} \\ &\quad + Ct^{\frac{\alpha-\varepsilon-\alpha^-}{\alpha}} \int_0^t (t-s)^{-\frac{\alpha-\varepsilon}{\alpha}} e^{-\lambda(t-s)} (\|\theta(s)\|^2_{W^{2\alpha^-,p}(\mathbb{R}^2)} + \|f\|_{L^p(\mathbb{R}^2)}) ds, \end{aligned} \tag{8.35}$$

thanks to the embeddings $W^{2\alpha^-,p}(\mathbb{R}^2) \subset W^{1,2p}(\mathbb{R}^2) \subset L^{2p}(\mathbb{R}^2)$. Using Remark 8.3.6, we see that if $\theta_0 \in D$, then by (8.35) and Lemma 8.3.4 we have

$$\|\theta(1)\|_{W^{2\alpha-2\varepsilon,p}(\mathbb{R}^2)} \leq C\|\theta_0\|_{W^{2\alpha^-,p}(\mathbb{R}^2)} + C\|f\|_{L^p(\mathbb{R}^2)} + C.$$

This implies that $D_1 = S(1)D$ is also an absorbing set, moreover bounded in $W^{2\alpha-2\varepsilon,p}(\mathbb{R}^2)$. □

We next prove the existence of the global attractor for $\{S(t)\}_{t\geq 0}$; the solution semigroup of problem (8.15) in $W^{2\alpha^-,p}(\mathbb{R}^2)$.

Theorem 8.3.11. *Assume that the hypotheses in Lemma 8.3.5 hold. Then the semigroup $\{S(t)\}_{t\geq 0}$ associated with problem (8.15) possesses a global attractor $\mathcal{A}$, in a sense of Definition 2.3.5, in $W^{2\alpha^-,p}(\mathbb{R}^2)$.*

Proof. The asymptotic compactness of semigroup $\{S(t)\}_{t\geq 0}$ in $W^{2\alpha^-,p}(\mathbb{R}^2)$ follows directly from Lemmas 8.3.9–8.3.10 and the Nirenberg–Gagliardo inequality. By Theorem 3.4.1 in [86], we see that $S(t)$ is continuous with respect to t and for fixed t in $W^{2\alpha^-,p}(\mathbb{R}^2)$, respectively. Thanks to Lemma 8.3.5, the existence of a global compact invariant attractor $\mathcal{A}$ follows immediately from the abstract results in [13, 83, 141, 172]. □

8.4 Solvability of subcritical (8.1), $\alpha \in (\frac{1}{2}, 1]$, in $H^{2\alpha^- +s}(\mathbb{R}^2)$

As was observed in [102] (compare the discussion in Section 9.2), it is also possible to choose $H^s(\mathbb{R}^2)$, $s > 1$, as a *base space* (where we consider the equation (8.1)). In that case, the phase space will be $H^{2\alpha^- +s}(\mathbb{R}^2)$. Since the considerations in that case are similar to the one presented in Section 8.2 in case of the base space $L^p(\mathbb{R}^2)$, we will shorten the presentation omitting a part of the technical proofs.

8.4.1 Local solvability

As in the previous case $X = L^p(\mathbb{R}^2)$ in Subsection 8.2.1, we need to check that the nonlinearity (8.5) is Lipschitz on bounded sets as a map from $H^{2\alpha^- +s}(\mathbb{R}^2)$ into $H^s(\mathbb{R}^2)$, $s > 1$. Indeed, as in (8.9), we have

$$\begin{aligned}\|F(\theta_1) - F(\theta_2)\|_{H^s(\mathbb{R}^2)} &\le \kappa\|\theta_1 - \theta_2\|_{H^s(\mathbb{R}^2)} + \Big\|R_2(\theta_1 - \theta_2)\frac{\partial\theta_1}{\partial x_1}\\ &+ R_2\theta_2\frac{\partial(\theta_1 - \theta_2)}{\partial x_1}\Big\|_{H^s(\mathbb{R}^2)} + \Big\|R_1(\theta_1 - \theta_2)\frac{\partial\theta_1}{\partial x_2} + R_1\theta_2\frac{\partial(\theta_1 - \theta_2)}{\partial x_2}\Big\|_{H^s(\mathbb{R}^2)}. \end{aligned} \tag{8.36}$$

We estimate the second term above using the property (e. g., [1, p. 115]), that *when Ω is a domain in $\mathbb{R}^N$ having the cone property, then $W^{m,r}(\Omega)$ is a Banach Algebra provided that $mr > N$*. In our case $2s > 2$, since $s > 1$. Hence

$$\begin{aligned}\Big\|R_2(\theta_1 - \theta_2)\frac{\partial\theta_1}{\partial x_1}\Big\|_{H^s(\mathbb{R}^2)} &\le C\|R_2(\theta_1 - \theta_2)\|_{H^s(\mathbb{R}^2)}\Big\|\frac{\partial\theta_1}{\partial x_1}\Big\|_{H^s(\mathbb{R}^2)}\\ &\le C\||u_1 - u_2|\|_{H^s(\mathbb{R}^2)}\|\theta_1\|_{H^{2\alpha^- +s}(\mathbb{R}^2)},\end{aligned} \tag{8.37}$$

where u_i, $i = 1, 2$ correspond to θ_i through the relation (8.3).

Applying the property (5.51) (see also [183, p. 12]) to the first term in (8.37) we get

$$\Big\|R_2(\theta_1 - \theta_2)\frac{\partial\theta_1}{\partial x_1}\Big\|_{H^s(\mathbb{R}^2)} \le C\|\theta_1 - \theta_2\|_{H^{2\alpha^- +s}(\mathbb{R}^2)}\|\theta_1\|_{H^{2\alpha^- +s}(\mathbb{R}^2)}.$$

The other components in (8.36) are estimated analogously. Consequently, we obtain

$$\|F(\theta_1) - F(\theta_2)\|_{H^s(\mathbb{R}^2)} \le \text{const}(\|\theta_1\|_{H^{2\alpha^- +s}(\mathbb{R}^2)}, \|\theta_2\|_{H^{2\alpha^- +s}(\mathbb{R}^2)})\|\theta_1 - \theta_2\|_{H^{2\alpha^- +s}(\mathbb{R}^2)}, \tag{8.38}$$

which proves local solvability of (8.1) in the phase space $H^{2\alpha^- +s}(\mathbb{R}^2)$. More precisely, following [32, 86], we formulate the corresponding result.

Theorem 8.4.1. *Let $s > 1$ be fixed. Then, for $f \in H^s(\mathbb{R}^2)$ and arbitrary $\theta_0 \in H^{2\alpha^- +s}(\mathbb{R}^2)$, there exists in the phase space $H^{2\alpha^- +s}(\mathbb{R}^2)$ a unique local in time mild solution $\theta(t)$ to the subcritical problem (8.1), $\alpha \in (\frac{1}{2}, 1]$, such that*

$$\theta \in C((0,\tau); H^{2\alpha+s}(\mathbb{R}^2)) \cap C([0,\tau); H^{2\alpha^- +s}(\mathbb{R}^2)), \quad \theta_t \in C((0,\tau); H^{2\gamma+s}(\mathbb{R}^2)),$$

with arbitrary $\gamma < \alpha$. Here, $\tau > 0$ is the "lifetime"' of that local solution. Moreover, the Cauchy formula is satisfied:

$$\theta(t) = e^{-A_\alpha t}\theta_0 + \int_0^t e^{-A_\alpha(t-s)}F(\theta(s))ds, \quad t \in [0,\tau),$$

where $e^{-A_\alpha t}$ denotes the linear semigroup corresponding to the operator $A_\alpha := \kappa[(-\Delta)^\alpha + I]$ in $H^s(\mathbb{R}^2)$, and F is given by formula (8.5).

8.4.2 Global solvability

To guarantee the global in time solvability of (8.1) in $H^{2\alpha^-+s}(\mathbb{R}^2)$, the a priori estimates (8.46), (8.47) below will be used. They show that $\|\theta(t,\cdot)\|_{H^{2\alpha^-+s}(\mathbb{R}^2)}$ is bounded for $t \geq 0$. Consequently, we have global Lipschitz continuity and boundedness of the nonlinear term F acting from $H^{2\alpha^-+s}(\mathbb{R}^2)$ to $H^s(\mathbb{R}^2)$.

Further a priori estimates

Following the calculations in [184, p. 1165], we are able to estimate higher Sobolev norms of the solutions to (8.1). Let $l \geq \alpha$ be fixed and $f \in H^{l-\alpha}(\mathbb{R}^2)$. Multiplying the equation (8.1) by $(-\Delta)^l\theta$, we obtain

$$\begin{aligned}\frac{1}{2}\frac{d}{dt}\int_{\mathbb{R}^2}[(-\Delta)^{\frac{l}{2}}\theta]^2 dx + \kappa\int_{\mathbb{R}^2}[(-\Delta)^{\frac{l+\alpha}{2}}\theta]^2 dx \\ = \int_{\mathbb{R}^2}(-\Delta)^{\frac{l-\alpha}{2}} f(-\Delta)^{\frac{l+\alpha}{2}}\theta dx - \int_{\mathbb{R}^2} u\cdot\nabla\theta(-\Delta)^l\theta dx.\end{aligned} \tag{8.39}$$

Using Hölder inequality, the nonlinear term is transformed as follows:

$$\begin{aligned}-\int_{\mathbb{R}^2} u\cdot\nabla\theta(-\Delta)^l\theta dx &\leq \left|\int_{\mathbb{R}^2}(-\Delta)^{\frac{l-\alpha}{2}}(u\cdot\nabla\theta)(-\Delta)^{\frac{l+\alpha}{2}}\theta dx\right| \\ &\leq \|(-\Delta)^{\frac{l+\alpha}{2}}\theta\|_{L^2(\mathbb{R}^2)}\|(-\Delta)^{\frac{l-\alpha}{2}}(u\cdot\nabla\theta)\|_{L^2(\mathbb{R}^2)}.\end{aligned} \tag{8.40}$$

Noting that $\nabla\cdot u = 0$ and the nontrivial estimate of Proposition 5.4.4 with $\frac{1}{q} + \frac{1}{r} = \frac{1}{2}$, we obtain

$$\begin{aligned}\|(-\Delta)^{\frac{l-\alpha}{2}}(u\cdot\nabla\theta)\|_{L^2(\mathbb{R}^2)} &= \|(-\Delta)^{\frac{l-\alpha}{2}}\nabla\cdot(u\theta)\|_{L^2(\mathbb{R}^2)} \\ &\leq C[\|(-\Delta)^{\frac{l-\alpha+1}{2}}u\|_{L^q(\mathbb{R}^2)}\|\theta\|_{L^r(\mathbb{R}^2)} + \|u\|_{L^r(\mathbb{R}^2)}\|(-\Delta)^{\frac{l-\alpha+1}{2}}\theta\|_{L^q(\mathbb{R}^2)}].\end{aligned} \tag{8.41}$$

Then, by (5.50) and (8.3), we infer from (8.41) that

$$\|(-\Delta)^{\frac{l-\alpha}{2}}(u\cdot\nabla\theta)\|_{L^2(\mathbb{R}^2)} \leq C\|\theta\|_{L^r(\mathbb{R}^2)}\|\theta\|_{W^{l+1-\alpha,q}(\mathbb{R}^2)}. \tag{8.42}$$

Select

$$q = \frac{1}{1-\alpha^-}, \quad r = \frac{2}{2\alpha^- - 1}, \tag{8.43}$$

and by the Nirenberg–Gagliardo inequality, we have

$$\|\theta\|_{W^{l+1-\alpha,q}(\mathbb{R}^2)} \leq C\|\theta\|^{\eta}_{H^{l+\alpha}(\mathbb{R}^2)}\|\theta\|^{1-\eta}_{L^2(\mathbb{R}^2)}, \tag{8.44}$$

with some $\eta \in [\frac{l+1-\alpha}{l+\alpha}, 1)$. Thus, it follows from (8.40)–(8.44) and the Young inequality that

$$\left|-\int_{\mathbb{R}^2} u\cdot\nabla\theta(-\Delta)^l\theta dx\right| \leq \frac{\kappa}{4}\|(-\Delta)^{\frac{l+\alpha}{2}}\theta\|^2_{L^2(\mathbb{R}^2)} + \text{const}(\|\theta\|_{L^2(\mathbb{R}^2)}, \|\theta\|_{L^r(\mathbb{R}^2)}). \tag{8.45}$$

Consequently, from (8.39) and (8.45) we obtain a differential inequality:

$$\begin{aligned}&\frac{d}{dt}\int_{\mathbb{R}^2}[(-\Delta)^{\frac{l}{2}}\theta]^2dx + \kappa\int_{\mathbb{R}^2}[(-\Delta)^{\frac{l+\alpha}{2}}\theta]^2dx\\ &\leq \text{const}(\|\theta\|_{L^2(\mathbb{R}^2)}, \|\theta\|_{L^r(\mathbb{R}^2)}) + C\|f\|^2_{H^{l-\alpha}(\mathbb{R}^2)},\end{aligned} \tag{8.46}$$

providing us, together with Lemma 8.2.4, the bound for $\|\theta\|_{H^l(\mathbb{R}^2)}$, $l \geq \alpha$, through $\|f\|_{H^{l-\alpha}(\mathbb{R}^2)}$.

From Theorem 8.4.1, we know that there exists a local solution of (8.1) in the phase space $H^{2\alpha^-+s}(\mathbb{R}^2)$. Such local solution will be extended globally in time provided that the nonlinearity F fulfills a subordination condition. We choose the space $Y = H^s(\mathbb{R}^2) \supset H^{2\alpha+s}(\mathbb{R}^2)$ for the problem (8.1). Then, by (8.5), the similar argument in (8.37), the property (5.51) and interpolation inequality, we obtain

$$\begin{aligned}\|F(\theta)\|_{H^s(\mathbb{R}^2)} &\leq \left\|R_2\theta\frac{\partial\theta}{\partial x_1} - R_1\theta\frac{\partial\theta}{\partial x_2}\right\|_{H^s(\mathbb{R}^2)} + \|f\|_{H^s(\mathbb{R}^2)} + \kappa\|\theta\|_{H^s(\mathbb{R}^2)}\\ &\leq \left\|R_2\theta\frac{\partial\theta}{\partial x_1}\right\|_{H^s(\mathbb{R}^2)} + \left\|R_1\theta\frac{\partial\theta}{\partial x_2}\right\|_{H^s(\mathbb{R}^2)} + \|f\|_{H^s(\mathbb{R}^2)} + \kappa\|\theta\|_{H^s(\mathbb{R}^2)}\\ &\leq C\|u\|_{H^s(\mathbb{R}^2)}\|\theta\|_{H^{s+1}(\mathbb{R}^2)} + \|f\|_{H^s(\mathbb{R}^2)} + \kappa\|\theta\|_{H^s(\mathbb{R}^2)}\\ &\leq \text{const}(\|\theta\|_{H^s(\mathbb{R}^2)})(1 + \|\theta\|^{\eta'}_{H^{2\alpha^-+s}(\mathbb{R}^2)}) + \|f\|_{H^s(\mathbb{R}^2)},\end{aligned} \tag{8.47}$$

where $0 < \eta' = \frac{1}{2\alpha^-} < 1$. Therefore, by the a priori estimate (8.46) with $l = s$ and Lemma 8.2.4, the solution of the problem (8.1) constructed in Theorem 8.4.1 will be extended globally in time.

8.5 Asymptotic behavior of solutions to (8.15) in $H^{2\alpha^-+s}(\mathbb{R}^2)$

In this section, we study the long time behavior of solution of equation (8.15) (containing linear damping term) in $H^{2\alpha^-+s}(\mathbb{R}^2)$. As for problem (8.1) in Section 8.4, we choose

$H^s(\mathbb{R}^2)$, $s > 1$, as the base space. By the standard approach of [32, 86] reported in Chapter 2, we have the following theorem.

Theorem 8.5.1. *Let $s > 1$ be fixed. Then, for $f \in H^s(\mathbb{R}^2)$ and for arbitrary $\theta_0 \in H^{2\alpha^-+s}(\mathbb{R}^2)$, there exists in the phase space $H^{2\alpha^-+s}(\mathbb{R}^2)$ a unique local in time mild solution $\theta(t)$ to the subcritical problem* (8.15), *$\alpha \in (\frac{1}{2}, 1]$. Moreover,*

$$\theta \in C((0,\tau); H^{2\alpha+s}(\mathbb{R}^2)) \cap C([0,\tau); H^{2\alpha^-+s}(\mathbb{R}^2)), \quad \theta_t \in C((0,\tau); H^{2\gamma+s}(\mathbb{R}^2)),$$

with arbitrary $\gamma < \alpha$. Here, τ is the "lifetime" of that local solution. Furthermore, the Cauchy formula is satisfied:

$$\theta(t) = e^{-A_{\alpha,\lambda}t}\theta_0 + \int_0^t e^{-A_{\alpha,\lambda}(t-s)}F_1(\theta(s))ds, \quad t \in [0,\tau),$$

where $e^{-A_{\alpha,\lambda}t}$ denotes the linear semigroup corresponding to the operator $A_{\alpha,\lambda} := \kappa(-\Delta)^\alpha + \lambda I$ in $H^s(\mathbb{R}^2)$, and the nonlinear term F_1 is given in (8.17).

Further, following [86, Theorem 3.4.1], one can show that $\theta(t,\theta_0)$ is continuous with respect to θ_0 in $H^{2\alpha^-+s}(\mathbb{R}^2)$. Thanks to Lemma 8.5.4, we see that every solution of problem (8.15) corresponding to $\theta_0 \in H^{2\alpha^-+s}(\mathbb{R}^2)$ can be globally defined. Hence we now define a semigroup $S : \mathbb{R}^+ \times H^{2\alpha^-+s}(\mathbb{R}^2) \to H^{2\alpha^-+s}(\mathbb{R}^2)$ by

$$S(t)\theta_0 = \theta(t,\theta_0) \quad \text{for all } (t,\theta_0) \in \mathbb{R}^+ \times H^{2\alpha^-+s}(\mathbb{R}^2).$$

Now, we derive some uniform estimates of solutions of problem (8.15), which will be used to prove the existence of a global attractor in $H^{2\alpha^-+s}(\mathbb{R}^2)$.

Lemma 8.5.2. *Fix $\alpha \in (\frac{1}{2}, 1]$ and $s > 1$. Assume that $f \in H^s(\mathbb{R}^2)$ and $\theta_0 \in H^{2\alpha^-+s}(\mathbb{R}^2)$. Then there exists $\mathcal{T}_0 = \mathcal{T}_0(\|\theta_0\|_{L^2(\mathbb{R}^2)}, \|\theta_0\|_{L^r(\mathbb{R}^2)}) > 0$ such that for all $t \geq \mathcal{T}_0$, the solution θ of problem* (8.15) *satisfies*

$$\begin{aligned}\|\theta(t)\|^2_{H^s(\mathbb{R}^2)} &\le Ce^{-\lambda t}\|\theta_0\|^2_{H^{2\alpha^-+s}(\mathbb{R}^2)} + C\|f\|^2_{H^s(\mathbb{R}^2)} + C \\ &\quad + \sup_{r'\in[0,\mathcal{T}_0]} \text{const}(\|\theta(r')\|_{L^2(\mathbb{R}^2)}, \|\theta(r')\|_{L^r(\mathbb{R}^2)})e^{-\lambda(t-\mathcal{T}_0)},\end{aligned} \tag{8.48}$$

$$\begin{aligned}\|\theta(t)\|^2_{H^{\alpha+s}(\mathbb{R}^2)} &\le Ce^{-\lambda t}\|\theta_0\|^2_{H^{2\alpha^-+s}(\mathbb{R}^2)} + C\|f\|^2_{H^s(\mathbb{R}^2)} + C \\ &\quad + \sup_{r'\in[0,\mathcal{T}_0]} \text{const}(\|\theta(r')\|_{L^2(\mathbb{R}^2)}, \|\theta(r')\|_{L^r(\mathbb{R}^2)})e^{-2\lambda(t-\mathcal{T}_0)},\end{aligned} \tag{8.49}$$

and

$$\kappa \int_{t}^{t+1} \|\theta(s')\|^2_{\dot{H}^{2\alpha+s}(\mathbb{R}^2)} ds' \le e^{-2\lambda t}\|\theta_0\|^2_{H^{2\alpha^- +s}(\mathbb{R}^2)} + C\|f\|^2_{H^s(\mathbb{R}^2)} + C$$
$$+ \sup_{r'\in[0,\mathcal{T}_0]} \operatorname{const}(\|\theta(r')\|_{L^2(\mathbb{R}^2)}, \|\theta(r')\|_{L^r(\mathbb{R}^2)})e^{-2\lambda(t-\mathcal{T}_0)}, \tag{8.50}$$

where $r > 2$ is given by (8.43).

Proof. The proof will be sketched only. We need to obtain three estimates. The first one follows analyzing the equation obtained multiplying (8.15) by $(-\Delta)^s\theta$:

$$\frac{1}{2}\frac{d}{dt}\int_{\mathbb{R}^2} [(-\Delta)^{\frac{s}{2}}\theta]^2 dx + \lambda \int_{\mathbb{R}^2} [(-\Delta)^{\frac{s}{2}}\theta]^2 dx + \kappa \int_{\mathbb{R}^2} [(-\Delta)^{\frac{s+\alpha}{2}}\theta]^2 dx$$
$$= \int_{\mathbb{R}^2} f(-\Delta)^s\theta dx - \int_{\mathbb{R}^2} u\cdot\nabla\theta(-\Delta)^s\theta dx.$$

Using the Young inequality and arguments similar as in (8.40)–(8.44), we obtain (8.48).

The remaining estimates, (8.49) and (8.50), are consequences of the equation obtained multiplying (8.15) by $(-\Delta)^{\alpha+s}\theta$ and then integrating over $\mathbb{R}^2$:

$$\frac{1}{2}\frac{d}{dt}\int_{\mathbb{R}^2} [(-\Delta)^{\frac{\alpha+s}{2}}\theta]^2 dx + \lambda \int_{\mathbb{R}^2} [(-\Delta)^{\frac{\alpha+s}{2}}\theta]^2 dx + \kappa \int_{\mathbb{R}^2} [(-\Delta)^{\frac{2\alpha+s}{2}}\theta]^2 dx$$
$$= \int_{\mathbb{R}^2} f(-\Delta)^{\alpha+s}\theta dx - \int_{\mathbb{R}^2} u\cdot\nabla\theta(-\Delta)^{\alpha+s}\theta dx, \tag{8.51}$$

and will be obtained as in (8.40)–(8.45). □

Lemma 8.5.3. *Fix $\alpha \in (\frac{1}{2}, 1]$ and $s > 1$. Assume that $f \in H^s(\mathbb{R}^2)$. Then for any bounded set $B \subset H^{2\alpha^- +s}(\mathbb{R}^2)$, there exists $\mathcal{T}_1 = \mathcal{T}_1(B) > 0$ such that for all $t \ge \mathcal{T}_1$, the solution θ of the problem* (8.15) *satisfies*

$$\|\theta_t(t,\theta_0)\|^2_{\dot{H}^s(\mathbb{R}^2)} \le \rho_0 + \rho_1,$$

where $\theta_0 \in B$, ρ_0 and ρ_1 are positive constants independent of B.

Proof. We know by Theorem 8.5.1 that $\lambda\theta$, $u\cdot\nabla\theta$ and $\kappa(-\Delta)^\alpha\theta$ are continuously differentiable with respect to $t > 0$. Then we can differentiate the first equation of (8.15) to get

$$\theta_{tt} + \lambda\theta_t + \frac{d}{dt}(u\cdot\nabla\theta) + \kappa(-\Delta)^\alpha\theta_t = 0. \tag{8.52}$$

Multiplying (8.52) by $(-\Delta)^s\theta_t$ and then integrating over $\mathbb{R}^2$, we obtain

$$\frac{1}{2}\frac{d}{dt}\int_{\mathbb{R}^2}[(-\Delta)^{\frac{s}{2}}\theta_t]^2dx+\lambda\int_{\mathbb{R}^2}[(-\Delta)^{\frac{s}{2}}\theta_t]^2dx+\kappa\int_{\mathbb{R}^2}[(-\Delta)^{\frac{s+\alpha}{2}}\theta_t]^2dx$$
$$=-\int_{\mathbb{R}^2}\frac{d}{dt}(u\cdot\nabla\theta)(-\Delta)^s\theta_t dx. \tag{8.53}$$

By the Hölder inequality, the nonlinear term is transformed as follows:

$$-\int_{\mathbb{R}^2}\frac{d}{dt}(u\cdot\nabla\theta)(-\Delta)^s\theta_t dx\le\left|(-\Delta)^{\frac{s-\alpha}{2}}\frac{d}{dt}(u\cdot\nabla\theta)(-\Delta)^{\frac{s+\alpha}{2}}\theta_t\right|$$
$$\le\left\|(-\Delta)^{\frac{s-\alpha}{2}}\frac{d}{dt}(u\cdot\nabla\theta)\right\|_{L^2(\mathbb{R}^2)}\|(-\Delta)^{\frac{s+\alpha}{2}}\theta_t\|_{L^2(\mathbb{R}^2)}. \tag{8.54}$$

Note that $(-\Delta)^\beta R_j f=R_j(-\Delta)^\beta f$ for $f\in D((-\Delta)^\beta)$, $\beta>0$, $j=1,2$ (see, e. g., [130, p. 300]). Then by the commutator estimate and (5.51), we obtain

$$\left\|(-\Delta)^{\frac{s-\alpha}{2}}\left(R_j\theta_t\frac{\partial\theta}{\partial x_i}\right)\right\|_{L^2(\mathbb{R}^2)}$$
$$\le C\|(-\Delta)^{\frac{s-\alpha}{2}}R_j\theta_t\|_{L^{p_1}(\mathbb{R}^2)}\left\|\frac{\partial\theta}{\partial x_i}\right\|_{L^{p_2}(\mathbb{R}^2)}+C\|R_j\theta_t\|_{L^{p_3}(\mathbb{R}^2)}\left\|(-\Delta)^{\frac{s-\alpha}{2}}\frac{\partial\theta}{\partial x_i}\right\|_{L^{p_4}(\mathbb{R}^2)}$$
$$\le C\|\theta_t\|_{W^{s-\alpha,p_1}(\mathbb{R}^2)}\|\theta\|_{W^{1,p_2}(\mathbb{R}^2)}+C\|\theta_t\|_{L^{p_3}(\mathbb{R}^2)}\|\theta\|_{W^{s-\alpha+1,p_4}(\mathbb{R}^2)}, \tag{8.55}$$

where $i=1,2$ and $p_1,p_2,p_3,p_4\in[1,\infty]$ satisfy

$$\frac{1}{p_1}+\frac{1}{p_2}=\frac{1}{p_3}+\frac{1}{p_4}=\frac{1}{2}.$$

Similarly, it follow from (8.3) that

$$\left\|(-\Delta)^{\frac{s-\alpha}{2}}\frac{d}{dt}(u\cdot\nabla\theta)\right\|_{L^2(\mathbb{R}^2)}$$
$$=\left\|(-\Delta)^{\frac{s-\alpha}{2}}\left(-R_2\theta_t\frac{\partial\theta}{\partial x_1}-R_2\theta\frac{\partial\theta_t}{\partial x_1}+R_1\theta_t\frac{\partial\theta}{\partial x_2}+R_1\theta\frac{\partial\theta_t}{\partial x_2}\right)\right\|_{L^2(\mathbb{R}^2)}$$
$$\le C\|\theta_t\|_{W^{s-\alpha,p_1}(\mathbb{R}^2)}\|\theta\|_{W^{1,p_2}(\mathbb{R}^2)}+C\|\theta\|_{W^{s-\alpha,p_1}(\mathbb{R}^2)}\|\theta_t\|_{W^{1,p_2}(\mathbb{R}^2)}$$
$$+C\|\theta_t\|_{L^{p_3}(\mathbb{R}^2)}\|\theta\|_{W^{s-\alpha+1,p_4}(\mathbb{R}^2)}+C\|\theta\|_{L^{p_3}(\mathbb{R}^2)}\|\theta_t\|_{W^{s-\alpha+1,p_4}(\mathbb{R}^2)}. \tag{8.56}$$

By the Sobolev embedding theorem (see Section 2.1.4), we can select p_i, $i=1,2,3,4$, such that $H^{\alpha+s}(\mathbb{R}^2)$ is embedded into $W^{s-\alpha,p_1}(\mathbb{R}^2)$, $W^{1,p_2}(\mathbb{R}^2)$, $L^{p_3}(\mathbb{R}^2)$ and $W^{s-\alpha+1,p_4}(\mathbb{R}^2)$, respectively. For example, we can take

$$p_1=\frac{2}{1-\alpha^-},\quad p_2=\frac{2}{\alpha^-},\quad p_3=\frac{2}{2\alpha^--1},\quad p_4=\frac{1}{1-\alpha^-}.$$

Furthermore, using the Nirenberg–Gagliardo inequality and the Young inequality, we deduce that

$$\left\|(-\Delta)^{\frac{s-\alpha}{2}}\frac{d}{dt}(u\cdot\nabla\theta)\right\|_{L^2(\mathbb{R}^2)}\|(-\Delta)^{\frac{s+\alpha}{2}}\theta_t\|_{L^2(\mathbb{R}^2)}$$

$$\begin{aligned}
&\leq C\|\theta_t\|_{H^{\alpha+s}(\mathbb{R}^2)}^{\eta_1+1}\|\theta_t\|_{L^2(\mathbb{R}^2)}^{1-\eta_1}\|\theta\|_{H^{\alpha+s}(\mathbb{R}^2)} + C\|\theta_t\|_{H^{\alpha+s}(\mathbb{R}^2)}^{\eta_2+1}\|\theta_t\|_{L^2(\mathbb{R}^2)}^{1-\eta_2}\|\theta\|_{H^{\alpha+s}(\mathbb{R}^2)} \\
&\quad + C\|\theta_t\|_{H^{\alpha+s}(\mathbb{R}^2)}^{\eta_3+1}\|\theta_t\|_{L^2(\mathbb{R}^2)}^{1-\eta_3}\|\theta\|_{H^{\alpha+s}(\mathbb{R}^2)} + C\|\theta_t\|_{H^{\alpha+s}(\mathbb{R}^2)}^{\eta_4+1}\|\theta_t\|_{L^2(\mathbb{R}^2)}^{1-\eta_4}\|\theta\|_{H^{\alpha+s}(\mathbb{R}^2)} \\
&\leq \frac{\kappa}{2}\|\theta_t\|_{H^{\alpha+s}(\mathbb{R}^2)}^2 + \|\theta_t\|_{L^2(\mathbb{R}^2)}^2\mathrm{const}(\|\theta\|_{H^{\alpha+s}(\mathbb{R}^2)}),
\end{aligned} \tag{8.57}$$

where $\eta_1, \eta_2, \eta_3, \eta_4 \in (0, 1)$. Collecting (8.53)–(8.54) and (8.57) we obtain

$$\frac{d}{dt}\|\theta_t\|_{\dot{H}^s(\mathbb{R}^2)}^2 + 2\lambda\|\theta_t\|_{\dot{H}^s(\mathbb{R}^2)}^2 + \kappa\|\theta_t\|_{H^{\alpha+s}(\mathbb{R}^2)}^2 \leq \|\theta_t\|_{L^2(\mathbb{R}^2)}^2(\mathrm{const}(\|\theta\|_{H^{\alpha+s}(\mathbb{R}^2)}) + C). \tag{8.58}$$

Using (5.51), Hölder and Sobolev inequalities, we have

$$\begin{aligned}
\|\theta_t\|_{L^2(\mathbb{R}^2)}^2 &\leq C\|\theta\|_{L^2(\mathbb{R}^2)}^2 + C\|u\cdot\nabla\theta\|_{L^2(\mathbb{R}^2)}^2 + C\|(-\Delta)^\alpha\theta\|_{L^2(\mathbb{R}^2)}^2 + C\|f\|_{L^2(\mathbb{R}^2)}^2 \\
&\leq C\|\theta\|_{H^{\alpha+s}(\mathbb{R}^2)}^2 + C\|\theta\|_{H^{\alpha+s}(\mathbb{R}^2)}^4 + C\|f\|_{L^2(\mathbb{R}^2)}^2.
\end{aligned} \tag{8.59}$$

On the other hand, multiplying (8.15) by $(-\Delta)^s\theta_t$ and then integrating over $\mathbb{R}^2$, we obtain

$$\begin{aligned}
&\int_{\mathbb{R}^2}[(-\Delta)^{\frac{s}{2}}\theta_t]^2dx + \frac{1}{2}\lambda\frac{d}{dt}\int_{\mathbb{R}^2}[(-\Delta)^{\frac{s}{2}}\theta]^2dx + \frac{1}{2}\kappa\frac{d}{dt}\int_{\mathbb{R}^2}[(-\Delta)^{\frac{s+\alpha}{2}}\theta]^2dx \\
&\quad = \int_{\mathbb{R}^2}f(-\Delta)^s\theta_t dx - \int_{\mathbb{R}^2}u\cdot\nabla\theta(-\Delta)^s\theta_t dx.
\end{aligned} \tag{8.60}$$

By the Young inequality, we easily obtain that

$$\left|\int_{\mathbb{R}^2}f(-\Delta)^s\theta_t dx\right| \leq \frac{1}{4}\|(-\Delta)^{\frac{s}{2}}\theta_t\|_{L^2(\mathbb{R}^2)}^2 + \|f\|_{H^s(\mathbb{R}^2)}^2, \tag{8.61}$$

and similar to the arguments of (8.40)–(8.44), we get

$$\begin{aligned}
&-\int_{\mathbb{R}^2}u\cdot\nabla\theta(-\Delta)^s\theta_t dx \leq \left|\int_{\mathbb{R}^2}(-\Delta)^{\frac{s}{2}}(u\cdot\nabla\theta)(-\Delta)^{\frac{s}{2}}\theta_t dx\right| \\
&\leq \frac{1}{4}\|(-\Delta)^{\frac{s}{2}}\theta_t\|_{L^2(\mathbb{R}^2)}^2 + \|(-\Delta)^{\frac{s}{2}}(u\cdot\nabla\theta)\|_{L^2(\mathbb{R}^2)}^2 \\
&\leq \frac{1}{4}\|(-\Delta)^{\frac{s}{2}}\theta_t\|_{L^2(\mathbb{R}^2)}^2 + \|(-\Delta)^{\frac{2\alpha+s}{2}}\theta\|_{L^2(\mathbb{R}^2)}^2 + \mathrm{const}(\|\theta\|_{L^2(\mathbb{R}^2)}, \|\theta\|_{L^r(\mathbb{R}^2)}).
\end{aligned} \tag{8.62}$$

Using (8.61)–(8.62) in (8.60), we get

$$\begin{aligned}
&\|\theta_t\|_{\dot{H}^s(\mathbb{R}^2)}^2 + \lambda\frac{d}{dt}\|\theta\|_{\dot{H}^s(\mathbb{R}^2)}^2 + \kappa\frac{d}{dt}\|\theta\|_{\dot{H}^{\alpha+s}(\mathbb{R}^2)}^2 \\
&\leq 2\|\theta\|_{\dot{H}^{2\alpha+s}(\mathbb{R}^2)}^2 + 2\|f\|_{H^s(\mathbb{R}^2)}^2 + \mathrm{const}(\|\theta\|_{L^2(\mathbb{R}^2)}, \|\theta\|_{L^r(\mathbb{R}^2)}).
\end{aligned} \tag{8.63}$$

Integrating (8.63) from t to $t+1$ with respect to s', we have that for all $t \geq \mathcal{T}_0$,

$$\int_t^{t+1} \|\theta_t(s')\|^2_{\dot{H}^s(\mathbb{R}^2)} ds' \leq \lambda \|\theta(t)\|^2_{\dot{H}^s(\mathbb{R}^2)} + \kappa \|\theta(t)\|^2_{\dot{H}^{\alpha+s}(\mathbb{R}^2)} + 2\|f\|^2_{H^s(\mathbb{R}^2)} + C$$
$$+ 2\int_t^{t+1} \|\theta(s')\|^2_{\dot{H}^{2\alpha+s}(\mathbb{R}^2)} ds'.$$

Let $B \subset H^{2\alpha^-+s}(\mathbb{R}^2)$ be bounded. Then by Lemma 8.5.2, there exists $\mathcal{T}_0' = \mathcal{T}_0'(B) > 0$ such that for all $\theta_0 \in B$,

$$\int_t^{t+1} \|\theta_t(s', \theta_0)\|^2_{\dot{H}^s(\mathbb{R}^2)} ds' \leq \rho_0, \quad \forall t > \mathcal{T}_0^*, \tag{8.64}$$

where $\mathcal{T}_0^* = \max\{\mathcal{T}_0, \mathcal{T}_0'\} > 0$ is sufficiently large, $\mathcal{T}_0 > 0$ is given in Lemma 8.5.2 and $\rho_0 > 0$ is independent of B. In addition, it follows from (8.49)–(8.50) and (8.59) that for all $t \geq \mathcal{T}_0^*$ and $\theta_0 \in B$,

$$\int_t^{t+1} \|\theta_t(s', \theta_0)\|^2_{L^2(\mathbb{R}^2)} (\mathrm{const}(\|\theta(s', \theta_0)\|_{H^{\alpha+s}(\mathbb{R}^2)}) + C) ds' \leq \rho_1, \tag{8.65}$$

where $\rho_1 > 0$ is independent of B. Combining (8.58) with (8.64)–(8.65), and using the uniform Gronwall lemma (see Chapter 2), we obtain that for all $t \geq \mathcal{T}_0^* + 1$,

$$\|\theta_t(t, \theta_0)\|^2_{\dot{H}^s(\mathbb{R}^2)} \leq \rho_0 + \rho_1.$$

Setting $\mathcal{T}_1 = \mathcal{T}_0^* + 1$, we complete the proof. □

Lemma 8.5.4. *Fix $\alpha \in (\frac{1}{2}, 1]$ and $s > 1$. Assume that $f \in H^s(\mathbb{R}^2)$. Then for any bounded set $B \subset H^{2\alpha^-+s}(\mathbb{R}^2)$, there exists $\mathcal{T}_2 = \mathcal{T}_2(B) > 0$ such that for all $t \geq \mathcal{T}_2$, the solution θ of problem* (8.15) *satisfies*

$$\|\theta(t, \theta_0)\|^2_{H^{2\alpha+s}(\mathbb{R}^2)} \leq \rho_2, \tag{8.66}$$

where $\theta_0 \in B$ and ρ_2 is a positive constant independent of B.

Proof. By the first equation of (8.15), we find that

$$\|(-\Delta)^{\frac{s}{2}}(-\Delta)^\alpha \theta\|^2_{L^2(\mathbb{R}^2)} \leq C\|(-\Delta)^{\frac{s}{2}}\theta_t\|^2_{L^2(\mathbb{R}^2)} + C\|(-\Delta)^{\frac{s}{2}}(u \cdot \nabla\theta)\|^2_{L^2(\mathbb{R}^2)}$$
$$+ C\|(-\Delta)^{\frac{s}{2}}\theta\|^2_{L^2(\mathbb{R}^2)} + C\|(-\Delta)^{\frac{s}{2}} f\|^2_{L^2(\mathbb{R}^2)}. \tag{8.67}$$

By similar arguments as in (8.42)–(8.44) and the Young inequality, we have

$$C\|(-\Delta)^{\frac{s}{2}}(u \cdot \nabla\theta)\|^2_{L^2(\mathbb{R}^2)} \leq C\|(-\Delta)^{\frac{2\alpha+s}{2}}\theta\|^{2\eta'}_{L^2(\mathbb{R}^2)} \|\theta\|^{2(1-\eta')}_{L^2(\mathbb{R}^2)} \|\theta\|^2_{L^r(\mathbb{R}^2)}$$
$$\leq \frac{1}{2}\|(-\Delta)^{\frac{2\alpha+s}{2}}\theta\|^2_{L^2(\mathbb{R}^2)} + \mathrm{const}(\|\theta\|_{L^r(\mathbb{R}^2)}, \|\theta\|_{L^2(\mathbb{R}^2)}). \tag{8.68}$$

Using (8.68) in (8.67), we get

$$\begin{aligned}\|\theta(t)\|^2_{\dot{H}^{2\alpha+s}(\mathbb{R}^2)} \le C\|\theta_t(t)\|^2_{\dot{H}^s(\mathbb{R}^2)} + C\|\theta(t)\|^2_{\dot{H}^s(\mathbb{R}^2)} + C\|f\|^2_{H^s(\mathbb{R}^2)} \\ + \operatorname{const}(\|\theta\|_{L^r(\mathbb{R}^2)}, \|\theta\|_{L^2(\mathbb{R}^2)}) \end{aligned} \tag{8.69}$$

Let $B \subset H^{2\alpha^-+s}(\mathbb{R}^2)$ be bounded. Then by (8.69) and Lemmas 8.3.4 and 8.5.2–8.5.3, there exists $\mathcal{T}_2 = \mathcal{T}_2(B) > \mathcal{T}_1 > 0$ such that for all $\theta_0 \in B$,

$$\|\theta(t,\theta_0)\|^2_{H^{2\alpha+s}(\mathbb{R}^2)} \le \rho_2, \quad \forall\, t \ge \mathcal{T}_2,$$

where $\mathcal{T}_1 > 0$ is given in Lemma 8.5.3 and $\rho_2 > 0$ is independent of B. The proof is thus completed. □

Next, we derive an estimate of the tails of solutions of (8.15) in $L^2(\mathbb{R}^2)$.

Lemma 8.5.5. *Fix $\alpha \in (\frac{1}{2}, 1]$ and $s > 1$. Assume that $f \in H^s(\mathbb{R}^2)$. Then for any $\varepsilon > 0$ and any bounded set $B \subset H^{2\alpha^-+s}(\mathbb{R}^2)$, there exist $\mathcal{T}_1^* = \mathcal{T}_1^*(B,\varepsilon) > 0$ and $\mathcal{K}_1^* = \mathcal{K}_1^*(\varepsilon) > 0$ such that for all $t \ge \mathcal{T}_1^*$ and $K \ge \mathcal{K}_1^*$, the solution θ of problem* (8.15) *satisfies*

$$\int_{\mathcal{O}_K} |\theta(t,\theta_0)|^2 dx \le \varepsilon,$$

where $\mathcal{O}_K = \{x \in \mathbb{R}^2 : |x| \ge K\}$ and $\theta_0 \in B$.

The proof is contained in that of Lemma 8.3.9 (when $p = 2$) and will be omitted.

Theorem 8.5.6. *Assume that the hypotheses in Lemma* 8.5.4 *hold. Then the semigroup $\{S(t)\}_{t\ge0}$ associated with problem* (8.15) *possesses a unique global attractor $\mathcal{A}$ in $H^{2\alpha^-+s}(\mathbb{R}^2)$.*

Proof. Asymptotic compactness of the semigroup $\{S(t)\}_{t\ge0}$ in $H^{2\alpha^-+s}(\mathbb{R}^2)$ follows directly from Lemmas 8.5.4, 8.5.5, and the Nirenberg–Gagliardo inequality. By Theorem 3.4.1 in [86], we see that $S(t)$ is continuous with respect to t and for fixed t in $H^{2\alpha^-+s}(\mathbb{R}^2)$, respectively. Thanks to Lemma 8.5.4, the existence of a global attractor $\mathcal{A}$ follows immediately from the abstract results in [13, 83, 141, 172, 180] or in subsection 2.3.2. □

8.6 Critical equation (8.1); $\alpha = \frac{1}{2}$

8.6.1 Passing to the limit

A precise description of letting $\alpha \to \frac{1}{2}^+$ in the equation (8.1) is given next. Technical details used in the main considerations below are discussed in the following subsection.

In this section, we deal with solutions θ^α (we add the superscript for clarity) constructed in subsection 8.4 on the base space $H^s(\mathbb{R}^2)$, $s > 1$. Such solutions, for any $\alpha \in (\frac{1}{2}, 1]$, are varying continuously in $H^{1+s}(\mathbb{R}^2)$, $s > 1$; hence also in each of the spaces $W^{1,p}(\mathbb{R}^2)$, $2 \le p \le +\infty$.

In particular, they fulfill the *uniform in* $\alpha \in (\frac{1}{2}, 1]$ estimates in $L^p(\mathbb{R}^2)$ of Lemma 8.2.4, the *main information* allowing us to let $\alpha \to \frac{1}{2}^+$ in the equation (8.1). More precisely, for such solutions, we have

$$\exists_{\text{const}>0} \forall_{\alpha \in (\frac{1}{2}, \frac{3}{4}]} \forall_{p \in [2,+\infty)} \|\theta^\alpha\|_{L^\infty(0,T;L^p(\mathbb{R}^2))} \le \text{const}, \tag{8.70}$$

where $T > 0$ is fixed. To work with *sectorial positive* operator $A_\alpha := \kappa[(-\Delta)^\alpha + I]$, we add to both sides of (8.1) the term $\kappa\theta$ to obtain

$$\begin{aligned} &\theta_t^\alpha + u \cdot \nabla\theta^\alpha + A_\alpha\theta^\alpha = f + \kappa\theta^\alpha, \quad x \in \mathbb{R}^2, t > 0, \\ &\theta^\alpha(0, x) = \theta_0(x). \end{aligned} \tag{8.71}$$

We look at (8.1) as an equation in $L^p(\mathbb{R}^2)$, $p \ge 2$, and multiply it by the test function $(A_\alpha^{-1})^*\phi$ where $\phi \in D((-\Delta_q)^{-1})$, $\frac{1}{p} + \frac{1}{q} = 1$ (see Lemma 5.3.3), to get

$$\begin{aligned} &\langle [\theta_t^\alpha + u \cdot \nabla\theta^\alpha], (A_\alpha^{-1})^*\phi\rangle_{L^p,L^q} \\ &\quad = -\langle A_\alpha\theta^\alpha, (A_\alpha^{-1})^*\phi\rangle_{L^p,L^q} + \langle [f + \kappa\theta^\alpha], (A_\alpha^{-1})^*\phi\rangle_{L^p,L^q}. \end{aligned} \tag{8.72}$$

We will discuss now the convergence of the terms in (8.72) one by one. By Lemma 5.3.3, when $\alpha \to \frac{1}{2}^+$, $(A_\alpha^{-1})^*\phi \to (A_{\frac{1}{2}}^{-1})^*\phi$ for $\phi \in D((-\Delta_q)^{-1})$. Thanks to uniform in $\alpha \in (\frac{1}{2}, 1]$ boundedness of θ^α in $L^p(\mathbb{R}^2)$, $p \ge 2$ (Lemma 8.2.4), we obtain

$$\begin{aligned} &\langle [f + \kappa\theta^\alpha], (A_\alpha^{-1})^*\phi\rangle_{L^p,L^q} \\ &\quad = \langle [f + \kappa\theta^\alpha], (A_\alpha^{-1})^*\phi - (A_{\frac{1}{2}}^{-1})^*\phi\rangle_{L^p,L^q} + \langle [f + \kappa\theta^\alpha], (A_{\frac{1}{2}}^{-1})^*\phi\rangle_{L^p,L^q} \\ &\quad \to \langle [f + \kappa\theta], (A_{\frac{1}{2}}^{-1})^*\phi\rangle_{L^p,L^q}, \end{aligned} \tag{8.73}$$

where θ is the weak limit of θ^α in $L^p(\mathbb{R}^2)$ as $\alpha \to \frac{1}{2}^+$ (over a sequence $\{\alpha_n\}$ convergent to $\frac{1}{2}^+$; various sequences may lead to various weak limits).

For the intermediate term, we have the equality:

$$\langle A_\alpha\theta^\alpha, (A_\alpha^{-1})^*\phi\rangle_{L^p,L^q} = \langle \theta^\alpha, \phi\rangle_{L^p,L^q}. \tag{8.74}$$

Moreover, since $\theta^\alpha \in W^{1,p}(\mathbb{R}^2)$, $p \ge 2$, we have $\theta^\alpha \in L^p(\mathbb{R}^2)$ and $A_{\frac{1}{2}}\theta^\alpha =: T^\alpha \in L^p(\mathbb{R}^2)$, which gives that $\theta^\alpha = A_{\frac{1}{2}}^{-1}T^\alpha$ for some $T^\alpha \in L^p(\mathbb{R}^2)$. Inserting to the equation above, we obtain

$$\langle \theta^\alpha, \phi\rangle_{L^p,L^q} = \langle A_{\frac{1}{2}}^{-1}T^\alpha, \phi\rangle_{L^p,L^q} = \langle T^\alpha, (A_{\frac{1}{2}}^{-1})^*\phi\rangle_{L^p,L^q}.$$

Since the left-hand side has a limit as $\alpha_n \to \frac{1}{2}^+$, the right-hand side also has the same limit:

$$\langle T^\alpha, (A_{\frac{1}{2}}^{-1})^* \phi \rangle_{L^p,L^q} \to \langle \Theta, (A_{\frac{1}{2}}^{-1})^* \phi \rangle_{L^p,L^q} = \langle \theta, \phi \rangle_{L^p,L^q} \quad \text{for all } \phi \in D((-\Delta_q)^{-1}); \tag{8.75}$$

note that $\mathcal{L}_2^q(\mathbb{R}^2)$ is dense in $L^q(\mathbb{R}^2)$ and that $\theta, \Theta \in L^p(\mathbb{R}^2)$.

Consequently, returning to (8.72), we see that the first term also has a limit:

$$\langle [\theta_t^\alpha + u \cdot \nabla \theta^\alpha], (A_\alpha^{-1})^* \phi \rangle_{L^p,L^q} \to \omega_\phi, \tag{8.76}$$

as $\alpha \to \frac{1}{2}^+$.

Considering the limits together, taken over the same sequence $\{\alpha_n\}$ convergent to $\frac{1}{2}^+$, we find a weak form of the limit equation:

$$\forall_{\phi \in \mathcal{L}_2^q(\mathbb{R}^2)} \, \omega_\phi = -\langle \theta, \phi \rangle_{L^p,L^q} + \langle f + \kappa\theta, (A_{\frac{1}{2}}^{-1})^* \phi \rangle_{L^p,L^q}. \tag{8.77}$$

Remembering that the set $\{(A_{\frac{1}{2}}^{-1})^* \phi; \phi \in D((-\Delta_q)^{-1})\}$ is dense in $L^q(\mathbb{R}^2)$, the right-hand side above defines a unique element in $L^p(\mathbb{R}^2)$.

Separation of terms

The time derivative will be separated from the term $[\theta_t^\alpha + u \cdot \nabla \theta^\alpha]$ when letting $\alpha \to \frac{1}{2}^+$. More precisely we have the following.

Remark 8.6.1. Since the approximating solutions θ^α satisfy

$$\theta^\alpha \in L^\infty(0, T; L^p(\mathbb{R}^2)), \theta_t^\alpha \in L^2(0, T; L^p(\mathbb{R}^2)), \quad \alpha \in \left(\frac{1}{2}, \frac{3}{4}\right], \tag{8.78}$$

then by [171, Lemma 1.1, Chapter III]

$$\forall_{\eta \in L^q(\mathbb{R}^2)} \langle \theta_t^\alpha, \eta \rangle_{L^p,L^q} = \frac{d}{dt} \langle \theta^\alpha, \eta \rangle_{L^p,L^q} \to \frac{d}{dt} \langle \theta, \eta \rangle_{L^p,L^q}, \tag{8.79}$$

(here $\frac{1}{p} + \frac{1}{q} = 1$) the derivative $\frac{d}{dt}$ and the convergence are in $D'(0, T)$ (space of the "scalar distributions"). Consequently,

$$\omega_\phi = \frac{d}{dt} \langle \theta, (A_{\frac{1}{2}}^{-1})^* \phi \rangle_{L^p,L^q} + \omega_\phi^1, \tag{8.80}$$

where ω_ϕ^1 is a limit in $D'(0, T)$ of $\langle u \cdot \nabla \theta^\alpha, (A_\alpha^{-1})^* \phi \rangle_{L^p,L^q}$ over a chosen sequence $\alpha_n \to \frac{1}{2}^+$.

Similar to (8.73), we obtain

$$\begin{aligned} &|\langle u^\alpha \cdot \nabla \theta^\alpha, (A_\alpha^{-1})^* \phi \rangle_{L^p,L^q} - \langle u^\alpha \cdot \nabla \theta^\alpha, (A_{\frac{1}{2}}^{-1})^* \phi \rangle_{L^p,L^q}| \\ &= |\langle u^\alpha \cdot \nabla \theta^\alpha, (A_\alpha^{-1})^* \phi - (A_{\frac{1}{2}}^{-1})^* \phi \rangle_{L^p,L^q}| \to 0 \quad \text{as} \quad \alpha \to \frac{1}{2}^+. \end{aligned} \tag{8.81}$$

Then we can deduce that ω_ϕ^1 is also a limit in $D'(0,T)$ of $\langle u \cdot \nabla\theta^\alpha, (A_{\frac{1}{2}}^{-1})^*\phi\rangle_{L^p,L^q}$ over a chosen sequence $\alpha_n \to \frac{1}{2}^+$.

We were unable to specify the exact form of the limit ω_ϕ^1 in case of equation (8.1). But, if we replace equation (8.1) with the damped equation (8.15), then using the "tail estimates" technique we are able to specify the form of the corresponding to it term ω_ϕ^1. By Lemmas 8.3.5 and 8.3.9, we have the uniform boundedness of θ^α and the tail estimates for solution θ^α for (8.15) in $L^p(\mathbb{R}^2)$, combining the following interpolation inequality:

$$\|\theta^\alpha(t)\|_{W^{1,p}(\mathcal{O}_k)} \le c\|\theta^\alpha(t)\|_{W^{2\alpha^-,p}(\mathbb{R}^2)}^{\frac{1}{2\alpha^-}} \|\theta^\alpha(t)\|_{L^p(\mathcal{O}_k)}^{\frac{2\alpha^- -1}{2\alpha^-}},$$

we can find the tail estimates in $W^{1,p}(\mathbb{R}^2)$. Thus, for any $\epsilon > 0$, we can choose $k > 0$ sufficiently large such that

$$\int_0^T \|\theta^\alpha(t)\|_{W^{1,p}(\mathcal{O}_k)}^2 dt < \frac{\epsilon}{8}. \tag{8.82}$$

On the other hand, using the Lions-Aubin compactness lemma (see section 5.4.2), we can find a subsequence $\{\theta^\alpha\}$ (after relabeling) that converges to θ strongly in $L^2(0,T; W^{1,p}(B_k))$, where B_k denotes a closed ball in $\mathbb{R}^2$ centered at zero with radius k, which implies that θ^α is a Cauchy sequence in $L^2(0,T; W^{1,p}(B_k))$, that is, for any $\epsilon > 0$ and $\theta^\alpha \in L^2(0,T; W^{1,p}(\mathbb{R}^2))$, there exists a positive constant $N = N(\epsilon)$ such that

$$\int_0^T \|\theta^{\alpha_n} - \theta^{\alpha_m}\|_{W^{1,p}(B_k)}^2 dt < \frac{\epsilon}{2}, \quad \text{for } n, m > N. \tag{8.83}$$

Using (8.82) and (8.83), for any $\epsilon > 0$ and $\theta^\alpha \in L^2(0,T; W^{1,p}(\mathbb{R}^2))$, we have

$$\begin{aligned}\int_0^T \|\theta^{\alpha_n} - \theta^{\alpha_m}\|_{W^{1,p}(\mathbb{R}^2)}^2 dt &\le \int_0^T \|\theta^{\alpha_n} - \theta^{\alpha_m}\|_{W^{1,p}(B_k)}^2 dt + \int_0^T \|\theta^{\alpha_n} - \theta^{\alpha_m}\|_{W^{1,p}(\mathcal{O}_k)}^2 dt \\ &\le \int_0^T \|\theta^{\alpha_n} - \theta^{\alpha_m}\|_{W^{1,p}(B_k)}^2 dt + 2\left(\int_0^T \|\theta^{\alpha_n}\|_{W^{1,p}(\mathcal{O}_k)}^2 dt + \int_0^T \|\theta^{\alpha_m}\|_{W^{1,p}(\mathcal{O}_k)}^2 dt\right) \\ &< \frac{\epsilon}{2} + \frac{\epsilon}{2} < \epsilon, \quad \text{for } n, m > N,\end{aligned}$$

which implies that θ^α is also a Cauchy sequence in $L^2(0,T; W^{1,p}(\mathbb{R}^2))$, therefore, $\{\theta^\alpha\}$ converges to θ strongly in $L^2(0,T; W^{1,p}(\mathbb{R}^2))$. Using boundedness of the Riesz operator, for $(A_{\frac{1}{2}}^{-1})^*\phi$ with $\phi \in \mathcal{L}_2^q(\mathbb{R}^2)$, we get

$$\left|\int_0^T \int_{\mathbb{R}^2} u^\alpha \cdot \nabla\theta^\alpha (A_{\frac{1}{2}}^{-1})^* \phi - u \cdot \nabla\theta (A_{\frac{1}{2}}^{-1})^* \phi dx dt\right|$$

$$\leq \left| \int_0^T \int_{\mathbb{R}^2} (u^\alpha - u) \cdot \nabla \theta^\alpha (A_{\frac{1}{2}}^{-1})^* \phi + u \cdot \nabla(\theta^\alpha - \theta)(A_{\frac{1}{2}}^{-1})^* \phi dx dt \right|$$

$$\leq \int_0^T (\|u^\alpha - u\|_{L^\infty(\mathbb{R}^2)} \|\theta^\alpha\|_{W^{1,p}(\mathbb{R}^2)} + \|u\|_{L^\infty(\mathbb{R}^2)} \|\theta^\alpha - \theta\|_{W^{1,p}(\mathbb{R}^2)}) \|(A_{\frac{1}{2}}^{-1})^* \phi\|_{L^q(\mathbb{R}^2)} dt$$

$$\leq c \int_0^T \|u^\alpha - u\|_{W^{1,p}(\mathbb{R}^2)} \|\theta^\alpha\|_{W^{1,p}(\mathbb{R}^2)} \|(A_{\frac{1}{2}}^{-1})^* \phi\|_{L^q(\mathbb{R}^2)} dt$$

$$+ c \int_0^T \|u\|_{W^{1,p}(\mathbb{R}^2)} \|\theta^\alpha - \theta\|_{W^{1,p}(\mathbb{R}^2)} \|(A_{\frac{1}{2}}^{-1})^* \phi\|_{L^q(\mathbb{R}^2)} dt$$

$$\leq c\|(A_{\frac{1}{2}}^{-1})^* \phi\|_{L^\infty(0,T;L^q(\mathbb{R}^2))} \|\theta^\alpha\|_{L^2(0,T;W^{1,p}(\mathbb{R}^2))} \|\theta^\alpha - \theta\|_{L^2(0,T;W^{1,p}(\mathbb{R}^2))}$$

$$+ c\|(A_{\frac{1}{2}}^{-1})^* \phi\|_{L^\infty(0,T;L^q(\mathbb{R}^2))} \|\theta\|_{L^2(0,T;W^{1,p}(\mathbb{R}^2))} \|\theta^\alpha - \theta\|_{L^2(0,T;W^{1,p}(\mathbb{R}^2))} \to 0$$

as $\alpha \to \frac{1}{2}^+$, which shows that $\omega_\phi^1 = \langle u \cdot \nabla\theta, (A_{\frac{1}{2}}^{-1})^* \phi \rangle_{L^p, L^q}$.

The construction presented above allows us to formulate the following theorem.

Theorem 8.6.2. *Let $\{\theta^\alpha\}_{\alpha \in (\frac{1}{2}, \frac{3}{4}]}$ be the set of regular $H^{s+2\alpha^-}(\mathbb{R}^2)$ solutions to subcritical equation (8.1). Such solutions are, in particular, bounded in each space $L^p(\mathbb{R}^2)$ for $p \in [2, +\infty)$, uniformly in α. As a consequence of that and the smoothness properties of regular solutions (they vary continuously in $W^{1,p}(\mathbb{R}^2)$), for arbitrary sequence $\{\alpha_n\} \subset (\frac{1}{2}, \frac{3}{4}]$ convergent to $\frac{1}{2}^+$ we can find a subsequence $\{\alpha_{n_k}\}$ that the corresponding sequence $\{\theta^{n_k}\}$ converges weakly in $L^p(\mathbb{R}^2)$ to a function θ fulfilling the equation:*

$$\forall_{\phi \in \mathcal{L}_2^q(\mathbb{R}^2)}\ \omega_\phi = -\langle \theta, \phi \rangle_{L^p, L^q} + \langle f + \kappa\theta, (A_{\frac{1}{2}}^{-1})^* \phi \rangle_{L^p, L^q}. \tag{8.84}$$

Due to denseness of the set $D((-\Delta_q)^{-1})$ in $L^q(\mathbb{R}^2)$, $\frac{1}{p} + \frac{1}{q} = 1$, the right-hand side of (8.84) defines a unique element in $L^p(\mathbb{R}^2)$. The left-hand side ω_ϕ is defined in (8.76) and discussed in Remark 8.6.1. We will call such θ a weak L^p solution to the critical equation (8.1), $\alpha = \frac{1}{2}$.

Bibliographical notes

The method of tail estimates used in that chapter was originated in [178] (see also [146]) and is now a standard tool for proving compactness (if available) for solutions of the Cauchy problem.

9 Dirichlet's problem for critical 2D quasi-geostrophic equation

9.1 Introduction

This chapter is devoted to the Dirichlet problem for subcritical and critical *Viscous Surface Quasi-geostrophic Equation* (Q-g equation, in short) [23, 37, 40, 53, 101, 117, 131, 183, 184, 188]:

$$\begin{aligned}
&\theta_t + u \cdot \nabla\theta + \kappa(-\Delta)^{\alpha}\theta = f, \quad x \in \Omega,\ t > 0,\\
&\theta(t,x) = 0, \quad x \in \partial\Omega,\ t > 0,\\
&\theta(0,x) = \theta_0(x),
\end{aligned} \tag{9.1}$$

where θ represents the potential temperature, $\kappa > 0$ is a diffusivity coefficient, $\alpha \in [\frac{1}{2}, 1]$ a fractional exponent, and $u = (u_1, u_2)$ is the *velocity field* determined by θ through the relation:

$$u = \left(-\frac{\partial\psi}{\partial x_2}, \frac{\partial\psi}{\partial x_1}\right), \quad \text{where } (-\Delta)^{\frac{1}{2}}\psi = -\theta, \tag{9.2}$$

or, in a more explicit way,

$$u = R^{\perp}\theta = (-R_2\theta, R_1\theta), \tag{9.3}$$

where R is the Riesz transform in a bounded domain $\Omega \subset \mathbb{R}^2$ defined as

$$R = \nabla(-\Delta)^{-\frac{1}{2}}, \quad \text{with components } R_j = \frac{\partial}{\partial x_j}(-\Delta)^{-\frac{1}{2}}, j = 1, 2; \tag{9.4}$$

with fractional Laplacian in the sense of Balakrishnan-Komatsu.

The Dirichlet boundary value problem with zero boundary conditions is considered in a bounded domain Ω with $\partial\Omega$ of class C^2. Definition 5.2.1 is used to introduce fractional powers of negative Dirichlet Laplacian.

As in the case of the Navier–Stokes equations, we analyze first the subcritical range of exponents $\alpha \in (\frac{1}{2}, 1]$, letting then the parameter $\alpha \to \frac{1}{2}^{+}$. See [131] for another regularization of the problem, where an extra Laplacian is added to join the existing dissipative term (see also [40, p. 521]). We remark here that in the subcritical case $\alpha \in (\frac{1}{2}, 1]$ the already existing dissipation $\kappa(-\Delta)^{\alpha}\theta$ alone is strong enough to guarantee good properties of solutions; an extra regularization is certainly needed when $\alpha \in (0, \frac{1}{2})$.

The technique used here is an extension of the classical *vanishing viscosity technique* that comes back to the 1950s and the studies of E. Hopf, O. A. Oleinik, P. D. Lax and J.-L. Lions dealing initially with the Burgers model and some hyperbolic problems

https://doi.org/10.1515/9783110599831-009

(e. g., [147]). It is applicable to critical and supercritical equations, in which the nonlinear term is "equivalent or more valid" than the main dissipative term in the equation (equivalent with first or higher power of the main part operator). Our idea is simple. We just strengthen that dissipative term, replacing it with its fractional power with sufficiently large exponent greater than 1. Next, using the strong and elegant semigroup technique we solve easily the modified problems. The final step is to pass to the limit over a sequence of solutions to such regularized problems, to obtain a weak solution of the original problem. Essential in that step are the *uniform with respect to the approximation parameter* estimates of solutions to the regularized problems. In Chapter 6, such technique was used to the 3-D Navier–Stokes equation, and the results obtained are fully comparable with those obtained within another approaches.

The advantage of our approach is that the approximations used are smooth (since they are solutions of the regular dissipative problems). Moreover, the approximating problems are very similar in nature to their critical or supercritical limit. Note that for regular functions the difference between $(-\Delta)\phi$ and $(-\Delta)^{1+\epsilon}\phi$ tends to zero as $\epsilon \to 0^+$ (e. g., [130, Proposition 3.1.1]). The solutions to our approximations exist globally in time, while for the limits they may be only local, with difficult estimation of the lifetime. This property helps when looking numerically for solutions, since one can be sure that the approximating solutions exist and have good properties for arbitrary positive time.

Let us discuss briefly the difference, in the light of the above described technique, between the Navier–Stokes equation (in 3-D) and the 2-D viscous quasi-geostrophic equation. The difference is significant, since the 3-D N-S equation is locally well posed in many phase spaces used in considerations of that problem. This property is a consequence of the fact that the nonlinearity of the 3-D N-S acts between elements of the fractional order scale corresponding to the Stokes operator, the difference of which (measured by the difference of exponents) is strictly less than one. Consequently, the standard semigroup approach (like in [32, 86, 138]) will be used to get *local in time* solutions. The weakness in case of the N-S equation is that the known in 3-D a priori estimates are too weak to guarantee, together with the action of the Stokes operator, global in time extendibility of that local solutions.

In case of the critical 2-D Q-g equation, the situation is essentially different. Since the nonlinearity is functionally equivalent (it contains ∇) with the main part operator $(-\Delta)^{\frac{1}{2}}$, the classical perturbation technique in the semigroup approach will not work. Consequently, we will not have following from general theory [32, 86] local in time solutions for that problem. To get some kind solutions, we are forced to obtain them by solving the regularized equations (with stronger viscosity exponent $\alpha > \frac{1}{2}$) and passing to the limiting critical equation.

In the present chapter, we will study the *Dirichlet problem* in a bounded domain $\Omega \subset \mathbb{R}^2$ having regular (C^2, say) boundary $\partial\Omega$. Let us mention, that the Cauchy problem in $\mathbb{R}^N$ was more often studied in the literature (compare Chapter 8).

9.2 Local in time solvability of subcritical quasi-geostrophic equation

We will discuss now the, possible, choices for the pairs base/phase spaces in which the local in time solvability of the critical Q-g equation in N-D will be verified. Moreover, we need to make that choice in such a way, that possible is the limit passage in the approximating subcritical problems to obtain in such limit a "weak form" solution for the critical equation. For the subcritical approximations, the possible set of phase spaces is larger, while the latter requirement we formulate limits that possibility, if we want to have a phase space suitable for the whole family of approximating problems with parameter $\alpha \in (\frac{1}{2}, \frac{3}{4}]$ including also the limiting critical problem.

The reasoning is as follows. Consider the nonlinear term in the limiting critical Q-g equation (exponent $\alpha = \frac{1}{2}$) that equals $u \cdot \nabla\theta$, and compare its action with the corresponding action of the main part operator $\kappa(-\Delta)^{\frac{1}{2}}\theta$. Evidently, the linear main part operator acts between the spaces $X^{a+\frac{1}{2}}$ and X^a, $a \in \mathbb{R}$ arbitrary, of the power scale of $(-\Delta)$ operator subjected to chosen boundary conditions. Consequently, the corresponding nonlinearity, even in critical case, also need to act (at most) between the mentioned spaces (perhaps with a certain, properly chosen and suitable for it $a \in \mathbb{R}$). But since the nonlinearity already contains first-order derivatives, multiplied by u, the phase space should be included in the space L^∞ to assure that such multiplication will not enlarge the distance of the spaces the nonlinearity acts between. Among the "usual Sobolev type spaces" such requirement is fulfilled in particular by the following pairs of the base/phase spaces:

- Base space $L^p(\Omega)$, phase space $W_0^{1,p}(\Omega)$ with $p > N$. Note $W_0^{1,p}(\Omega)$ is a Banach Algebra contained in $L^\infty(\Omega)$ when $p > N$,
- Base space $D((-\Delta)^{\frac{r}{2}}) \subset W^{r,p}(\Omega)$ with $r > 0$, phase space $D((-\Delta)^{\frac{r+1}{2}}) \subset W^{r+1,p}(\Omega)$, again with $p > N$,
- Base space $D((-\Delta)^{\frac{s}{2}}) \subset H^s(\Omega)$ with $s > \frac{N}{2}$, phase space $D((-\Delta)^{\frac{s+1}{2}})$ contained in $H^{s+1}(\Omega)$, again a Banach Algebra contained in $L^\infty(\Omega)$,
- The borderline case: base space $L^2(\Omega)$ with the phase space $H_0^1(\Omega) \cap L^\infty(\Omega)$, $N = 2$, as used in Remark 9.5.4. Compare Remark 5.4.11 or directly [96] for details.

Characterization of the domains $D((-\Delta)^\gamma)$, $\gamma > 0$, (incorporating boundary conditions) is given, for example, in [187, pp. 558–561], or in [118]. In all of the discussed above pairs, the nonlinearity acts between elements of the fractional power scale $X^\beta = D((-\Delta)^\beta)$ whose distance (measured by the difference of exponents) is not larger than $\frac{1}{2}$, same as the main part operator exponent in (9.1). Such situation will still be handled in our approach. *Further in the text, we will just call the phase space, for example, $H^s(\Omega)$, not including the required boundary conditions in that notation.*

We will consider here in details the case of the $H^{s+1}(\Omega)$ phase space (more precisely $D((-\Delta)^{\frac{s+1}{2}})$), for arbitrary exponent $\alpha \in (\frac{1}{2}, 1]$. Note that since $u = (-R_2\theta, R_1\theta)$ then

$\operatorname{div} u = 0$, and the nonlinear term in (9.1) will be rewritten as

$$u \cdot \nabla\theta = \nabla \cdot (u\theta) - \theta \operatorname{div} u = \nabla \cdot (u\theta). \tag{9.5}$$

To justify local in time solvability of (9.1), we need to check that the nonlinearity $u \cdot \nabla\theta = \nabla \cdot (u\theta)$ is Lipschitz on bounded sets as a map from $D((-\Delta)^{\frac{s+1}{2}}) \subset H^{s+1}(\Omega)$ into $D((-\Delta)^{\frac{s}{2}}) \subset H^s(\Omega)$, $s \in (\frac{3}{2}, \frac{5}{2})$ fixed arbitrarily (where we will use equivalent norms of the $H^s(\Omega)$ spaces). Indeed, we have

$$\begin{aligned} \|F(\theta_1) - F(\theta_2)\|_{H^s(\Omega)} \le & \left\| R_2(\theta_1 - \theta_2)\frac{\partial\theta_1}{\partial x_1} + R_2\theta_2\frac{\partial(\theta_1 - \theta_2)}{\partial x_1} \right\|_{H^s(\Omega)} \\ & + \left\| R_1(\theta_1 - \theta_2)\frac{\partial\theta_1}{\partial x_2} + R_1\theta_2\frac{\partial(\theta_1 - \theta_2)}{\partial x_2} \right\|_{H^s(\Omega)}. \end{aligned} \tag{9.6}$$

Estimating the second term above we are using the property (e. g., [1, p. 115]), that *when Ω is a domain in $\mathbb{R}^N$ then $W^{m,r}(\Omega)$ is a Banach algebra provided that $mr > N$*. Hence

$$\begin{aligned} \left\| R_2(\theta_1 - \theta_2)\frac{\partial\theta_1}{\partial x_1} \right\|_{H^s(\Omega)} & \le c\|R_2(\theta_1 - \theta_2)\|_{H^s(\Omega)} \left\| \frac{\partial\theta_1}{\partial x_1} \right\|_{H^s(\Omega)} \\ & \le c\||u_1 - u_2|\|_{H^s(\Omega)} \|\theta_1\|_{H^{s+1}(\Omega)}, \end{aligned} \tag{9.7}$$

where u_i, $i = 1, \ldots, N$, correspond to θ_i through relation (9.3).

Applying the property as in [183, p. 12] (see also (5.58)):

$$\||D^j u(t,\cdot)|\|_{L^q(\Omega)} \le c\|D^j\theta(t,\cdot)\|_{L^q(\Omega)}, \quad q \in (1,\infty),\ |j| \le k, \tag{9.8}$$

to the first term in (9.7), we get

$$\left\| R_2(\theta_1 - \theta_2)\frac{\partial\theta_1}{\partial x_1} \right\|_{H^s(\Omega)} \le c\|\theta_1 - \theta_2\|_{H^s(\Omega)} \|\theta_1\|_{H^{s+1}(\Omega)}.$$

The other components in (9.6) are estimated analogously. Consequently, we obtain that for each $s > 1$,

$$\|F(\theta_1) - F(\theta_2)\|_{H^s(\Omega)} \le c'(\|\theta_1\|_{H^{s+1}(\Omega)}, \|\theta_2\|_{H^{s+1}(\Omega)})\|\theta_1 - \theta_2\|_{H^{s+1}(\Omega)}. \tag{9.9}$$

If additionally $s \in (\frac{3}{2}, \frac{5}{2})$ then, due to the characterization of the domains of fractional powers

$$D((-\Delta_2)^{\frac{s}{2}}) = \begin{cases} H^s(\Omega) & \text{for } 0 \le s < \frac{1}{2}, \\ H_0^s(\Omega) & \text{if } \frac{1}{2} < s < 1, \\ H^s(\Omega) \cap H_0^1(\Omega) & \text{if } 1 \le s < \frac{5}{2}, s \ne \frac{3}{2} \\ H^s(\Omega) \cap H_0^2(\Omega) & \text{if } \frac{5}{2} < s < \frac{9}{2}, s \ne \frac{7}{2}, \end{cases} \tag{9.10}$$

and since $F : H_0^2(\Omega) \to H_0^1(\Omega)$, we have that $F : D((-\Delta)^{\frac{s+1}{2}}) \to D((-\Delta)^{\frac{s}{2}})$. Moreover, as a consequence of (9.9) F is Lipschitz on bounded sets of $D((-\Delta)^{\frac{s+1}{2}})$ (under equivalent norm of $H^{s+1}(\Omega)$) which, following [32, 86], proves local solvability of (9.1) in the $D((-\Delta)^{\frac{s+1}{2}})$ phase space. More precisely, we formulate the following.

Theorem 9.2.1. *Let $s \in (\frac{3}{2}, \frac{5}{2})$ be fixed. Then, for $f \in D((-\Delta)^{\frac{s}{2}}) \subset H^s(\Omega)$ and for arbitrary $\theta_0 \in D((-\Delta)^{\frac{s+1}{2}}) \subset H^{s+1}(\Omega)$, there exists in the phase space $D((-\Delta)^{\frac{s+1}{2}})$ a unique local in time mild solution θ^α to the problem (9.1), $\alpha \in (\frac{1}{2}, 1]$. Moreover,*

$$\theta^\alpha \in C((0,\tau); H^{s+2\alpha}(\Omega)) \cap C([0,\tau); H^{s+1}(\Omega)), \quad \theta^\alpha_t \in C((0,\tau); H^{s+2\gamma}(\Omega)),$$

with arbitrary $\gamma < \alpha$. Here, $\tau > 0$ is the "lifetime" of that local in time solution. Moreover, the Cauchy formula is satisfied:

$$\theta^\alpha(t) = e^{-A_\alpha t}\theta_0 + \int_0^t e^{-A_\alpha(t-s)}F(\theta^\alpha(s))ds, \quad t \in [0,\tau),$$

where $e^{-A_\alpha t}$ denotes the linear semigroup corresponding to the operator $A_\alpha := (-\Delta)^\alpha$ in $D((-\Delta)^{\frac{s+1}{2}}) \subset H^{s+1}(\Omega)$, and $F(\theta^\alpha) = -u^\alpha \cdot \nabla\theta^\alpha + f$.

Optional phase space

As an option, a phase space contained in the space $W^{1,p}(\Omega)$, $p > N$, will be considered. Let $p > N$ be given, and set $D((-\Delta)^{\frac{1}{2}}) = W^{1,p}_0(\Omega)$ (e. g., [187, p. 561]) as a phase space, with $L^p(\Omega)$ as a base space, in which the equation will be fulfilled. A simple calculation below will show Lipschitz continuity on bounded sets of the nonlinear term $F : W^{1,p}_0(\Omega) \to L^p(\Omega)$. Indeed,

$$\begin{aligned}\|F(\theta_1) - F(\theta_2)\|_{L^p(\Omega)} \le{}& \left\| R_2(\theta_1 - \theta_2)\frac{\partial\theta_1}{\partial x_1} + R_2\theta_2\frac{\partial(\theta_1 - \theta_2)}{\partial x_1}\right\|_{L^p(\Omega)} \\ &+ \left\| R_1(\theta_1 - \theta_2)\frac{\partial\theta_1}{\partial x_2} + R_1\theta_2\frac{\partial(\theta_1 - \theta_2)}{\partial x_2}\right\|_{L^p(\Omega)}. \end{aligned} \tag{9.11}$$

Extending that estimate, for a chosen term, using boundedness of the Riesz operator we get

$$\begin{aligned}\left\| R_2(\theta_1 - \theta_2)\frac{\partial\theta_1}{\partial x_1}\right\|_{L^p(\Omega)} &\le \|R_2(\theta_1 - \theta_2)\|_{L^\infty(\Omega)}\left\|\frac{\partial\theta_1}{\partial x_1}\right\|_{L^p(\Omega)} \\ &\le c\|R_2(\theta_1 - \theta_2)\|_{W^{1,p}(\Omega)}\|\theta_1\|_{W^{1,p}(\Omega)} \\ &\le C\|\theta_1 - \theta_2\|_{W^{1,p}(\Omega)}\|\theta_1\|_{W^{1,p}(\Omega)}, \end{aligned} \tag{9.12}$$

with similar result for another terms. Note that also $F : W^{1,p}_0(\Omega) \to L^p(\Omega)$, and consequently, it follows from the abstract theory of [32, 86] or Chapter 3 that

Theorem 9.2.2. *Let $p > N$ be fixed. Then, for $f \in L^p(\Omega)$ and for arbitrary $\theta_0 \in W^{1,p}_0(\Omega)$, there exists a unique local in time mild solution θ to the subcritical problem (9.1), $\alpha \in (\frac{1}{2}, 1]$. Moreover,*

$$\theta \in C((0,\tau); W^{2\alpha,p}(\Omega)) \cap C([0,\tau); W^{1,p}_0(\Omega)), \quad \theta_t \in C((0,\tau); W^{2\gamma,p}(\Omega))$$

with arbitrary $\gamma < \alpha$. Here, $\tau > 0$ is the "lifetime" of that local in time solution. Moreover, the Cauchy formula is satisfied:

$$\theta(t) = e^{-A_\alpha t}\theta_0 + \int_0^t e^{-A_\alpha(t-s)}F(\theta(s))ds, \quad t \in [0,\tau),$$

where $e^{-A_\alpha t}$ denotes the linear semigroup corresponding to the operator $A_\alpha := (-\Delta)^\alpha$ in $W_0^{1,p}(\Omega)$, and $F(\theta) = -u \cdot \nabla\theta + f$.

Remark 9.2.3. It is a general observation (compare, for example, [32, Corollary 2.1.2]) that, as a consequence of Theorem 9.2.2, for each $\beta \in (\frac{1}{2}, \alpha)$ there exists a local in time solution to the subcritical problem (9.1) in the phase space $X^\beta \subset W^{2\beta,p}(\Omega)$. This follows from the property, that if the local Lipschitz continuity (9.11) holds, similar condition is guaranteed if we look at nonlinearity as a map $F : X^\beta \to L^p(\Omega)$, where $p > N$.

9.2.1 Natural a priori estimate

We recall now the *natural a priori estimate* known for solutions of the Q-g equation, that is a variant of the maximum principle, as formulated below.

Lemma 9.2.4. *For arbitrary $q \in [2,\infty)$, and a sufficiently regular solution of (9.1), if f is nonzero and in $L^q(\Omega)$ then*

$$\|\theta(t,\cdot)\|^q_{L^q(\Omega)} \le \|\theta_0\|^q_{L^q(\Omega)} e^{(q-1)t} + te^{(q-1)t}\|f\|^q_{L^q(\Omega)}. \tag{9.13}$$

Moreover, when $q = 2$, the following estimates are satisfied:

$$\|\theta\|^2_{L^2(\Omega)} \le c\max\left\{\|\theta_0\|^2_{L^2(\Omega)}; \frac{2C}{\kappa\lambda_1^\alpha}\|f\|^2_{L^2(\Omega)}\right\},$$

$$\kappa\int_0^T \|(-\Delta)^{\frac{\alpha}{2}}\theta(t)\|^2_{L^2(\Omega)}dt \le c(\|\theta_0\|^2_{L^2(\Omega)} + T\|f\|^2_{L^2(\Omega)}), \tag{9.14}$$

where λ_1 is the first positive eigenvalue of the negative Dirichlet Laplacian in Ω

$$\lambda_1^\alpha\|\phi\|^2_{L^2(\Omega)} \le \|(-\Delta)^{\frac{\alpha}{2}}\phi\|^2_{L^2(\Omega)}. \tag{9.15}$$

When $f = 0$, the corresponding to (9.13) estimate takes the form:

$$\|\theta(t,\cdot)\|_{L^q(\Omega)} \le \|\theta_0\|_{L^q(\Omega)}. \tag{9.16}$$

Proof. Note that for smooth solutions after multiplying the nonlinear term $u \cdot \nabla\theta$ by $|\theta|^{q-1}\operatorname{sgn}(\theta)$ and integrating over Ω, the resulting term will vanish

$$\int_\Omega \left(\frac{\partial\theta}{\partial x_1}\frac{\partial}{\partial x_2}[(-\Delta)^{-\frac{1}{2}}\theta] - \frac{\partial\theta}{\partial x_2}\frac{\partial}{\partial x_1}[(-\Delta)^{-\frac{1}{2}}\theta]\right)|\theta|^{q-1}\operatorname{sgn}(\theta)dx$$
$$= \frac{1}{q}\int_\Omega \frac{\partial(|\theta|^q)}{\partial x_1}\frac{\partial}{\partial x_2}[(-\Delta)^{-\frac{1}{2}}\theta] - \frac{\partial(|\theta|^q)}{\partial x_2}\frac{\partial}{\partial x_1}[(-\Delta)^{-\frac{1}{2}}\theta]dx = 0,$$

thanks to integration by parts. Consequently, multiplying (9.1) by $|\theta|^{q-1}\operatorname{sgn}(\theta)$, we obtain

$$\int_\Omega \theta_t|\theta|^{q-1}\operatorname{sgn}(\theta)dx + \kappa\int_\Omega (-\Delta)^\alpha\theta|\theta|^{q-1}\operatorname{sgn}(\theta)dx = \int_\Omega f|\theta|^{q-1}\operatorname{sgn}(\theta)dx.$$

When $q = 2$, we get in a standard way estimates (9.14). Indeed, multiplying (9.1) by θ^α, observing as above that the income from the nonlinear term will vanish, we obtain

$$\frac{1}{2}\frac{d}{dt}\|\theta^\alpha\|^2_{L^2(\Omega)} + \kappa\|(-\Delta)^{\frac{\alpha}{2}}\theta^\alpha\|^2_{L^2(\Omega)} = \int_\Omega f\theta^\alpha dx. \tag{9.17}$$

Recall next the *generalized Poincaré inequality* (5.37):

$$\lambda_1^\alpha\int_\Omega \phi^2 dx \le \int_\Omega [(-\Delta)^{\frac{\alpha}{2}}\phi]^2 dx, \quad \text{or more general}$$
$$\lambda_1^{\beta-\gamma}\int_\Omega [(-\Delta)^{\frac{\gamma}{2}}\phi]^2 dx \le \int_\Omega [(-\Delta)^{\frac{\beta}{2}}\phi]^2 dx, \quad \forall_{\beta>\gamma}, \tag{9.18}$$

where λ_1 denotes the first eigenvalue of negative Laplacian in Ω.

By the first inequality in (9.18), estimate (9.17) will be also extended to a uniform in $\alpha \in (\frac{1}{2}, 1]$ estimate

$$\frac{1}{2}\frac{d}{dt}\|\theta^\alpha\|^2_{L^2(\Omega)} + \kappa\lambda_1^\alpha\|\theta^\alpha\|^2_{L^2(\Omega)} \le \int_\Omega f\theta^\alpha dx,$$

which, together with the Cauchy inequality, gives us (9.14).

By the second inequality in (9.18), estimate (9.17) will be also extended to a uniform in $\alpha \in (\frac{1}{2}, 1]$ estimate

$$\frac{1}{2}\frac{d}{dt}\|\theta^\alpha\|^2_{L^2(\Omega)} + \kappa\lambda_1^{\alpha-\frac{1}{2}}\|(-\Delta)^{\frac{1}{4}}\theta^\alpha\|^2_{L^2(\Omega)} \le \int_\Omega f\theta^\alpha dx.$$

If additionally $f = 0$ in that case, using the Poincaré inequality, we get the decay estimate

$$\frac{1}{2}\frac{d}{dt}\int_\Omega \theta^2 dx \le -\kappa\lambda_1^\alpha\int_\Omega \theta^2 dx. \tag{9.19}$$

Since, for all the values $2 \le q < \infty$ and $0 \le \alpha \le 1$, as a consequence of the Kato–Beurling–Deny inequality in a bounded domain Ω (e. g., [52, Corollary 3.2] or Corollary 5.4.23),

$$\int_\Omega (-\Delta)^\alpha \theta |\theta|^{q-1} \operatorname{sgn}(\theta) dx \ge \frac{4(q-1)}{q^2} \int_\Omega [(-\Delta)^{\frac{\alpha}{2}} (|\theta|^{\frac{q}{2}})]^2 dx \ge 0, \tag{9.20}$$

using Hölder's and Young's inequalities, we obtain

$$\frac{1}{q}\frac{d}{dt}\int_\Omega |\theta|^q dx \le \int_\Omega f|\theta|^{q-1} \operatorname{sgn}(\theta) dx \le \frac{1}{q}\|f\|^q_{L^q(\Omega)} + \frac{q-1}{q}\|\theta\|^q_{L^q(\Omega)}. \tag{9.21}$$

Solving the above differential inequality, we get (9.13) when $f \neq 0$ and (9.16) otherwise.

Note that the estimate (9.20) is especially simple when $q = 4p$, $p \in \mathbb{N}$. We need not use absolute value $|\theta|$ at the right-hand side of it.

Let us observe also that we are able to pass to the limit $q \to +\infty$ in the estimate (9.13). We recall first, for completeness, the known property of the norms of L^p spaces.

Proposition 9.2.5. *Let $\Omega \subset \mathbb{R}^N$ be a bounded domain and $1 \le p \le \infty$. If $\phi \in L^\infty(\Omega)$, then*

$$\lim_{p\to\infty} \|\phi\|_{L^p(\Omega)} = \|\phi\|_{L^\infty(\Omega)}.$$

Moreover, if $\phi \in L^p(\Omega)$ for $1 \le p < \infty$ and there is a constant K such that for all such p holds $\|\phi\|_{L^p(\Omega)} \le K$, then

$$\phi \in L^\infty(\Omega) \quad \text{and} \quad \|\phi\|_{L^\infty(\Omega)} \le K.$$

See [1, p. 26] for the proof.

Taking the qth root and using the property that $\sqrt[q]{a+b} \le \sqrt[q]{a} + \sqrt[q]{b}$, $a, b \ge 0$, letting next $q \to +\infty$ one gets an estimate:

$$\|\theta(t,\cdot)\|_{L^\infty(\Omega)} \le e^t (\|\theta_0\|_{L^\infty(\Omega)} + \|f\|_{L^\infty(\Omega)}). \tag{9.22}$$

Note further that, when $f = 0$, we have from (9.27) a simpler estimate

$$\|\theta(t,\cdot)\|_{L^\infty(\Omega)} \le \|\theta_0\|_{L^\infty(\Omega)}. \tag{9.23}$$

□

A stronger property, that solutions decay to zero in $L^\infty(\Omega)$ norm as $t \to \infty$, when $f = 0$, will be discussed later in Lemma 9.5.7 using the Moser–Alikakos iteration technique.

9.3 Weak solutions of critical (9.1) with arbitrary data

From now on, to avoid confusion, the solutions of the approximating problems (9.1) will be denoted by θ^α. The limit of functions θ^α will be denoted θ, with the corresponding function $u = R^\perp\theta = (-R_2\theta, R_1\theta)$. We introduce now a *weak solution* to (9.1).

Definition 9.3.1. A weak solution to (9.1) is a function

$$\theta^\alpha \in L^2_{\mathrm{loc}}(0,T;H^\alpha(\Omega)) \cap C_w([0,T];L^2(\Omega)), \tag{9.24}$$

such that for any test function

$$\phi \in C_0^\infty((0,T);C_0^\infty(\Omega))$$

the following condition holds:

$$\begin{aligned}
&-\int_0^T \langle\theta^\alpha(t),\phi_t(t)\rangle dt - \int_0^T \langle\theta^\alpha(t),u(t)\cdot\nabla\phi(t)\rangle dt \\
&+\int_0^T \langle(-\Delta)^{\frac{\alpha}{2}}\theta^\alpha(t),(-\Delta)^{\frac{\alpha}{2}}\phi(t)\rangle dt = \int_0^T \langle f,\phi(t)\rangle dt.
\end{aligned} \tag{9.25}$$

Note that the regular solutions reported in Theorem 9.2.1 are evidently weak solutions in the specified above sense, so that for arbitrary $\alpha \in (\frac{1}{2},1]$ the problem (9.1) has a weak solution (at least for regular data).

Through the described above natural a priori estimate (9.14), we have a, uniform in $\alpha \in (\frac{1}{2},1]$, estimate of the approximating solutions θ^α of the form

$$\|\theta^\alpha\|_{L^\infty(0,T;L^2(\Omega))} + \|\theta^\alpha\|_{L^2(0,T;H^{\frac{1}{2}}(\Omega))} \le \text{const}, \tag{9.26}$$

and also

$$\|\theta^\alpha\|_{L^2(0,T;H^\alpha(\Omega))} \le \text{const}. \tag{9.27}$$

The last information allows us to let $\alpha_n \to \frac{1}{2}^+$, passing several times to a subsequences, $\{\theta^{\alpha_{n_k}}\}$ in a weakly formulated approximating equations (9.1).

Note first that the uniform boundedness of θ^α in $L^\infty(0,T;L^2(\Omega))$ allows to pass to the limit in most linear components in (9.25). Indeed, thanks to the weak* $L^\infty(0,T;L^2(\Omega))$ compactness

$$-\int_0^T \langle\theta^\alpha(t),\phi_t(t)\rangle dt \to -\int_0^T \langle\theta(t),\phi_t(t)\rangle dt$$

as $\alpha \to \frac{1}{2}^+$. Further, we can rewrite the last term as

$$\int_0^T \langle f, \phi(t)\rangle dt = \int_0^T \langle (-\Delta)^{-\frac{\alpha}{2}} f, (-\Delta)^{\frac{\alpha}{2}}\phi(t)\rangle dt$$
$$\to \int_0^T \langle (-\Delta)^{-\frac{1}{4}} f, (-\Delta)^{\frac{1}{4}}\phi(t)\rangle dt, \tag{9.28}$$

compare, for example, [130, Proposition 3.1.1] to get the convergence $(-\Delta)^{\frac{\alpha}{2}}\phi \to (-\Delta)^{\frac{1}{4}}\phi$ in $L^2(\Omega)$ for $\phi \in D(-\Delta)$ (see Section 5.3).

The only difficulty is connected with the nonlinear term, for which we will use an estimate

$$\begin{aligned}\left| -\int_0^T \langle \theta^\alpha(t), u^\alpha(t)\cdot\nabla\phi(t)\rangle dt\right| &\le \int_0^T |\langle \theta^\alpha(t), u^\alpha(t)\cdot\nabla\phi(t)\rangle| dt\\ &\le \|\theta^\alpha\|_{L^\infty(0,T;L^2(\Omega))}\|u^\alpha\|_{L^2(0,T;L^3(\Omega))}\|\nabla\phi\|_{L^2(0,T;L^6(\Omega))}\\ &\le c\|\theta^\alpha\|_{L^\infty(0,T;L^2(\Omega))}\|u^\alpha\|_{L^2(0,T;H^{\frac{1}{2}^-}(\Omega))}\|\phi\|_{L^2(0,T;W^{1,6}(\Omega))}.\end{aligned} \tag{9.29}$$

Adding and subtracting, we will estimate as above the resulting terms

$$\begin{aligned}&\left| -\int_0^T \langle \theta^\alpha(t), u^\alpha(t)\cdot\nabla\phi(t)\rangle dt + \int_0^T \langle \theta(t), u(t)\cdot\nabla\phi(t)\rangle dt\right|\\ &\quad\le \left|\int_0^T \langle (\theta^\alpha(t)-\theta(t)), u^\alpha(t)\cdot\nabla\phi(t)\rangle dt\right|\\ &\quad\quad + \left|\int_0^T \langle \theta(t), (u(t)-u^\alpha(t))\cdot\nabla\phi(t)\rangle dt\right|.\end{aligned} \tag{9.30}$$

Note that since $u^\alpha = (-R_2\theta^\alpha, R_1\theta^\alpha)$ then $\operatorname{div} u^\alpha = 0$, and the nonlinear term in (9.1) will be rewritten as

$$u^\alpha\cdot\nabla\theta^\alpha = \nabla\cdot(u^\alpha\theta^\alpha) - \theta^\alpha \operatorname{div} u^\alpha = \nabla\cdot(u^\alpha\theta^\alpha).$$

Further, thanks to the uniform in $\alpha \in (\frac{1}{2}, 1]$ estimates of θ^α ((9.26), (9.27), and (8.12)), for nonlinearity given in that form, we have

$$\begin{aligned}\|\nabla\cdot(u^\alpha\theta^\alpha)\|_{H^{-1}(\Omega)} &= \|(-\Delta)^{-\frac{1}{2}}\nabla\cdot(u^\alpha\theta^\alpha)\|_{L^2(\Omega)} \le c\||u^\alpha\theta^\alpha|\|_{L^2(\Omega)}\\ &\le c\|u^\alpha\|_{L^4(\Omega)}\|\theta^\alpha\|_{L^4(\Omega)} \le c\|\theta^\alpha\|_{L^4(\Omega)}\|\theta^\alpha\|_{H^{\frac{1}{2}}(\Omega)},\end{aligned} \tag{9.31}$$

and thus it will be seen from equation (9.1) that θ_t^α are bounded in $L^2(0,T;H^{-1}(\Omega))$ uniformly in $\alpha \in (\frac{1}{2}, 1]$.

Therefore, due to (9.14) and the Lions-Aubin compactness lemma [124, Chapter 1] with $H^{\frac{1}{2}}(\Omega) \subset H^{\frac{1}{2}^-}(\Omega) \subset L^2(\Omega)$ with first compact embedding (a consequence of interpolation inequality), the following strong convergence holds:

$$\|u - u^\alpha\|_{L^2(0,T;L^3(\Omega))} \leq \|u - u^\alpha\|_{L^2(0,T;H^{\frac{1}{2}^-}(\Omega))} \to 0,$$

as $\alpha \to \frac{1}{2}^+$, consequently the last term in (9.30) goes to 0 since θ^α are bounded in $L^\infty(0,T;L^2(\Omega))$. Since also u^α are bounded in $L^\infty(0,T;L^2(\Omega))$ uniformly in $\alpha \in (\frac{1}{2},1]$, then

$$\left|\int_0^T \langle(\theta^\alpha(t) - \theta(t)), u^\alpha(t)\cdot\nabla\phi(t)\rangle dt\right|$$
$$\leq c\|u^\alpha\|_{L^\infty(0,T;L^2(\Omega))}\|\theta^\alpha - \theta\|_{L^2(0,T;L^3(\Omega))}\|\phi\|_{L^2(0,T;W^{1,6}(\Omega))} \to 0. \tag{9.32}$$

Weak continuity, $\theta \in C_w([0,T];L^2(\Omega))$, follows from [162, Corollary 2.1] since $\theta \in L^\infty(0,T;L^2(\Omega))$ with $\theta_t \in L^2(0,T;H^{-1}(\Omega))$. Consequently, the limit function $\theta \in L^\infty(0,T;L^2(\Omega)) \cap L^2(0,T;H^{\frac{1}{2}}(\Omega))$ fulfills the Definition 9.3.1 with $\alpha = \frac{1}{2}$, therefore, it is a weak solution of the critical equation.

We are thus able to conclude the following theorem.

Theorem 9.3.2. *When $\theta_0 \in D((-\Delta)^{\frac{s+1}{2}}) \subset H^{s+1}(\Omega)$ and $f \in H^s(\Omega)$, $s \in (\frac{3}{2},\frac{5}{2})$, then there exists a (not necessary unique which is connected with passing to the limits over different sequences) weak solution θ of the critical quasi-geostrophic equation.*

Remark 9.3.3. The latter theorem will be generalized to cover a larger class of initial data and right-hand side. Inspecting the proof, it is evident that in the process of approximating weak solution of the critical problem (9.1), $\alpha = \frac{1}{2}$, we can choose a sequence of approximating solutions $\{\theta^{\alpha_n}\}$, $\alpha_n \to \frac{1}{2}$, in such a way that the corresponding initial data $\theta_0^{\alpha_n}$ will satisfy

$$\theta_0^{\alpha_n} \in D((-\Delta)^{\frac{s+1}{2}}) \subset H^{s+1}(\Omega), \quad \theta_0^{\alpha_n} \to \theta_0 \quad \text{in } L^2(\Omega) \text{ as } \alpha_n \to \frac{1}{2}^+.$$

This is possible thanks to density of $D((-\Delta)^{\frac{s+1}{2}})$ in $L^2(\Omega)$ (e. g., [86, p. 29]). For such solutions θ^{α_n} of (9.1), $\alpha = \alpha_n$, the uniform estimates (9.26) and (9.27) are still valid. While the whole range of initial data $\theta_0 \in L^2(\Omega)$ will be reached in such a construction (see similar consideration in [55, p. 54]).

Further, it will be seen that due to (9.25) and (9.28) we can even take $f \in H^{-\frac{1}{2}}(\Omega)$ in the critical problem (9.1). Indeed, in the approximating process we will consider a sequence of subcritical problems (9.1) with right-hand sides $f_n \in H^s(\Omega)$, $s \in (\frac{3}{2},\frac{5}{2})$, which are convergent to a limit function f in $H^{-\frac{1}{2}}(\Omega)$ only. We will still obtain a weak solution θ of the critical problem in the limit; note that the uniform estimates (9.26), (9.27) remain valid also in that case.

Further, if we want a more regular solution θ satisfying estimate (9.22), we need to take $f \in L^\infty(\Omega)$ approximated by a sequence $f_n \in H^s(\Omega)$, $s \in (\frac{3}{2}, \frac{5}{2})$, in the $L^\infty(\Omega)$ norm.

In the light of the latter observation, we will sharpen Theorem 9.3.2 to the form which will be used below:

Corollary 9.3.4. *When $\theta_0 \in L^2(\Omega)$ and $f \in L^2(\Omega)$, then there exists a (not necessary unique) weak solution θ of the critical quasi-geostrophic equation.*

9.4 Regular solutions of critical (9.1) with small data

Our next task is to extend the constructed above local in time solution globally (that means, onto arbitrary large time interval $[0, T]$, $T > 0$). This is typical situation, since the solutions corresponding to small data obey better a priori estimates than the one connected with large data (which is a consequence of the two properties of nonlinear term; $F(0) = 0$ and quadratic, or faster, growth of $F(\theta)$).

Consider a slightly more regular local solution to subcritical problem (9.1) with $\alpha > \frac{1}{2}$, as reported in Remark 9.2.3; let the phase space equal $W_0^{1+\epsilon,p}(\Omega)$ with small positive $\epsilon < \frac{N}{p}$ (depending on α). It now follows from the estimate

$$\begin{aligned}\|u \cdot \nabla\theta\|_{L^p(\Omega)} &\leq \left\| -R_2\theta\frac{\partial\theta}{\partial x_1} + R_1\theta\frac{\partial\theta}{\partial x_2}\right\|_{L^p(\Omega)} \leq c\|\theta\|_{L^{\frac{N}{\epsilon}}(\Omega)}\|\theta\|_{W_0^{1,\frac{Np}{N-p\epsilon}}(\Omega)} \\ &\leq c\|\theta\|_{L^{\frac{N}{\epsilon}}(\Omega)}\|\theta\|_{W_0^{1+\epsilon,p}(\Omega)},\end{aligned} \tag{9.33}$$

that, due to Lemma 9.2.4 with $q = \frac{N}{\epsilon}$, the above is a form of the *subordination condition* for the local solution in $W_0^{1+\epsilon,p}(\Omega)$. Consequently, for arbitrary $\alpha > \frac{1}{2}$ the local solution to subcritical problem (9.1) will be extended globally in time.

Note that, due to Lemma 9.2.4, the introduced above global in time solutions are bounded, uniformly in $\alpha > \frac{1}{2}$, in all the spaces $L^q(\Omega)$, $1 \leq q \leq \infty$. This information will be used in passing to the limiting critical equation in its weak formulation.

9.4.1 Uniform in $\alpha > \frac{1}{2}$ estimates for subcritical problems with small data

When the data θ_0, f are small, we will obtain a *uniform in $\alpha > \frac{1}{2}$ estimate* of solutions to approximating problems (9.1), first in a weak norms. Multiplying in $L^2(\Omega)$ equation (9.1) by $(-\Delta)^{-\frac{1}{2}}\theta^\alpha$, using (9.5), we get

$$\begin{aligned}&\frac{1}{2}\frac{d}{dt}\int_\Omega [(-\Delta)^{-\frac{1}{4}}\theta^\alpha]^2 dx + \int_\Omega \nabla\cdot(u^\alpha\theta^\alpha)(-\Delta)^{-\frac{1}{2}}\theta^\alpha dx + \int_\Omega [(-\Delta)^{\frac{\alpha-\frac{1}{2}}{2}}\theta^\alpha]^2 dx \\ &= \int_\Omega f(-\Delta)^{-\frac{1}{2}}\theta^\alpha dx.\end{aligned} \tag{9.34}$$

The nonlinear term is next estimated using boundedness of the operator $(-\Delta)^{-\frac{1}{2}}\nabla$ and (9.22),

$$\begin{aligned}\|(-\Delta)^{-\frac{1}{2}}\nabla\cdot(u^\alpha\theta^\alpha)\|_{L^2(\Omega)} &\le c\||u^\alpha\theta^\alpha|\|_{L^2(\Omega)} \le c\|u^\alpha\|_{L^2(\Omega)}\|\theta^\alpha\|_{L^\infty(\Omega)}\\ &\le c(\|\theta_0\|_{L^\infty(\Omega)},\|f\|_{L^\infty(\Omega)})\|\theta^\alpha\|_{L^2(\Omega)}.\end{aligned} \tag{9.35}$$

Noting that the exponent $\frac{\alpha-\frac{1}{2}}{2}$ is positive, we get from (5.37) that

$$\lambda_1^{\alpha-\frac{1}{2}}\int_\Omega[\theta^\alpha]^2dx \le \int_\Omega[(-\Delta)^{\frac{\alpha-\frac{1}{2}}{2}}\theta^\alpha]^2dx.$$

Inserting the above two estimates into (9.34), using the Cauchy inequality and (9.22), we obtain

$$\begin{aligned}&\frac{d}{dt}\int_\Omega[(-\Delta)^{-\frac{1}{4}}\theta^\alpha]^2dx + 2\kappa C\int_\Omega[\theta^\alpha]^2dx\\ &\quad\le 2c(\|\theta_0\|_{L^\infty(\Omega)},\|f\|_{L^\infty(\Omega)})\|\theta^\alpha\|^2_{L^2(\Omega)} + \|(-\Delta)^{-\frac{1}{2}}f\|^2_{L^2(\Omega)},\end{aligned} \tag{9.36}$$

$t\in[0,T]$, where the constant $C := \inf_{\alpha\in(\frac{1}{2},1]}\lambda_1^{\alpha-\frac{1}{2}}$ will be chosen uniformly ($\lambda_1^{\alpha-\frac{1}{2}}\to 1$ when $\alpha\to\frac{1}{2}^+$). Consequently, *when the data are small*,

$$c(\|\theta_0\|_{L^\infty(\Omega)},\|f\|_{L^\infty(\Omega)}) < \kappa C, \tag{9.37}$$

we get a uniform in α estimate

$$\|\theta^\alpha\|_{L^\infty(0,T;H^{-\frac{1}{2}}(\Omega))} + \|\theta^\alpha\|_{L^2(0,T;L^2(\Omega))} \le \text{const}, \tag{9.38}$$

with a general constant, independent of $\alpha\in(\frac{1}{2},1]$.

Remark 9.4.1. Let us explain here the possibility of choosing the constant above *uniform* for $\alpha\in(\frac{1}{2},1]$ (or, more general, in bounded subsets of $(\frac{1}{2},\infty)$). It is known from the spectral resolution of self-adjoint positive definite operators in a Hilbert space, that the Poincaré constant equals $\lambda_1^{\alpha-\frac{1}{2}}$ in that case with λ_1 being the first eigenvalue of the negative Dirichlet Laplacian in Ω. In order (e. g., [78, p. 164]), when $u\in W_0^{1,p}(\Omega)$, $1\le p<\infty$, then the classical *Poincaré inequality* reads

$$\|u\|_{L^p(\Omega)} \le \left(\frac{1}{\omega_N}|\Omega|\right)^{\frac{1}{N}}\|\nabla u\|_{L^p(\Omega)}, \tag{9.39}$$

where ω_N stands for the volume of the unite ball in $\mathbb{R}^N$ and $|\Omega|$ for the Lebesgue measure of Ω in $\mathbb{R}^N$. Evidently, when $\alpha>\frac{1}{2}$ then the constant will be estimated uniformly, since

$$0 < \lambda_1^{\frac{1}{2}} \le \left(\frac{1}{\omega_N}|\Omega|\right)^{-\frac{1}{N}}; \quad 0 < \lambda_1 \le \frac{\pi}{|\Omega|} \quad \text{when } N = 2,$$

moreover $\lambda_1^{\frac{\alpha}{2}-\frac{1}{4}} \to 1$ as $\alpha \to \frac{1}{2}^+$.

In the case of regular solutions, $f = 0$ and for small $\|\theta_0\|_{L^\infty(\Omega)}$ we will present now uniform in $\alpha \in (\frac{1}{2}, 1]$ estimates of solutions to (9.1) in better norms. First, following the calculations, borrowed from [37], we will obtain uniform in $\alpha \in (\frac{1}{2}, 1]$ estimate of the $H^1(\Omega)$ norm.

Multiplying (9.1) in $L^2(\Omega)$ by $(-\Delta)\theta^\alpha$, we obtain

$$\frac{1}{2}\frac{d}{dt}\int_\Omega |\nabla\theta^\alpha|^2 dx + \kappa \int_\Omega [(-\Delta)^{\frac{\alpha+1}{2}}\theta^\alpha]^2 dx \le \left|\int_\Omega u^\alpha \nabla \cdot \theta^\alpha (-\Delta)\theta^\alpha dx\right|. \tag{9.40}$$

Thanks to integration by parts, the nonlinear term is transformed as follows:

$$\left|\int_\Omega u^\alpha \nabla \cdot \theta^\alpha (-\Delta)\theta^\alpha dx\right| = \left|\int_\Omega \nabla(u^\alpha \cdot \nabla\theta^\alpha) \cdot \nabla\theta^\alpha dx\right|. \tag{9.41}$$

Then we arrived at

$$\left|\int_\Omega \nabla(u^\alpha \cdot \nabla\theta^\alpha) \cdot \nabla\theta^\alpha dx\right| \le C \int_\Omega |\nabla\theta^\alpha|^2 |\nabla u^\alpha| dx \le C\|\nabla\theta^\alpha\|^3_{L^3(\Omega)}. \tag{9.42}$$

The above estimate is smart, since the second-order term from the original expression is divided between u^α and θ^α.

Using a borderline version of the Nirenberg–Gagliardo inequality (as in [37]; see Remark 5.4.11),

$$\|\psi\|_{W^{1,3}(\Omega)} \le c\|\psi\|^{\frac{1}{3}}_{L^\infty(\Omega)} \|\psi\|^{\frac{2}{3}}_{H^{\frac{3}{2}}(\Omega)}, \tag{9.43}$$

estimate (9.42) extends to

$$\left|\int_\Omega \nabla(u^\alpha \nabla \cdot \theta^\alpha) \cdot \nabla\theta^\alpha dx\right| \le c\|\theta^\alpha\|_{L^\infty(\Omega)} \|\theta^\alpha\|^2_{H^{\frac{3}{2}}(\Omega)}. \tag{9.44}$$

Thanks to the Poincaré inequality (5.37), we have an estimate

$$\lambda_1^{\alpha-\frac{1}{2}} \|\phi\|^2_{H^{\frac{3}{2}}(\Omega)} \le \|\phi\|^2_{H^{\alpha+1}(\Omega)},$$

and the constant $C = \inf_{\alpha\in(\frac{1}{2},1]} \lambda_1^{\alpha-\frac{1}{2}}$ is uniform for $\alpha \in (\frac{1}{2}, 1]$; see Remark 9.4.1. The last observation together with the maximum principle allows us to close the estimate (9.40)

$$\frac{1}{2}\frac{d}{dt}\int_\Omega |\nabla\theta^\alpha|^2 dx + \kappa C \int_\Omega [(-\Delta)^{\frac{3}{4}}\theta^\alpha]^2 dx \le c\|\theta^\alpha\|_{L^\infty(\Omega)} \|\theta^\alpha\|^2_{H^{\frac{3}{2}}(\Omega)}. \tag{9.45}$$

Under the *smallness restriction* connected with the estimate (9.16),

$$c\|\theta_0\|_{L^\infty(\Omega)} < \kappa C, \tag{9.46}$$

we thus obtain, in a standard way, an estimate

$$\|\theta^\alpha\|_{L^\infty(0,T;H^1(\Omega))} + \|\theta^\alpha\|_{L^2(0,T;H^{\frac{3}{2}}(\Omega))} \le \text{const}, \tag{9.47}$$

with the constant *uniform in* $\alpha \in (\frac{1}{2}, 1]$.

Multiplying equation (9.1) in $L^2(\Omega)$ by $(-\Delta)^2\theta^\alpha$, again as in [37], using the expression for the nonlinear term

$$\int_\Omega \Delta^2\theta^\alpha u^\alpha \cdot \nabla\theta^\alpha dx = 2\int_\Omega \nabla u^\alpha \cdot (\nabla\nabla\theta^\alpha)\Delta\theta^\alpha dx + \int_\Omega (\Delta u^\alpha \cdot \nabla\theta^\alpha)\Delta\theta^\alpha dx, \tag{9.48}$$

estimates similar as above lead to a uniform, for small $\|\theta_0\|_{L^\infty(\Omega)}$ and $f = 0$, bound

$$\|\theta^\alpha\|_{L^\infty(0,T;H^2(\Omega))} + \|\theta^\alpha\|_{L^2(0,T;H^{\frac{5}{2}}(\Omega))} \le \text{const}, \tag{9.49}$$

with the constant *uniform in* $\alpha \in (\frac{1}{2}, 1]$.

The two estimates above with the use of interpolation provide us the bound

$$\|\theta^\alpha\|_{L^\infty(0,T;H^\beta(\Omega))} + \|\theta^\alpha\|_{L^2(0,T;H^{\beta+\frac{1}{2}}(\Omega))} \le \text{const}, \quad \beta \in (1,2), \tag{9.50}$$

with the constant uniform in $\alpha \in (\frac{1}{2}, 1]$ (valid under the smallness restriction).

Remark 9.4.2. Since the described above estimates (9.47), valid for small $\|\theta_0\|_{L^\infty(\Omega)}$ (9.46) and $f = 0$, are uniform for $\alpha \in (\frac{1}{2}, 1]$, they are inherited by the weak solution of the critical Q-g equation when letting $\alpha \to \frac{1}{2}^+$ (under the same smallness restriction (9.46)).

9.5 Critical quasi-geostrophic equation. Small data

We will construct now the global regular solution in the critical case $\alpha = \frac{1}{2}$ as a limit when $\alpha \to \frac{1}{2}^+$ of solutions to the subcritical approximations in case of the *small data.*

9.5.1 Critical equation; passing to the limit $\alpha \to \frac{1}{2}^+$ in subcritical approximations in case of small data

Our task now is to pass to the limit over a sequence $\alpha_n \to \frac{1}{2}^+$, extracting a subsequence of approximating solutions $\theta^{\alpha_n}(t)$ convergent for $t \in [0, T]$, $T > 0$ arbitrary large, to certain (weak type) solution of the limit critical problem ($\alpha = \frac{1}{2}$). A uniform in $\alpha > \frac{1}{2}$ estimate of the solutions θ^α of Lemma 9.2.4 will be used in that process.

We will describe next the way of passing to the limit in (9.1) when the data are small. Remembering that $\frac{1}{2} \le \alpha \le 1$, we apply the operator $A^{-\alpha}$ to (9.1) and multiply the result by a "test function" $\phi \in H_0^1(\Omega) \subset L^q(\Omega)$, $q \in [1, \infty)$ ($N = 2$), to obtain

$$\begin{aligned}&\langle A^{-\alpha}\theta_t^\alpha, \phi\rangle_{L^p,L^q}\\ &\quad = -\langle \theta^\alpha, \phi\rangle_{L^p,L^q} + \langle A^{-\alpha}F(\theta^\alpha), \phi\rangle_{L^p,L^q} + \langle A^{-\alpha}f, \phi\rangle_{L^p,L^q}.\end{aligned} \tag{9.51}$$

We will discuss convergence of components one by one. Note that, by [130, Theorems 3.1.6, and 7.1.1], $A^{-\alpha}\psi \to A^{-\frac{1}{2}}\psi$ as $\alpha \to \frac{1}{2}^+$. Thanks to the uniform in $\alpha \in (\frac{1}{2}, 1]$ boundedness of θ^α in $L^\infty(0, T; L^p(\Omega))$, $1 < p \le \infty$, by weak compactness $\theta^\alpha \to \theta$ weakly in $L^p(\Omega)$ (weak* compactness for the case $p = \infty$), we have next

$$\langle \theta^\alpha, \phi\rangle \to \langle \theta, \phi\rangle. \tag{9.52}$$

For the nonlinear term, we have an estimate: (in Lemma 9.2.4 we will let $q \to +\infty$, first taking qth roots)

$$\begin{aligned}\|A^{-\alpha}F(\theta^\alpha)\|_{L^p(\Omega)} &= \|A^{-(\alpha-\frac{1}{2})}A^{-\frac{1}{2}}F(\theta^\alpha)\|_{L^p(\Omega)}\\ &\le c\|A^{-\frac{1}{2}}F(\theta^\alpha)\|_{L^p(\Omega)} \le c\|\theta^\alpha\|_{L^\infty(\Omega)}\|u^\alpha\|_{L^p(\Omega)}\\ &\le c(\|\theta_0\|_{L^\infty(\Omega)}, \|f\|_{L^\infty(\Omega)})\|\theta^\alpha\|_{L^p(\Omega)},\end{aligned} \tag{9.53}$$

the estimate being uniform with respect to $\alpha \in (\frac{1}{2}, 1]$. Consequently, using equation (9.51),

$$\|\theta_t^\alpha\|_{L^\infty(0,T;D(A_p^{-1}))} \le \text{const}, \tag{9.54}$$

uniformly in $\alpha \in (\frac{1}{2}, 1]$.

Using next a generalization of the Lions–Aubin compactness lemma in [153], thanks to (9.54) and the uniform in $\alpha \in (\frac{1}{2}, 1]$ estimates (9.14):

$$\kappa\int_0^T \|(-\Delta)^{\frac{1}{4}}\theta^\alpha(t)\|^2_{L^2(\Omega)}dt \le c(\|\theta_0\|^2_{L^2(\Omega)} + T\|f\|^2_{L^2(\Omega)}), \tag{9.55}$$

we claim that the family $\{\theta^\alpha; \alpha \in (\frac{1}{2}, 1]\}$ is precompact in the space $L^2(0, T; H^{\frac{1}{2}^-}(\Omega))$. Consequently, denoting $\theta^\alpha \to \theta$, we will pass to the limit, over a sequence $\alpha_n \to \frac{1}{2}^+$, in the nonlinear term in the above mentioned space; as $\alpha_n \to \frac{1}{2}^+$,

$$\begin{aligned}\int_0^T \langle A^{-\alpha}F(\theta^\alpha), \phi\rangle_{[L^2(\Omega)]^2}dt &= \int_0^T \langle A^{-\frac{1}{2}}F(\theta^\alpha), A^{\frac{1}{2}-\alpha}\phi\rangle_{[L^2(\Omega)]^2}dt\\ &\to \int_0^T \langle A^{-\frac{1}{2}}F(\theta), \phi\rangle_{[L^2(\Omega)]^2}dt.\end{aligned} \tag{9.56}$$

Indeed, the convergence above holds thanks to Lemma 9.2.4 and a consequence of (5.60):

$$\begin{aligned}\|A^{-\frac{1}{2}}(F(\theta^\alpha)-F(\theta))\|_{L^2(\Omega)} &\le c\|u^\alpha\theta^\alpha - u\theta\|_{L^2(\Omega)}\\ &\le c\|\theta^\alpha-\theta\|_{L^{\frac{4+2\epsilon}{1+\epsilon}}(\Omega)}(\|\theta^\alpha\|_{L^{4+2\epsilon}(\Omega)}+\|\theta\|_{L^{4+2\epsilon}(\Omega)})\\ &\le c\|\theta^\alpha-\theta\|_{H^{\frac{1}{2}^-}(\Omega)}(\|\theta^\alpha\|_{L^{4+2\epsilon}(\Omega)}+\|\theta\|_{L^{4+2\epsilon}(\Omega)}),\end{aligned} \tag{9.57}$$

using the embedding $H^{\frac{1}{2}^-}(\Omega)\subset L^{\frac{4+2\epsilon}{1+\epsilon}}(\Omega)$ when $N=2$ (valid when $\frac{1}{2}^- \ge \frac{1}{2+\epsilon}$). Note that for arbitrary $\phi\in L^2(\Omega)$, $A^{\frac{1}{2}-\alpha}\phi\to\phi$ in $L^2(\Omega)$ (see, e. g., [130], [53, Lemma 7.2], or Lemma 5.4.21).

Finally, in the term containing time derivative, we can pass to the limit in the sense of distributions with values in Banach spaces (e. g., [124, Chapter 1]) to obtain

$$\langle A^{-\alpha}\theta^\alpha_t,\phi\rangle = \frac{d}{dt}\langle A^{-\alpha}\theta^\alpha,\phi\rangle \to \frac{d}{dt}\langle A^{-\frac{1}{2}}\theta,\phi\rangle, \tag{9.58}$$

with distributional derivative $\frac{d}{dt}$. We thus described passing to the limit, over a sequence $\alpha_n\to\frac{1}{2}^+$ for a given $\phi\in L^2(\Omega)$. But the space is separable, hence choosing countable many times a subsequence we will pass to the limit for a dense set of "test functions" $\phi\in L^2(\Omega)$ (taken from $H^1_0(\Omega)$, say). Since the limit *may depend on the chosen subsequence*, it must not be unique.

Consequently, we have justified the possibility of passing to the limit in weakly formulated approximating equation (9.51). In the limit, we get a weak solution to the critical Q-g equation:

$$\frac{d}{dt}\langle A^{-\frac{1}{2}}\theta,\phi\rangle = -\langle\theta,\phi\rangle + \langle A^{-\frac{1}{2}}F(\theta),\phi\rangle + \langle A^{-\frac{1}{2}}f,\phi\rangle \tag{9.59}$$

valid for each "test function" $\phi\in L^2(\Omega)$.

We will formulate the corresponding theorem on global solvability of critical Q-g equation.

Theorem 9.5.1. *Let $s\in(\frac{3}{2},\frac{5}{2})$ be fixed. Then, for $f\in D((-\Delta)^{\frac{s}{2}})\subset H^s(\Omega)$ and for arbitrary $\theta_0\in D((-\Delta)^{\frac{s+1}{2}})\subset H^{s+1}(\Omega)$ there exists a solution θ^α (constructed in Theorem 9.2.1) of the subcritical problem (9.1) with $\alpha\in(\frac{1}{2},1]$. Letting $\alpha\to\frac{1}{2}^+$, over a sequence, we get a weak solution θ to the critical problem with $\alpha=\frac{1}{2}$ (not necessary unique), fulfilling*

$$\theta\in L^2(0,T;H^{\frac{1}{2}}(\Omega)),\quad \theta_t\in L^\infty(0,T;D((-\Delta)^{-1})) \tag{9.60}$$

satisfying, for each "test function" $\phi\in L^2(\Omega)$, the equality

$$\frac{d}{dt}\langle A^{-\frac{1}{2}}\theta,\phi\rangle = -\langle\theta,\phi\rangle + \langle A^{-\frac{1}{2}}F(\theta),\phi\rangle + \langle A^{-\frac{1}{2}}f,\phi\rangle.$$

9.5.2 Properties of the weak solution to the critical Q-g equation

We start with collecting the properties inherited by the solution θ of the critical Q-g equation in the process of passing to the limit ($\alpha \to \frac{1}{2}^+$). Directly, such limit possesses the following properties:

$$\begin{aligned}
&\forall_{1<p\leq\infty} \quad \theta \in L^\infty(0,\infty;L^p(\Omega)),\\
&\theta \in L^2(0,T;H^{\frac{1}{2}}(\Omega)),\\
&\theta_t \in L^\infty(0,T;D(A^{-1})), \quad \text{here } D(A^{-1}) = (H^2(\Omega)\cap H_0^1(\Omega))^*.
\end{aligned} \tag{9.61}$$

We will describe now briefly further properties of the constructed above weak solution of (9.59).

Our first task will be to show that the local solutions of the critical 2-D quasi-geostrophic equation, varying in the base space $H^s(\Omega)$, $s > 1$, are *locally unique*. The last means that, if there are two different local solutions of that kind, they must coincide up to the existence time of the shorter existing one.

Lemma 9.5.2. *The solution of the critical 2-D quasi-geostrophic equation satisfying*

$$\theta \in L^\infty(0,\tau;H^s(\Omega)), s > 1, \quad \theta_t \in L^2(0,\tau;H^{-1}(\Omega)),$$

is locally unique. Note that, by the well-known Lions lemma (e. g., [124, Lemma 1.2]), the above function is continuous as a map $\theta : [0,\tau) \to L^2(\Omega)$.

Proof. Let θ_1, θ_2 be the two local in time solutions in the above class. We denote the difference of the two solutions as: $w := \theta_1 - \theta_2$, and write an equation for the difference:

$$\begin{aligned}
&w_t + u_1 \cdot \nabla w + (u_1 - u_2)\cdot \nabla\theta_2 + \kappa(-\Delta)^{\frac{1}{2}} w = 0,\\
&w(0) = 0, \quad w = 0 \text{ on } \partial\Omega.
\end{aligned} \tag{9.62}$$

We multiply next the equation by w to get

$$\frac{1}{2}\frac{d}{dt}\int_\Omega w^2 dx + \int_\Omega u_1\cdot\nabla w w dx + \int_\Omega (u_1-u_2)\cdot\nabla\theta_2 w dx + \kappa\int_\Omega [(-\Delta)^{\frac{1}{4}} w]^2 dx = 0. \tag{9.63}$$

Now, for the nonlinear terms we notice that

$$\int_\Omega u_1\cdot\nabla(w) w dx = \frac{1}{2}\int_\Omega u_1\cdot\nabla(w^2) dx = 0, \tag{9.64}$$

where integration by parts formula and the definition of u_1 (see (9.3)) were used. Further, by the Hölder inequality,

$$\begin{aligned}
\left|\int_\Omega (u_1-u_2)\cdot\nabla\theta_2 w dx\right| &\leq \|u_1-u_2\|_{L^{4-2\epsilon}(\Omega)}\|w\|_{L^{4-2\epsilon}(\Omega)}\|\nabla\theta_2\|_{L^{\frac{2-\epsilon}{1-\epsilon}}(\Omega)}\\
&\leq c\|\theta_2\|_{W^{1,\frac{2-\epsilon}{1-\epsilon}}(\Omega)}\|w\|^2_{L^{4-2\epsilon}(\Omega)},
\end{aligned}$$

where $0 < \epsilon < 1$ and the estimate (9.8) was used. Note a Nirenberg–Gagliardo-type inequality, $N = 2$,

$$\|\phi\|_{L^{4-2\epsilon}(\Omega)} \le C\|\phi\|^{\frac{2-2\epsilon}{2-\epsilon}}_{H^{\frac{1}{2}}(\Omega)} \|\phi\|^{\frac{\epsilon}{2-\epsilon}}_{L^2(\Omega)}, \tag{9.65}$$

and an embedding

$$\|\theta_2\|_{W^{1,\frac{2-\epsilon}{1-\epsilon}}(\Omega)} \le c\|\theta_2\|_{H^s(\Omega)}, \tag{9.66}$$

valid whenever $s \ge 1 + \frac{\epsilon}{2-\epsilon}$. For arbitrary fixed $s > 1$, there is thus an $0 < \epsilon < 1$, the nonlinear term above is estimated, using the Young inequality, as follows:

$$\begin{aligned}\left|\int\limits_\Omega (u_1 - u_2)\cdot\nabla\theta_2 w dx\right| &\le c\|\theta_2\|_{H^s(\Omega)}\|w\|^{\frac{4-4\epsilon}{2-\epsilon}}_{H^{\frac{1}{2}}(\Omega)} \|w\|^{\frac{2\epsilon}{2-\epsilon}}_{L^2(\Omega)} \\ &\le \frac{\kappa}{2}\|(-\Delta)^{\frac{1}{4}} w\|^2_{L^2(\Omega)} + c'\|\theta_2\|^{\frac{2-\epsilon}{\epsilon}}_{H^s(\Omega)}\|w\|^2_{L^2(\Omega)}.\end{aligned} \tag{9.67}$$

Inserting the above estimate to (9.63), remembering that $\|\theta_2\|_{H^s(\Omega)}$ is bounded, we get a differential inequality for the function $y(t) := \|w(t,\cdot)\|^2_{L^2(\Omega)}$:

$$\begin{aligned}&\frac{d}{dt}y(t) + \kappa\int\limits_\Omega [(-\Delta)^{\frac{1}{4}} w]^2 dx \le 2c'\|\theta_2\|^{\frac{2-\epsilon}{\epsilon}}_{H^s(\Omega)} y(t), \\ &y(0) = 0,\end{aligned} \tag{9.68}$$

having only zero solution on $[0,\tau)$. Consequently, the solutions as specified in the lemma are unique, as long as they exist (property we called *local uniqueness*). □

9.5.3 The case $f = 0$

In a particular case $f = 0$ when no forcing term is present in the system, the behavior of solutions to critical equation (9.1) will be described better (see, e. g., [40, Section 5]). Using the observations of Subsection 9.4.1, we will formulate the following.

Proposition 9.5.3. *Let $s \in (1,2]$, $f = 0$ and θ_0 satisfies the smallness restriction* (9.46). *Then the constructed in Theorem* 9.5.1 *weak solution is regular, more precisely*

$$\theta \in L^\infty(0,T;H^{2^-}(\Omega)) \cap L^2(0,T;H^{\frac{5}{2}^-}(\Omega)). \tag{9.69}$$

The above solution is also unique, as a consequence of Lemma 9.5.2.

Proof. Using [153], this follows directly from the uniform in $\alpha \in (\frac{1}{2},1]$ estimate (9.50). □

If θ_0 is not necessary small, observe that from Lemma 9.2.4 the approximating solutions θ^α fulfill, uniformly in $\alpha > \frac{1}{2}$, the following estimates:

$$\|\theta^\alpha(t,\cdot)\|_{L^q(\Omega)} \le \|\theta_0\|_{L^q(\Omega)} \quad 1 \le q \le \infty,$$
$$\kappa \int_0^\infty \int_\Omega [(-\Delta)^{\frac{\alpha}{2}}\theta^\alpha]^2 dxdt \le \frac{1}{2}\int_\Omega \theta_0^2 dx. \tag{9.70}$$

This gives the estimate inherited by the limiting solution θ of the critical equation

$$\kappa \int_0^\infty \int_\Omega [(-\Delta)^{\frac{1}{4}}\theta]^2 dxdt \le \frac{1}{2}\int_\Omega \theta_0^2 dx. \tag{9.71}$$

Such estimate will be used to determine the asymptotic behavior of solution θ when time goes to infinity (compare, e. g., [40, Section 5]).

Remark 9.5.4. For the purpose of that subsection, we will extend now a bit the local/global solvability result *for approximations* (9.1), $\alpha \in (\frac{1}{2}, 1]$ of the critical Q-g equation. We want to have, for $\alpha \in (\frac{1}{2}, 1]$, all the solutions in the (larger than previously) phase space $H_0^1(\Omega)$. In addition, we will take $\theta_0 \in H_0^1(\Omega) \cap L^\infty(\Omega)$ to have global boundedness of the solution. For this, we need to choose the base space as a negative (but small, $\epsilon < \alpha - \frac{1}{2}$) order Sobolev space $H^{-\epsilon}(\Omega)$. We will check, to justify local solvability, that the nonlinear term acts as a Lipschitz map on bounded sets from $H_0^1(\Omega)$ to $H^{-\epsilon}(\Omega)$. Indeed, calculating for one component, by duality argument (since $H^{-\epsilon}(\Omega) \supset L^{\frac{2}{\epsilon+1}}(\Omega)$),

$$\begin{aligned}
&\|F(\theta_1^\alpha) - F(\theta_2^\alpha)\|_{H^{-\epsilon}(\Omega)} \\
&\quad \le c\|F(\theta_1^\alpha) - F(\theta_2^\alpha)\|_{L^{\frac{2}{1+\epsilon}}(\Omega)} \text{(calculating for one component)} \\
&\quad \le c\left\|R_2(\theta_1^\alpha - \theta_2^\alpha)\frac{\partial\theta_1^\alpha}{\partial x_1}\right\|_{L^{\frac{2}{1+\epsilon}}(\Omega)} \le c\|\theta_1^\alpha - \theta_2^\alpha\|_{L^{\frac{2}{\epsilon}}(\Omega)}\|\theta_1^\alpha\|_{H^1(\Omega)} \\
&\quad \le c\|\theta_1^\alpha - \theta_2^\alpha\|_{H_0^1(\Omega)}\|\theta_1^\alpha\|_{H_0^1(\Omega)}.
\end{aligned} \tag{9.72}$$

Consequently, [32, 86], local in time solvability for arbitrary $\alpha \in (\frac{1}{2}, 1]$ follows. With the aid of the maximum principle, this will be extended to the global in time solvability. In a similar way as above, using further interpolation, we obtain a *subordination type* estimate:

$$\begin{aligned}
\|F(\theta^\alpha)\|_{H^{-\epsilon}(\Omega)} &\le c\|F(\theta^\alpha)\|_{L^{\frac{2}{1+\epsilon}}(\Omega)} \text{(calculating for one component)} \\
&\le \left\|R_2(\theta^\alpha)\frac{\partial\theta^\alpha}{\partial x_1}\right\|_{L^{\frac{2}{1+\epsilon}}(\Omega)} \le c\|\theta^\alpha\|_{L^{\frac{2}{\epsilon}^+}(\Omega)}\|\theta^\alpha\|_{W^{1,2^-}(\Omega)} \\
&\le c(\|\theta^\alpha\|_{L^\infty(\Omega)})\|\theta^\alpha\|_{H_0^1(\Omega)}^{1^-}.
\end{aligned} \tag{9.73}$$

Therefore, the local solution θ^α varying in the phase space $H_0^1(\Omega)$ (in fact, it varies in $H_0^{-\epsilon+2\alpha}(\Omega)$ for $t > 0$; note that $-\epsilon + 2\alpha > \frac{1}{2} + \alpha > 1$) will be extended globally in time in that phase space.

We thus have a family, parameter $\alpha > \frac{1}{2}$, of the global in time solutions $\theta^\alpha(t,\cdot) \in H^1_0(\Omega)$ of the approximating problems. The next task will be to let $\alpha \to \frac{1}{2}$ and obtain a weak solution for the limit critical problem.

Now we recall the definition of the, so called, *Leray–Hopf weak solutions* (e. g., [37, p. 3]) to the Q-g equation with $\alpha > \frac{1}{2}$ and then pass to the limit in that weak formulation to get a weak solution of the critical problem.

Definition 9.5.5. A function $\theta^\alpha \in L^\infty(\mathbb{R}^+; L^2(\Omega)) \cap L^2(\mathbb{R}^+; H^\alpha(\Omega))$, $\alpha \geq \frac{1}{2}$, is called *a Leray–Hopf weak solution* of the problem (9.1) with $f = 0$, if:

- θ^α is weakly continuous from $\mathbb{R}^+$ to $L^2(\Omega)$,
- For arbitrary test function $\phi \in C_0^\infty(\mathbb{R}^+ \times \Omega)$, the following equality holds:

$$-\int_0^\infty \left\langle \theta^\alpha, \frac{d}{dt}\phi \right\rangle d\tau - \int_0^\infty \langle u^\alpha \theta^\alpha, \nabla\phi \rangle d\tau + \kappa \int_0^\infty \langle \theta^\alpha, (-\Delta)^\alpha \phi \rangle d\tau = 0.$$

We need to modify slightly the way we define the approximating sequence of solutions of the regularizations to get the above weak solution of the critical problem in the limit. For fixed $\alpha \in (\frac{1}{2}, 1]$, following Remark 9.5.4, we find $n \in \mathbb{N}$ such that $\frac{1}{n} < \alpha - \frac{1}{2}$ and chose $H^{-\frac{1}{n}}(\Omega)$ as a base space. Then $X_n := H_0^{-\frac{1}{n}+2\alpha}(\Omega)$ will be the corresponding phase space (note that $-\frac{1}{n} + 2\alpha > 1$, so that $X_n \subset H^1_0(\Omega)$). To have the maximum principle valid for the limiting solution θ of critical problem, we chose $H^1_0(\Omega) \cap L^\infty(\Omega)$ as a phase space from which the initial data θ_0 in Definition 9.5.5 will be taken. Let $\{\theta_0^n;\ \theta_0^n \in X_n\}_{n\in\mathbb{N}}$ be a sequence, convergent to θ_0 in $H^1_0(\Omega)$ and bounded uniformly in $L^\infty(\Omega)$, of initial data corresponding to approximating problems (9.1) with $\alpha = \alpha_n > \frac{1}{2} + \frac{1}{n} \to \frac{1}{2}$ as $n \to \infty$. As reported in Remark 9.5.4, we construct the corresponding to θ_0^n sequence of approximating solutions θ^n. Similarly, as described in Theorem 9.5.1, we obtain its weak limit θ when $\alpha_n \to \frac{1}{2}^+$. Using the inherited by the solution θ in the limiting process estimates (9.61), (9.71), we have the following theorem

$$\|\theta\|_{L^2(\mathbb{R}^+; H^{\frac{1}{2}}(\Omega))} + \|\theta\|_{L^\infty(\mathbb{R}^+; L^\infty(\Omega))} \leq \text{const.} \tag{9.74}$$

Consequently, we are able to formulate, with evident proof (since weak convergence is stronger than the distributional one), the following theorem.

Theorem 9.5.6. *When $f = 0$ then, for arbitrary $\theta_0 \in H^1_0(\Omega) \cap L^\infty(\Omega)$, there exists a weak Leray–Hopf solution of the critical Q-g equation. Moreover, it satisfies the estimate* (9.74).

It was shown in Theorem 4.1 of [40], using E. De Giorgi technique, that the $L^\infty(\mathbb{R}^2)$ norm of solutions of critical Q-g equation decays to zero as $t \to \infty$ when $f = 0$. So, for a sufficiently large time condition (9.46) is fulfilled.

We will propose here another than in [40] reasoning showing that when $f = 0$ the $L^\infty(\Omega)$ norm of θ^α decays to zero. It is based on *the Moser–Alikakos iteration technique* (which is known to be equivalent with the De Giorgi estimates) and was used to second-order parabolic semilinear equations in [46]. More precisely, the following property holds.

Lemma 9.5.7. *For the solutions θ^α of (9.1), $f = 0$, the following implication holds:*

$$\|\theta^\alpha(t,\cdot)\|_{L^2(\Omega)} \to 0 \Rightarrow \|\theta^\alpha(t,\cdot)\|_{L^\infty(\Omega)} \to 0, \quad t \to \infty, \tag{9.75}$$

and the above property is uniform in $\alpha \in (\frac{1}{2}, 1]$.

Proof. The proof is standard while technical, so we recall only its main points. Multiplying equation (9.1) in $L^2(\Omega)$ by $(\theta^\alpha)^{2^m-1}$, $m \in \mathbb{N}$, we obtain

$$\int_\Omega \theta^\alpha_t (\theta^\alpha)^{2^m-1} + \int_\Omega u^\alpha \cdot \nabla\theta^\alpha (\theta^\alpha)^{2^m-1} dx + \kappa \int_\Omega (-\Delta)^\alpha \theta^\alpha (\theta^\alpha)^{2^m-1} dx = 0. \tag{9.76}$$

Next, as in Lemma 9.2.4, we observe that the intermediate nonlinear component is vanishing. Therefore, the estimate runs as for the corresponding linear equation. Consequently, using the Kato–Beurling–Deny inequality (5.112), we obtain

$$\frac{1}{2^m}\frac{d}{dt}\int_\Omega (\theta^\alpha)^{2^m} dx + \kappa\frac{4(2^m-1)}{2^{2m}}\int_\Omega [(-\Delta)^{\frac{\alpha}{2}}((\theta^\alpha)^{2^{m-1}})]^2 dx \le 0. \tag{9.77}$$

We notify that, when $m = 1$, the above with (9.15) provides us an exponential type decay of the $L^2(\Omega)$ norm of $\theta^\alpha(t,\cdot)$ to zero, $t \to \infty$. Using the tiring inductive estimates as in [46] based on the Nirenberg–Gagliardo inequality,

$$\|\phi\|^2_{L^2(\Omega)} \le c\|\phi\|^{\frac{4}{3}}_{H^{\frac{1}{2}}(\Omega)}\|\phi\|^{\frac{2}{3}}_{L^1(\Omega)} \le C(\|(-\Delta)^{\frac{1}{4}}\phi\|^2_{L^2(\Omega)} + \|\phi\|^2_{L^1(\Omega)}), \quad N = 2,$$

this information will be extended to the $L^\infty(\Omega)$ convergence of $\theta^\alpha(t,\cdot)$ to 0 as $t \to \infty$. Moreover, the estimates are seen to be uniform in $\alpha \in (\frac{1}{2}, 1]$, since in the calculations we will replace $(-\Delta)^{\frac{\alpha}{2}}$ with $(-\Delta)^{\frac{1}{4}}$, thanks to uniform embedding constant

$$\lambda_1^{\frac{\alpha}{2}-\frac{1}{4}}\|(-\Delta)^{\frac{1}{4}}\phi\|_{L^2(\Omega)} \le \|(-\Delta)^{\frac{\alpha}{2}}\phi\|_{L^2(\Omega)}, \quad \alpha \in \left(\frac{1}{2}, 1\right].$$

Indeed, multiplying (9.77) by 2^m and inserting the estimate above, one has

$$\frac{d}{dt}\int_\Omega (\theta^\alpha)^{2^m} dx \le 2\kappa[\|(\theta^\alpha)^{2^{m-1}}\|^2_{L^1(\Omega)} - C^{-1}\|(\theta^\alpha)^{2^{m-1}}\|^2_{L^2(\Omega)}]. \tag{9.78}$$

Solving the above differential inequality, we obtain

$$\|\theta^\alpha(t,\cdot)\|^{2^m}_{L^{2^m}(\Omega)} \le \|\theta_0\|^{2^m}_{L^{2^m}(\Omega)} e^{-2\kappa C^{-1}t} + \int_0^t 2\kappa\|\theta^\alpha(\tau,\cdot)\|^{2^m}_{L^{2^{m-1}}(\Omega)} e^{2\kappa C^{-1}(\tau-t)} d\tau. \tag{9.79}$$

We will next show inductively the convergence of the $L^{2^m}(\Omega)$ norms of $\theta^\alpha(t,\cdot)$ to zero; compare (9.19). Assuming that $\|\theta^\alpha(t,\cdot)\|_{L^{2^{m-1}}(\Omega)} \leq \epsilon$ for $t \geq t_0$ $(0 < \epsilon < \max\{1; C^{-1}\})$, enlarging if necessary the value t_0 to t_0' we will let the first right-hand side term in (9.79) less than $\frac{C\epsilon^{2^m}}{2}$ for $t \geq t_0'$. Note that θ^α are bounded in $L^\infty(\Omega)$ uniformly in α. We split the integral into $(0, t_0')$ and (t_0', t), the second part estimate through $C\epsilon^{2^m}$, the first integral through $\frac{C\epsilon^{2^m}}{2}$ for large t. It will be seen that t_0' will be chosen the same for all $\alpha \in (\frac{1}{2}, 1]$. Taking the $2m$-th roots we carefully put together the obtained estimates and letting $m \to \infty$ we obtain the required bound:

$$\lim_{m\to\infty} \|\theta^\alpha(t,\cdot)\|_{L^{2^m}(\Omega)} = \|\theta^\alpha(t,\cdot)\|_{L^\infty(\Omega)} \leq (2C)^{\sum_{k=1}^{\infty} 2^{-k}} \epsilon = 2C\epsilon, \tag{9.80}$$

valid for sufficiently large t. See [46] for more details.

Since the estimate above is uniform in $\alpha \in (\frac{1}{2}, 1]$, it is inherited by the weak solutions of the critical Q-g equation. □

Remark 9.5.8. An interesting regularity result for homogeneous ($f = 0$) critical Q-g equation (9.1) was reported recently in [38]. Namely, for smooth solutions of (9.1) with $\alpha = \frac{1}{2}$, existing on a time interval $[0, T)$, $T \leq \infty$ and with initial data θ_0 there exists a constant Γ depending only on Ω, such that an interior gradient bound holds

$$\sup_{[0,T)\times\Omega} d(x)|\nabla_x \theta(t,x)| \leq \Gamma_1 \Big[\sup_{\Omega} d(x)|\nabla_x \theta_0(x)| + (1 + \|\theta_0\|_{L^\infty(\Omega)})^4\Big],$$

where $d(x) = \operatorname{dist}(x, \partial\Omega)$.

Bibliographical notes

The Cauchy problem for the quasi-geostrophic equation was studied earlier than the Dirichlet one. As was observed, for example, in [109], most of the existing definitions of fractional powers are equivalent in case of negative Laplacian in $\mathbb{R}^N$ (of course, on suitable classes of functions). The situation is different in the Dirichlet case in bounded regular domain (compare Section 5) and that problem seems more involved. Recently, however, more authors are extending their studies to problems in bounded domains; see, for example, [22, 38].

10 Dirichlet's problem for critical Hamilton–Jacobi fractional equation

10.1 Introduction

Our task is to investigate the Hamilton–Jacobi problem with *critical exponent* $\alpha = \frac{1}{2}$,

$$\begin{cases} u_t + (-\Delta)^\alpha u + H(u, \nabla u) = 0, & t > 0,\ x \in \Omega \subset \mathbb{R}^N, \\ u(t,x) = 0, \quad t > 0,\ x \in \partial\Omega, \\ u(0,x) = u_0(x), \quad x \in \Omega, \end{cases} \tag{10.1}$$

where Ω is a bounded domain in $\mathbb{R}^N$ with $\partial\Omega \in C^2$. The Hamilton–Jacobi equation has attracted attention of many scholars, for example, [14, 49, 93, 191, 192]. Regularity of viscosity solutions to (10.1) was discussed recently in [151, 152], which motivates our studies reported below.

The present chapter is devoted to solvability of Dirichlet problem for the critical Hamilton–Jacobi equation (10.1) with $\alpha = \frac{1}{2}$. Such problem is *essentially critical* since the action of the ∇ in the nonlinear term H is equivalent to that of $(-\Delta)^{\frac{1}{2}}$ in the main part operator. It is thus impossible to use directly the *perturbation techniques* to treat that problem, since the nonlinearity is anymore *a perturbation* of the linear main part in that case. In particular, the Dan Henry approach in classical Sobolev spaces or its generalizations (e. g., [32, 86]) is not directly applicable and we need to modify the latter technique to treat the quasi-linear problem (10.1) with $\alpha = \frac{1}{2}$. A similar situation we face when considering the critical quasi-geostrophic equation as described in the previous chapters.

Section 10.2 is devoted to solutions of the subcritical equation (10.1), when $\alpha \in (\frac{1}{2}, 1]$. Such solutions will serve as approximations of the weak solutions to (10.1) in critical case $\alpha = \frac{1}{2}$. In subsection 10.2.1, several possible *phase spaces* are listed suitable for the local in time solvability of the subcritical problem (10.1) with $\alpha \in (\frac{1}{2}, 1]$. In Theorem 10.2.3, we prove existence of a local in time regular solution u_α of subcritical (10.1), which is varying continuously in $W^{2,q}(\Omega)$, $q > N$. We further discuss and prove a version of the maximum principle valid for (10.1); see Theorem 10.2.7. The following Theorem 10.2.14 concludes the possibility of extending globally the local solutions of subcritical (10.1) obtained in Theorem 10.2.3.

In subsections 10.2.3 and 10.2.5, we obtain, under suitably chosen assumptions, *uniform in $\alpha \in (\frac{1}{2}, 1]$ a priori estimates* of solutions to (10.1) in subcritical case. Such uniform estimates are used later in Section 10.3 when we construct a weak solution of the critical equation with $\alpha = \frac{1}{2}$, passing to the limit in subcritical equations (10.1).

The existence results for critical (10.1) found in the literature, as [90, 99, 151, 152], are based on the approach of *viscosity solutions* due to P.-L. Lions, and Perron's construction of lower/upper solutions to get a solution.

https://doi.org/10.1515/9783110599831-010

10.2 Subcritical equations (10.1) in $H^s(\Omega)$ type spaces

This section is devoted to construction, in classical Sobolev spaces, of solution to (10.1) in *subcritical case* $\alpha \in (\frac{1}{2}, 1]$ and in space dimension $N \geq 2$. We will also formulate assumptions required for the local in time solvability of equation (10.1).

We recall first, for completeness, description of the domain of fractional powers $(-\Delta)^\alpha$, $\alpha \in (0,1)$, of the negative Dirichlet Laplacian defined on $X = L^p(\Omega)$ in a C^2 domain Ω. In Hilbert case $X = L^2$, such description (for $\alpha < \frac{3}{4}$) was discussed in details in [118, p. 303], and also in [187, p. 559]

$$D((-\Delta)^\alpha_{L^2}) = \begin{cases} H^{2\alpha}(\Omega) & \text{if } 0 \leq \alpha < \frac{1}{4}, \\ H^{2\alpha}_0(\Omega) & \text{if } \frac{1}{4} < \alpha < \frac{1}{2}, \\ H^{2\alpha}(\Omega) \cap H^1_0(\Omega) & \text{if } \frac{1}{2} \leq \alpha < \frac{5}{4}, \alpha \neq \frac{3}{4} \\ H^{2\alpha}(\Omega) \cap H^2_0(\Omega) & \text{if } \frac{5}{4} < \alpha < \frac{9}{4}, \alpha \neq \frac{7}{4}. \end{cases} \tag{10.2}$$

In the case of Banach spaces (build over $L^p(\Omega)$), such description can be found in [86, p. 39] when $\partial\Omega \in C^2$, or in more details in [187, pp. 81, 558]. In particular for elliptic operators in the $L^p(\Omega)$ setting under the assumptions of [187, Theorem 16.14] (which are satisfied for negative Laplacian in part regarding boundedness of imaginary powers thanks to [140, p. 188]), we have[1]:

$$\begin{matrix} D((-\Delta)^\alpha_{L^p}) \\ \| \\ W^{2\alpha,p}_D(\Omega) \end{matrix} = \begin{cases} W^{2\alpha,p}(\Omega) & \text{if } 0 \leq \alpha < \frac{1}{2p}, \\ W^{2\alpha,p}_0(\Omega) & \text{if } \frac{1}{2p} < \alpha < \frac{1}{2}, \\ W^{2\alpha,p}(\Omega) \cap W^{1,p}_0(\Omega) & \text{if } \frac{1}{2} \leq \alpha < 1 + \frac{1}{2p}, \alpha \neq \frac{p+1}{2p}, \\ W^{2\alpha,p}(\Omega) \cap W^{2,p}_0(\Omega) & \text{if } 1 + \frac{1}{2p} < \alpha < 2 + \frac{1}{2p}, \alpha \neq \frac{3p+1}{2p}. \end{cases} \tag{10.3}$$

An even more complete description of the domains of fractional powers of negative Dirichlet Laplacian (and more general elliptic operators) was known much earlier; see [173, Sections 4.9.2, 5.5.1]. However, those results were formulated under the assumption of C^∞ regularity of the domain Ω and the coefficients of the differential operator involved (the latter satisfied in case of the Laplacian). A lot of effort was needed to weaken that high regularity assumptions (see in particular [187, Chapter 16], [133]).

10.2.1 Admissible phase spaces and local solvability

We will discuss first the *admissible* pairs of the base/phase spaces, in which the subcritical approximating problems will be considered together with the limiting critical

1 It follows from [43] that in bounded C^2 domains the negative Dirichlet Laplacian admits bounded $\mathcal{H}^\infty$ calculus (note that Hilbert spaces are Banach spaces of class $\mathcal{HT}$). Together with [187, Theorem 16.15], this allows to write equality here.

problem. The situation here is very similar as for the *critical quasi-geostrophic equation*, as reported recently in [56], or in Chapter 9. To simplify the considerations, we will use some "ready made" estimates from [187, p. 50].

1. Base space $X^{\frac{1}{2}} = W_0^{1,q}(\Omega)$, phase space $X^1 = W^{2,q}(\Omega) \cap W_0^{1,q}(\Omega)$

Let $H : \mathbb{R}\times\mathbb{R}^N \to \mathbb{R}$ fulfilling $H(0,p) = 0$, $p \in \mathbb{R}^N$, be a C^2 function. We will choose $N < q < \infty$ and $\alpha > \frac{1}{2}$. As a direct consequence of [187, p. 50] (just "adding and subtracting" changing only one argument of H, and using the original estimate of [187]) we have the following two estimates:

$$\begin{aligned}
&\|H(u,\nabla u)\|_{W^{1,q}(\Omega)} \le q(\|u\|_{W^{2,q}(\Omega)}) \quad \text{for every } u \in W^{2,q}(\Omega),\\
&\|H(u,\nabla u) - H(v,\nabla v)\|_{W^{1,q}(\Omega)} \le \sup_{0\le\theta\le1} p_1(\|\theta u + (1-\theta)v\|_{W^{2,q}(\Omega)})\|u - v\|_{W^{2,q}(\Omega)}\\
&\qquad \text{for every } u, v \in W^{2,q}(\Omega),
\end{aligned} \tag{10.4}$$

with some increasing functions q, p_1. Also since $u \in W^{2,q}(\Omega)\cap W_0^{1,q}(\Omega)$ then $H(u,\nabla u) \in W_0^{1,q}(\Omega)$ due to the assumption that $H \in C^2$ with $H(0,p) = 0$ for each $p \in \mathbb{R}^N$. Indeed, the nonlinearity acts from $W^{2,q}(\Omega) \cap W_0^{1,q}(\Omega)$ into $W_0^{1,q}(\Omega)$, as a map Lipschitz continuous on bounded subsets of $W^{2,q}(\Omega)$. In the light of Dan Henry's theory, the last properties together with the sectoriality of the main part operators $(-\Delta)^\alpha$, are sufficient for the local in time solvability of the problems (10.1) in $W^{2,q}(\Omega) \cap W_0^{1,q}(\Omega)$ for $\alpha > \frac{1}{2}$.

2. Base space $X^{\frac{s}{2}} \subset H^s(\Omega)$, $s \ge 1$ and $s > \frac{N}{2}$, phase space $X^{\frac{s+1}{2}} \subset H^{s+1}(\Omega)$

Let $H : \mathbb{R}\times\mathbb{R}^N \to \mathbb{R}$ be a C^{m+1} function with $m \ge s$. We will choose a base space for subcritical problems (10.1) to be included in $H^s(\Omega)$, that space is a Banach Algebra of functions. For larger values of s, to guarantee that the nonlinear term acts from $X^{\frac{s+1}{2}}$ to $X^{\frac{s}{2}}$ we need to impose the *compatibility conditions* on H:

$$H(u,\nabla u) \in X^{\frac{s}{2}} \quad \text{whenever } u \in X^{\frac{s+1}{2}}, \tag{10.5}$$

which includes *boundary conditions* for $H(u,\nabla u)$ on $\partial\Omega$. Compare [118, p. 326] for more details.

Then the following estimates are valid with increasing functions p, p_1 determined for H:

$$\begin{aligned}
&\|H(u,\nabla u)\|_{H^s(\Omega)} \le p(\|u\|_{H^{s+1}(\Omega)}) \quad \text{for every } u \in H^{s+1}(\Omega),\\
&\|H(u,\nabla u) - H(v,\nabla v)\|_{H^s(\Omega)} \le \sup_{0\le\theta\le1} p_1(\|\theta u + (1-\theta)v\|_{H^{s+1}(\Omega)})\|u - v\|_{H^{s+1}(\Omega)}\\
&\qquad \text{for every } u, v \in H^{s+1}(\Omega).
\end{aligned} \tag{10.6}$$

Again, the last two estimates show that the nonlinearity is Lipschitz continuous on bounded subsets as a map from $H^{s+1}(\Omega)$ to $H^s(\Omega)$ hence, in the subcritical case of exponent $\alpha \in (\frac{1}{2}, 1]$, the local in time solvability is guaranteed.

3. Base space included in $H^{\frac{N}{2}}(\Omega) \cap L^\infty(\Omega)$
This is a borderline case and the largest possible base space in our collection. However, this is still an *algebra of functions* (see, e. g., [96]) with a useful estimate:

$$\begin{aligned}&\|fg\|_{H^{\frac{N}{2}}(\Omega)} \leq c(\|f\|_{L^\infty(\Omega)}\|g\|_{H^{\frac{N}{2}}(\Omega)} + \|f\|_{H^{\frac{N}{2}}(\Omega)}\|g\|_{L^\infty(\Omega)})\\ &f,g \in H^{\frac{N}{2}}(\Omega) \cap L^\infty(\Omega).\end{aligned} \tag{10.7}$$

Remark 10.2.1. One must be aware that the abstract existence result (see Chapter 3) stays valid under the main Assumption A: "Let X be a Banach space, $A : D(A) \to X$ sectorial and positive operator in X and for some $\alpha' \in [0,1)$, $F : X^{\alpha'} \to X$ be Lipschitz continuous on bounded subsets of $X^{\alpha'}$."

In particular, the nonlinear term needs to act between $X^{\alpha'}$ and X. We are allowed to choose X to be located at any level of the fractional power scale associated with A (see [6, p. 294]), but then the requirement $F : X^{\alpha'} \to X$ includes the fact that eventually certain boundary conditions must be fulfilled both by the solution u and by the composite function $F(u)$ (the so-called, "compatibility conditions"; compare [116]). We must take care to fulfill such additional conditions, especially when dealing with more regular solutions to our problem.

Remark 10.2.2. When the solutions u are varying in the phase space $W^{2,q}(\Omega) \cap W^{1,q}_0(\Omega)$, $q > N$, H is a C^2 function and $\partial\Omega$ is C^2, thanks to the Sobolev embedding $W^{2,q}(\Omega) \subset C^{1+\mu}(\overline{\Omega})$ with $1 - \frac{N}{q} > \mu > 0$, the function u and its first derivatives have values on $\partial\Omega$.

Considering the problem (10.1) for subcritical range of exponents $\alpha \in (\frac{1}{2}, 1]$, we will settle it in case 1. of the base space $W^{1,q}_0(\Omega)$ with the corresponding to it phase space $W^{2,q}(\Omega) \cap W^{1,q}_0(\Omega)$. As known (see Remark 2.2.27), the nonlinear term H acts as a bounded operator between the above spaces; moreover, it is Lipschitz continuous on bounded subsets of $W^{2,q}(\Omega) \cap W^{1,q}_0(\Omega)$.

This suffices to formulate the following *local existence result* for the subcritical equation (10.1) in case 1. Similar theorems are valid in the two remaining cases without essential differences in the proofs.

Theorem 10.2.3. *Let $\alpha \in (\frac{1}{2}, 1]$ and $H : \mathbb{R} \times \mathbb{R}^N \to \mathbb{R}$ be a C^2 smooth real function, such that $H(0,p) = 0$, $p \in \mathbb{R}^N$. Then for arbitrary $u_0 \in X^1 = W^{2,q}(\Omega) \cap W^{1,q}_0(\Omega)$, $q > N$, there exists a unique local in time solution of* (10.1), *such that*

$$\begin{aligned}&u \in C^0([0,\tau); W^{2,q}(\Omega) \cap W^{1,q}_0(\Omega)) \cap C^0((0,\tau); D((-\Delta)^{\frac{1}{2}+\alpha})),\\ &u_t \in C^0((0,\tau); W^{1+2\alpha^-,q}(\Omega)),\end{aligned} \tag{10.8}$$

with $\tau = \tau(u_0)$ being its lifetime.

Furthermore, as stated in Corollary 3.3.1, if $u_0 \in D((-\Delta)^{1+\frac{\epsilon}{2}}) \subset W^{2+\epsilon,q}(\Omega)$, *where* $0 < \epsilon < 2\alpha - 1$, *then the above solution fulfills* (10.8) *and*

$$u \in C^0([0,\tau); D((-\Delta)^{1+\frac{\epsilon}{2}})). \tag{10.9}$$

See Chapter 3 for the proof.

10.2.2 A priori estimates and the global solvability

We formulate next, for the use of the *global in time solvability*, a generalization (for functions of several variables) of the remark [187, (7), p. 51]. The latter is, in order, quite a direct extension of [78, Theorem 7.8]. We have the following.

Proposition 10.2.4. *Let* $H : \mathbb{R} \times \mathbb{R}^N \to \mathbb{R}$ *be* C^2 *smooth with bounded all first partial derivatives:* $|D_i H| \le C$, $i = 1, \dots, N+1$, *and such that* $H(0,\dots,0) = 0$. *Then, for arbitrary* $N < q < \infty$, *the mapping* $u \to H(u, \nabla u)$ *is an operator from* $W^{2,q}(\Omega)$ *into* $W^{1,q}(\Omega)$ *(note that* $W^{2,q}(\Omega) \subset W^{1,\infty}(\Omega)$*), and the following estimate is valid:*

$$\|H(u,\nabla u)\|_{W^{1,q}(\Omega)} \le C\|\nabla H\|_{L^\infty}\|u\|_{W^{2,q}(\Omega)}, \quad u \in W^{2,q}(\Omega), \tag{10.10}$$

where, for a real function $H = H(u, v_1, \dots, v_N)$, $\nabla H = (D_1 H, D_2 H, \dots, D_{N+1} H)$ *denotes a vector of all its first-order partial derivatives* $(\frac{\partial H}{\partial u}, \frac{\partial H}{\partial v_1}, \dots, \frac{\partial H}{\partial v_N})$.

Proof. The proof is a rather direct extension of that in [187, p. 51]. While we will sketch it for completeness of the presentation. We will use the following two equalities:

$$\begin{aligned} H(u,\nabla u) &= H(u,\nabla u) - H(0,\nabla u) + H(0,\nabla u) - H\left(0, 0, \frac{\partial u}{\partial x_2}, \dots, \frac{\partial u}{\partial x_N}\right) \\ &\quad + \cdots - H(0,\dots,0) + H(0,\dots,0), \end{aligned} \tag{10.11}$$

and, for arbitrary $1 \le i \le N$,

$$\frac{\partial}{\partial x_i} H(u,\nabla u) = D_1 H \frac{\partial u}{\partial x_i} + \sum_{j=1}^{N} D_{j+1} H \frac{\partial^2 u}{\partial x_i \partial x_j}. \tag{10.12}$$

From (10.11), we obtain

$$\|H(u,\nabla u)\|_{L^q(\Omega)} \le c\|\nabla H\|_{L^\infty}\|u\|_{W^{1,q}(\Omega)} + \|H(0,\dots,0)\|_{L^q(\Omega)}. \tag{10.13}$$

From (10.12), we get

$$\left\|\frac{\partial}{\partial x_i} H(u,\nabla u)\right\|_{L^q(\Omega)} \le c'\|\nabla H\|_{L^\infty}\|u\|_{W^{2,q}(\Omega)}, \tag{10.14}$$

which completes the proof. □

Remark 10.2.5. The assumptions of the latter proposition are rather restrictive, especially the one on boundedness of all the first-order partial derivatives. This requires a special form of that nonlinearity, and we will consider here the real functions $H : \mathbb{R} \times \mathbb{R}^N \to \mathbb{R}$ like:

$$H = H(u, v_1, \dots, v_N) = f_1(u) + \sum_{i=1}^{N} v_i,$$

$$H(u, v_1, \dots, v_N) = f_2(u) + f_3(u) \sum_{i=1}^{N} \cos v_i, \tag{10.15}$$

with suitable functions f_1, f_2, f_3 (vanishing at $u = 0$, with bounded first derivatives).

Consider the first choice of the base/phase space in our collection corresponding to the base space $W_0^{1,q}(\Omega)$, $q > N$. As an immediate consequence of the proposition above, we have a *subordination condition* provided that u is bounded a priori in $W^{1,\infty}(\Omega)$. More precisely, if the real function H has locally bounded all first-order derivatives and $H(0, \dots, 0) = 0$, the nonlinearity in subcritical H-J equation (10.1) (or the Nemytskiĭ (composition) operator corresponding to H) acting from $W^{2,q}(\Omega)$ to $W^{1,q}(\Omega)$ ($N < q < \infty$, $\alpha \in (\frac{1}{2}, 1]$) fulfills (see [166, p. 90] for the Nirenberg–Gagliardo inequality):

$$\begin{aligned}\|H(u,\nabla u)\|_{W^{1,q}(\Omega)} &\le C\|\nabla H\|_{L^\infty}\|u\|_{W^{2,q}(\Omega)} \le c\|\nabla H\|_{L^\infty}\|u\|_{W^{2+\epsilon,q}(\Omega)}^{\frac{2}{2+\epsilon}}\|u\|_{L^q(\Omega)}^{\frac{\epsilon}{2+\epsilon}} \\ &\le c\|\nabla H\|_{L^\infty}\|u\|_{W^{2+\epsilon,q}(\Omega)}^{\frac{2}{2+\epsilon}}\|u\|_{L^\infty(\Omega)}^{\frac{\epsilon}{2+\epsilon}} \\ &\le c(\|u\|_{W^{1,\infty}(\Omega)}, H)\|u\|_{W^{2+\epsilon,q}(\Omega)}^{\frac{2}{2+\epsilon}},\end{aligned} \tag{10.16}$$

where $0 < \epsilon < 2\alpha - 1$. Moreover, if $H(0,p) = 0$ for $p \in \mathbb{R}^N$ then $u \in W^{2,q}(\Omega) \cap W_0^{1,q}(\Omega)$ implies $H(u, \nabla u) \in W_0^{1,q}(\Omega)$, $q > N$.

Remark 10.2.6. We need to mention here, that the condition "$H(0,p) = 0$ for $p \in \mathbb{R}^N$" imposed above is restrictive indeed. It is however necessary when we study regular solutions to (10.1), to guarantee the *compatibility conditions* are satisfied by the nonlinear term $H(u, \nabla u)$ at $\partial\Omega$. Note that the abstract theory of local existence for (10.1) requires that the nonlinearity acts between specified domains of fractional powers of the operator A; in particular the nonlinearity taken on the solution; $H(u, \nabla u)$ needs to satisfy the prescribed for the base space boundary conditions at $\partial\Omega$. We are thus limited to nonlinearities similar to the second example in (10.15).

If we study the Cauchy problem in $\mathbb{R}^N$, such type assumption on the nonlinearity is not needed (we *do not have* boundary conditions) except some integrability assumption for the solution over $\mathbb{R}^N$.

The estimate (10.16) provides us the subordination type condition (consequently, global in time extendibility) for regular solutions of subcritical approximations (10.1) with $\alpha \in (\frac{1}{2}, 1]$ and $u_0 \in D((-\Delta)^{1+\frac{\epsilon}{2}}) \subset W^{2+\epsilon,q}(\Omega)$ introduced in the second part of Theorem 10.2.3.

10.2.3 Maximum principle and another L^∞ a priori estimates

Our next task is to specify assumptions on the nonlinearity H, such that the local solutions obtained in Theorem 10.2.3 fulfill a $L^\infty(\Omega)$ a priori estimate. In case of the Condition (A) below, such a priori estimate has a from of the maximum principle (as for second-order parabolic equations). Condition (B) is connected with the "method of invariant regions" reported in monograph [154]. This assumption is not leading to the maximum principle, while it provides another a priori estimate of solutions in $L^\infty(\Omega)$. We mention finally, for "hydrodynamic type nonlinearities," another way of getting the maximum principle type estimates.

Assumption I. *We will assume that the real function $H : \mathbb{R} \times \mathbb{R}^N \to \mathbb{R}$, $H = H(u,p)$ is C^2 smooth, in particular locally Lipschitz with respect to (u,p),*

$$|H(u,p) - H(v,q)| \le l(|u - v| + |p - q|)$$

where $|\cdot|$ denotes also the Euclidean norm, $l = l(K) \ge 0$ and $(u,p),(v,q) \in K \subset \mathbb{R} \times \mathbb{R}^N$, where K denotes a compact set.

We additionally assume one of the following conditions (A) or (B).

Condition (A)

$$\begin{aligned} &\exists_{D>0} \forall_{u\in\mathbb{R}, p\in\mathbb{R}^N} -[H(u,p) - H(0,p)] \operatorname{sgn}(u) \le D|u|, \\ &\exists_{L>0} \forall_{p\in\mathbb{R}^N} |H(0,p)| \le L|p|, \end{aligned} \tag{10.17}$$

(the final condition is satisfied in particular when $H(0,p) = 0$) where 0 denotes *zero vector* (see [116, Theorem 2.9], [164, p. 143]).

Condition (B)

$$-H(r,0)r \le 0, \quad |r| \ge r_0 > 0, \tag{10.18}$$

and

$$|H(r,p)| \le c(|r|)(1 + |p|), \quad (r,p) \in \mathbb{R} \times \mathbb{R}^N, \tag{10.19}$$

where $c : [0,\infty) \to [0,\infty)$ is a continuous increasing function.

For completeness of the presentation, we give next a proof of the maximum principle for smooth solutions of the Dirichlet problem (10.1).

Theorem 10.2.7. *Let the assumptions of Theorem 10.2.3 hold and u be the corresponding local solution. Assume additionally the Assumption I above with Condition (A) is satisfied. Then the following estimate holds:*

$$\|u(t,\cdot)\|_{L^\infty(\Omega)} \le \|u_0\|_{L^\infty(\Omega)} e^{Dt}. \tag{10.20}$$

Proof. The proof will follow in several steps.

Step 1.

Remark 10.2.8. Let $\phi \in C(\overline{\Omega}) \cap C^2(\Omega)$. Let the sequence $\{x_m\}_{m\ge1} \subset \Omega$ satisfy $\phi(x_m) \to \sup_{x\in\Omega} \phi(x)$. Then

$$\lim_{m\to\infty} \nabla\phi(x_m) = 0 \quad \text{and} \quad \limsup_{m\to\infty}(-\Delta)^\alpha \phi(x_m) \ge 0. \tag{10.21}$$

See [98, p. 37] for the proof.

Step 2.

Lemma 10.2.9. *Let $x_0 \in \Omega$ be the point of the positive maximum of the regular function ϕ. Then the following inequality holds:*

$$-(-\Delta)^\alpha \phi(x_0) \le 0, \quad \alpha \in (0,1]. \tag{10.22}$$

The idea of the proof is as follows.

Working with the Dirichlet Laplacian in a bounded, regular domain $\Omega \subset \mathbb{R}^N$, the standard Balakrishnan definition of $(-\Delta)^\alpha$ is equivalent to the *Bochner's definition* (see Chapter 5):

$$((-\Delta)^\alpha \phi)(x) = c_\alpha \int_0^\infty [\phi(x) - e^{t\Delta}\phi(x)] t^{-\alpha-1} dt, \tag{10.23}$$

where $\phi \in D((-\Delta)^\alpha)$. Together with the expression for the heat semigroup,

$$(e^{t\Delta}\phi)(x) = \int_\Omega K_\Omega(t,x,y)\phi(y)dy, \tag{10.24}$$

where K_Ω denotes the corresponding *heat kernel* the last definition was used, for example, in [36]. Since the latter kernel is nonnegative and normalized [44, p. 62, Section 3], property (10.22) follows easily from (10.23).

Step 3. Following [98], Theorem 1.18, under the condition (10.17), we prove next a version of the maximum principle for solutions u of (10.1) such that $u \in C([0,T] \times \overline{\Omega})$ with $u_{tt} \in L^\infty(\Omega)([t_1,t_2] \times \overline{\Omega})$ in arbitrary closed interval $[t_1,t_2] \in (0,T)$.

We transform in a standard way equation (10.1) multiplying it by e^{-Dt} and denoting $v(t,x) = u(t,x)e^{-Dt}$, to obtain

$$v_t + (-\Delta)^\alpha v = -Dv - H(u,\nabla u)e^{-Dt}, \tag{10.25}$$

with the same initial-boundary conditions as for u. We have the following.

Proposition 10.2.10. *Define, for smooth solution v of (10.25), the function:*

$$\Phi(t) = \sup_{x\in\Omega} v(t,x) = \sup_{x\in\Omega} u(t,x)e^{-Dt}. \tag{10.26}$$

Such Φ is locally Lipschitz (hence differentiable almost everywhere) and $\Phi'(t) \le 0$ a. e. in $(0,T)$.

To show Lipschitz continuity of Φ, for small $\epsilon > 0$ we choose $x_\epsilon \in \Omega$ such that

$$\sup_{x\in\Omega} v(t,x) = v(t,x_\epsilon) + \epsilon.$$

Let $t,s \in I$ where $(0,T) \supset I$ is a closed interval. Suppose that $\Phi(t) \ge \Phi(s)$, then

$$\begin{aligned} 0 \le \Phi(t) - \Phi(s) &= \sup_{x\in\Omega} v(t,x) - \sup_{x\in\Omega} v(s,x) = v(t,x_\epsilon) + \epsilon - \sup_{x\in\Omega} v(s,x) \\ &\le \epsilon + v(t,x_\epsilon) - v(s,x_\epsilon) \le \epsilon + \sup_{x\in\Omega}|v(t,x) - v(s,x)| \le \epsilon + |t-s| \sup_{x\in\Omega,t\in I} |v_t(t,x)|. \end{aligned} \tag{10.27}$$

Since $t,s \in I$, $\epsilon > 0$, by letting $\epsilon \to 0$, this gives local Lipschitz continuity of Φ.

Now we show that $\Phi'(t) \le 0$.

Note that due to the boundary condition always $\Phi(t) \ge 0$, therefore, it suffices to consider the case when $\Phi(t) > 0$. As a consequence of Taylor's formula, for $s,t \in I$, $0 < s < t$:

$$v(t-s,x) = v(t,x) - sv_t(t,x) + \frac{s^2}{2}v_{tt}(\bar{t},x) \ge v(t,x) - sv_t(t,x) - Cs^2 \tag{10.28}$$

($\bar{t}$ denotes a point intermediate between $t-s$ and t), since the second time derivative is bounded.

Writing (10.28) in an equivalent form and using equation (10.25), we obtain

$$\begin{aligned} v(t,x) &\le v(t-s,x) + sv_t(t,x) + Cs^2 \\ &\le \sup_{x\in\Omega} v(t-s,x) + s[-(-\Delta)^\alpha v(t,x) - Dv(t,x) - H(u(t,x),\nabla u(t,x))e^{-Dt}] + Cs^2 \\ &\le \sup_{x\in\Omega} v(t-s,x) + Cs^2 + s[-(-\Delta)^\alpha v(t,x) - Dv(t,x) - [H(u(t,x),\nabla u(t,x)) \\ &\quad - H(0,\nabla u(t,x)) + H(0,\nabla u(t,x))]e^{-Dt}]. \end{aligned} \tag{10.29}$$

Setting $x = x_m$ in (10.29), we obtain

$$\begin{aligned} v(t,x_m) \le \sup_{x\in\Omega} v(t-s,x) + Cs^2 + s[-(-\Delta)^\alpha v(t,x_m) - Dv(t,x_m) \\ - [H(u(t,x_m),\nabla u(t,x_m)) - H(0,\nabla u(t,x_m)) + H(0,\nabla u(t,x_m))]e^{-Dt}]. \end{aligned}$$

Using Observation 10.2.8 and condition (10.17), letting $m \to \infty$, we see that the whole square bracket is nonpositive in that limit and, consequently,

$$\Phi(t) \le \Phi(t-s) + Cs^2, \tag{10.30}$$

or equivalently

$$\frac{\Phi(t) - \Phi(t-s)}{s} \leq Cs.$$

Letting $s \to 0$, we have: $\Phi'(t) \leq 0$, where $t \in I$ is a point where Φ is differentiable. Consequently Φ is nonincreasing in I. Symmetric property holds for the function $\Psi(t) = \inf_{x\in\Omega} v(t,x)$.

We notify that $\sup_{x\in\Omega} u(t,x)e^{-Dt}$ is nonincreasing, and $\inf_{x\in\Omega} u(t,x)e^{-Dt}$ is nondecreasing, thus we have a $L^\infty(\Omega)$ a priori estimate for the solution $u(t,\cdot)$:

$$\|u(t,\cdot)\|_{L^\infty(\Omega)} \leq \|u_0\|_{L^\infty(\Omega)} e^{Dt}.$$

Step 4. While the proof of the maximum principle above was given for smooth solutions, in $C(\overline{\Omega}) \cap C^2(\Omega)$, such property remains valid for less regular solutions to (10.1) through density argument. Indeed, when building in Theorem 10.2.3 the local solutions to (10.1) with $\alpha > \frac{1}{2}$, we will just approximate the initial value $u_0 \in W^{2,q}(\Omega)$ with a sequence $u_0^n \in W^{2,q}(\Omega) \cap C^2(\Omega)$ (when $q > N$, $W^{2,q}(\Omega) \subset L^\infty(\Omega)$), and noting validity of the above maximum principle for regular approximations and uniformity of that estimate, the same estimate will stay valid for the limit problem.

These complete the proof of the theorem. □

We will present now the proof of an *a priori* $L^\infty(\Omega)$ *estimate* of solution obtained under condition (B). Only small modifications (compare to the proof of the latter theorem) are needed to show an a priori $L^\infty(\Omega)$ estimate

$$\|u(t,\cdot)\|_{L^\infty(\Omega)} \leq \max\{\|u_0\|_{L^\infty(\Omega)}, r_0\}, \tag{10.31}$$

is valid under the condition (B) of Assumption I. The detailed calculations will be left for the reader. We only state the following.

Remark 10.2.11. The proof of Theorem 10.2.7 was given under the assumption (10.17). It is easy to see that replacing (10.17) with the condition (10.18) we can show that the following implication holds:

$$\Theta(t) = \sup_{x\in\Omega} u(t,x) > r_0 \Rightarrow \Theta(t) \text{ is nonincreasing.}$$

We give only the main estimate of the proof, which starts with an extension of (10.29) written for the solution u:

$$\begin{aligned} u(t,x) &\leq u(t-s,x) + su_t(t,x) + Cs^2 \leq \sup_{x\in\Omega} u(t-s,x) \\ &+ s\big[-(-\Delta)^\alpha u(t,x) - H(u(t,x),0) + \big|H(u(t,x),0) - H(u(t,x),\nabla u(t,x))\big|\big] + Cs^2. \end{aligned} \tag{10.32}$$

We set $x = x_m$ in that estimate, where $u(t,x_m) \to \sup_{x\in\Omega} u(t,x) > r_0$. Due to Remark 10.2.8, we have that $\limsup_{m\to\infty} -(-\Delta)^\alpha u(t,x_m) \leq 0$. Further, as a consequence of the condition (10.18); $-H(u(t,x_m),0) \le 0$ for m sufficiently large (i. e., whenever $u(t,x_m) > r_0$). Finally, thanks to the Lipschitz continuity of H with respect to the second argument and Remark 10.2.8, the whole square bracket above is nonpositive in the limit, hence we get

$$\Theta(t) \le \Theta(t-s) + Cs^2,$$

as long as $\Theta(t) > r_0$. Equivalently, $\Theta'(t) \le 0$ a. e. as long as $\Theta(t) > r_0$. A symmetric property holds for $\inf_{x\in\Omega} u(t,x)$:

$$\Psi(t) = \inf_{x\in\Omega} u(t,x) < -r_0 \Rightarrow \Psi(t) \text{ is nondecreasing.}$$

The two properties together show that the a priori $L^\infty(\Omega)$ estimate (10.31) holds.

Following [32, Appendix] we will finally show an iteration proof of a version of the maximum principle. The proof will be given for a "nonhomogeneous" extension of the equation (10.1):

$$u_t + (-\Delta)^\alpha u + H(u,\nabla u) = f(x), \quad t > 0,\ x \in \Omega \subset \mathbb{R}^N, \tag{10.33}$$

with the same initial-boundary condition as in (10.1).

Lemma 10.2.12. *Under the assumption* $-H(u,\nabla u)u \le A_0 u^2$, $A_0 \ge 0$, $f \in L^\infty(\Omega)$, *all sufficiently smooth solutions of* (10.33) *fulfill a version of the maximum principle:*

$$\|u(t)\|_{L^\infty(\Omega)} \le (\|u_0\|_{L^\infty(\Omega)} + \|f\|_{L^\infty(\Omega)} t)e^{A_0 t}. \tag{10.34}$$

Proof. It suffices to consider the case $A_0 = 0$, elsewhere we can use the transformation $v(t) = u(t)e^{-A_0 t}$ and study the equation for v, which reads:

$$v_t + A_0 v + (-\Delta)^\alpha v + \tilde{H}(v,\nabla v) = f(x)e^{-A_0 t},$$

where $\tilde{H}(v,\nabla v) := H(ve^{A_0 t}, \nabla v e^{A_0 t})e^{-A_0 t} = H(u,\nabla u)e^{-A_0 t}$, and in the further calculations only an estimate $\|fe^{-A_0 t}\|_{L^\infty(\Omega)} \le \|f\|_{L^\infty(\Omega)}$ is needed.

Multiplying (10.33) by u^{2^k-1}, integrating and using the above assumption (with $A_0 = 0$), we obtain

$$2^{-k}\frac{d}{dt}\int_\Omega u^{2^k}\,dx \le -\int_\Omega (-\Delta)^\alpha u u^{2^k-1}\,dx + \int_\Omega f u^{2^k-1}\,dx. \tag{10.35}$$

Applying next the Kato–Beurling–Deny inequality, and the Hölder inequality (with $p = \frac{2^k}{2^k-1}$, $q = 2^k$), we have further

$$2^{-k}\frac{d}{dt}\int_\Omega u^{2^k}\,dx \le -\frac{4(2^k-1)}{2^{2k}}\int_\Omega [(-\Delta)^{\frac{\alpha}{2}}(u^{2^{k-1}})]^2\,dx + \Big(\int_\Omega f^{2^k}\,dx\Big)^{2^{-k}}\Big(\int_\Omega u^{2^k}\,dx\Big)^{1-2^{-k}}. \tag{10.36}$$

For $w = u^{2^{k-1}}$, using the generalized Poincaré inequality (10.54),

$$\lambda_1^{\beta-\alpha}\|\phi\|_{X^\alpha} \le \|\phi\|_{X^\beta}, \quad \beta > \alpha,$$

we get

$$\begin{aligned}\frac{d}{dt}\int_\Omega w^2 dx &\le -\lambda_1^\alpha(2^k-1)2^{2-k}\int_\Omega w^2 dx + \|f\|_{L^\infty(\Omega)}2^k|\Omega|^{2^{-k}}\Big(\int_\Omega w^2 dx\Big)^{1-2^{-k}}\\ &\le -\alpha_k\|w\|^2_{L^2(\Omega)} + \beta_k 2^k\big(\|w\|^2_{L^2(\Omega)}\big)^{1-2^{-k}},\end{aligned} \tag{10.37}$$

where $\alpha_k = \lambda_1^\alpha\frac{2^k-1}{2^{k-2}}$, $\beta_k = \|f\|_{L^\infty(\Omega)}|\Omega|^{2^{-k}}$. A Bernoulli-type differential inequality (10.37) for the function $z(t) := \|w(t)\|^2_{L^2(\Omega)}$:

$$z'(t) \le -\alpha_k z(t) + \beta_k 2^k (z(t))^{1-2^{-k}},$$

will be solved using the substitution $y(t) = z(t)^{2^{-k}}$, to obtain

$$\|w(t)\|^2_{L^2(\Omega)} \le \Big[\|w(0)\|^{2^{1-k}}_{L^2(\Omega)} + \int_0^t \beta_k e^{\alpha_k 2^{-k} z}dz\Big]^{2^k} e^{-\alpha_k t}.$$

After taking the 2^k roots and returning to function u, we thus have

$$\|u(t)\|_{L^{2^k}(\Omega)} \le \Big[\|u_0\|_{L^{2^k}(\Omega)} + \beta_k\frac{2^k}{\alpha_k}(e^{\frac{\alpha_k t}{2^k}}-1)\Big]e^{-\frac{\alpha_k t}{2^k}}. \tag{10.38}$$

Finally, passing with k to $+\infty$, we get an estimate

$$\|u(t)\|_{L^\infty(\Omega)} \le \|u_0\|_{L^\infty(\Omega)} + \|f\|_{L^\infty(\Omega)}t. \tag{10.39}$$

The proof is complete. □

A similar lemma, for second-order equations and with a classical proof, can be found in [91].

Hydrodynamic type nonlinearity

Another way of proving the maximum principle is offered by the Moser–Alikakos iteration technique (e. g., [32, Section 9.3]). A $L^\infty(\Omega)$ estimate is obtained within that technique as a limit of the $L^{2^k}(\Omega)$ estimates, when $k \to \infty$, and such estimates will be obtained directly multiplying the equation by u^{2^k-1}. Consider a particular case of the nonlinearity in (10.1), the so-called *hydrodynamical type nonlinearity*:

$$u_t + (-\Delta)^\alpha u + b(x)\cdot\nabla(h(u)) = 0, \quad x \in \Omega,\ t > 0, \tag{10.40}$$

where $b : \Omega \to \mathbb{R}^N$ is a differentiable, bounded vector function and $h(u)$ was initially equal to $u|u|^{r-1}$ with certain $r > 1$. Here, $h : \mathbb{R} \to \mathbb{R}$ is C^1 with h' locally Lipschitz, and we require that

$$\operatorname{div} b(x) = 0, \quad x \in \Omega.$$

When calculating the $L^\infty(\Omega)$ estimate, as described above, the nonlinear term is transformed as follows:

$$-\int_\Omega b(x) \cdot \nabla(h(u))u^{2^k-1}dx = -\int_\Omega b(x) \cdot \nabla(h_k(u))dx = 0, \tag{10.41}$$

where $h_k(u) = \int_0^u h'(z)z^{2^k-1}dz$ (note that $h_k(0) = 0$). Hence, the $L^\infty(\Omega)$ estimate of solutions to (10.40) is the same as for the solutions of the corresponding to (10.40) *linear equation.*

Remark 10.2.13. Once we have obtained a $L^\infty(\Omega)$ a priori estimate for solutions to (10.1) with $\alpha \in (\frac{1}{2}, 1]$, and since we are working in the phase space contained in $L^\infty(\Omega)$ (compare Theorem 10.2.3), when the initial data u_0 is given varying in $[-M, M]$, we can *change eventually* the nonlinearity H outside the set $[-M, M] \times \mathbb{R}^N$ to obtain a globally Lipschitz real function. Evidently, such modification will not change the solution corresponding to initial data u_0.

10.2.4 Global in time solutions to (10.1) with $\alpha \in (\frac{1}{2}, 1]$

In Theorem 10.2.3, a local in time solution to (10.1), $\alpha \in (\frac{1}{2}, 1]$, was constructed satisfying:

$$u \in C^0([0,\tau); W^{2,q}(\Omega) \cap W_0^{1,q}(\Omega)) \cap C^0((0,\tau); D((-\Delta)^{\frac{1}{2}+\alpha})),$$
$$u_t \in C^0((0,\tau); W^{1+2\alpha^-,q}(\Omega)),$$

$q > N$ arbitrary, in the *base space* $W_0^{1,q}(\Omega)$. Moreover, it satisfies $u \in C^0([0,\tau); D((-\Delta)^{1+\frac{\epsilon}{2}}))$ with $0 < \epsilon < 2\alpha - 1$, provided that $u_0 \in D((-\Delta)^{1+\frac{\epsilon}{2}}) \subset W^{2+\epsilon,q}(\Omega)$.

To extend such solution globally in time, according to [32, p. 79], a *subordination condition* is needed. To guarantee it, for arbitrary fixed $\alpha > \frac{1}{2}$, the following assumption is formulated.

Assumption II. *Let Assumption I be satisfied. Moreover, the real function $H = H(u, v_1, \ldots, v_N)$ restricted to any set of the form $[-M, M] \times \mathbb{R}^N$ ($M > 0$ arbitrary) is globally Lipschitz continuous with bounded all first partial derivatives $\nabla H = (\frac{\partial H}{\partial u}, \frac{\partial H}{\partial v_1}, \ldots, \frac{\partial H}{\partial v_N})$, more precisely,*

$$\forall_{M>0} \exists_{C=C(M)>0} \forall_{(u,v_1,\ldots,v_N)\in[-M,M]\times\mathbb{R}^N} \quad \left|\left(\frac{\partial H}{\partial u}, \frac{\partial H}{\partial v_1}, \ldots, \frac{\partial H}{\partial v_N}\right)\right| \le C.$$

With the use of (10.14) the nonlinear term in (10.1) will be estimated as follows:

$$\begin{aligned} c\|(-\Delta)^{\frac{1}{2}}H(u,\nabla u)\|_{L^q(\Omega)} &\leq \|H(u,\nabla u)\|_{W^{1,q}(\Omega)} \\ &\leq \|H(u,\nabla u)\|_{L^q(\Omega)} + \sum_{i=1}^{N}\left\|\frac{\partial}{\partial x_i}H(u,\nabla u)\right\|_{L^q(\Omega)} \\ &\leq \|\nabla H\|_{L^\infty}\|u\|_{W^{1,q}(\Omega)} + c'\|\nabla H\|_{L^\infty}\|u\|_{W^{2,q}(\Omega)} \\ &\leq c''\|\nabla H\|_{L^\infty}\|u\|_{W^{2,q}(\Omega)}. \end{aligned} \tag{10.42}$$

The above is a form of the *subordination condition* for solutions of the problem (10.1) with $\alpha > \frac{1}{2}$, since for such problem we will choose $D((-\Delta)^{1+\frac{\epsilon}{2}}) \subset W^{2+\epsilon,q}(\Omega)$ $(0 < \frac{\epsilon}{2} < \alpha - \frac{1}{2})$ to be the phase space. Estimate (10.42) extends then to

$$\begin{aligned} \|H(u,\nabla u)\|_{W^{1,q}(\Omega)} \leq c''\|\nabla H\|_{L^\infty}\|u\|_{W^{2,q}(\Omega)} &\leq C\|\nabla H\|_{L^\infty}\|u\|^{\frac{2}{2+\epsilon}}_{W^{2+\epsilon,q}(\Omega)}\|u\|^{\frac{\epsilon}{2+\epsilon}}_{L^q(\Omega)} \\ &\leq c(\|u\|_{L^\infty(\Omega)})\|u\|^{\frac{2}{2+\epsilon}}_{W^{2+\epsilon,q}(\Omega)}, \end{aligned} \tag{10.43}$$

with $\|\nabla H\|_{L^\infty}$ bounded by the Assumption II while $\|u\|_{L^\infty(\Omega)}$ bounded by the maximum principle. As a consequence of the general result of Chapter 4 validity of such type subordination condition is sufficient to extend the local solutions varying continuously in $W^{2+\epsilon,q}(\Omega)$ globally in time.

We thus have the following conclusion.

Theorem 10.2.14. *Let Assumption* II *be satisfied. Whenever* $u_0 \in D((-\Delta)^{1+\frac{\epsilon}{2}}) \subset W^{2+\epsilon,q}(\Omega)$ *where* $0 < \epsilon < 2\alpha - 1$, *the local solution constructed in Theorem* 10.2.3 *will be extended globally in time,*

$$\begin{aligned} u &\in C^0([0,T]; D((-\Delta)^{1+\frac{\epsilon}{2}})) \cap C^0((0,T); ((-\Delta)^{\frac{1}{2}+\alpha})), \\ u_t &\in C^0((0,T); W^{1+2\alpha^-,q}(\Omega)), \end{aligned} \tag{10.44}$$

where $T > 0$ *is arbitrarily large.*

Remark 10.2.15. We need to note here, that since the a priori estimate coming from certain version of the maximum principle is of exponential type

$$\|u(t,\cdot)\|_{L^\infty(\Omega)} \leq \|u_0\|_{L^\infty(\Omega)}e^{Dt},$$

and we are eventually modifying the nonlinear term outside a sufficiently large set $[-M,M] \times \mathbb{R}^N$, in that chapter a "global solution" is understood as a solution, which is defined on an arbitrarily large time interval $[0,T]$ (as in the monograph [115]).

Thus, in our construction, we choose first suitable $T > 0$, then modify the nonlinearity outside the sufficiently large set $[-M,M] \times \mathbb{R}^N$ where $M = M(T) = \|u_0\|_{L^\infty(\Omega)}e^{DT}$, and get a solution on $[0,T]$. It will still exist beyond T, but using eventually the modified part of the nonlinearity, need not be a solution of the original equation for $t > T$.

10.2.5 A better norm a priori estimate in Hilbert spaces based on nonlinear interpolation theory

We will provide now an $L^\infty(0,T;H_0^1(\Omega))$ a priori estimate of the solution to (10.1) that uses an interesting *nonlinear interpolation theory* (see [168, p. 137]), a generalization of interpolation theory to Lipschitz operators we owe to J.-L. Lions and F. Browder.

From now on, we will consider the nonlinearity H satisfying the following *structural condition.*

Assumption III. *Let Assumption* II *be satisfied. Moreover,*

$$H(u,\nabla u) = H_1(u,\nabla u) + h_2(u). \tag{10.45}$$

Recall that both H_1 and h_2 are C^2 functions and such that restricted to any set of the form $[-M,M]\times\mathbb{R}^N$ or $[-M,M]$ ($M>0$ arbitrary) are globally Lipschitz continuous. We additionally assume that $H_1(0,\dots,0)$ and $h_2(0)=0$.

Multiplying (10.1) by $(-\Delta u)$, we obtain

$$\int_\Omega u_t(-\Delta u)dx + \int_\Omega [(-\Delta)^{\frac{\alpha+1}{2}}u]^2 dx + \int_\Omega (H_1(u,\nabla u) + h_2(u))(-\Delta u)dx = 0. \tag{10.46}$$

Now we rewrite the nonlinear term in the following way:

$$\begin{aligned}\int_\Omega H_1(u,\nabla u)(-\Delta u)dx &= \int_\Omega (-\Delta)^{\frac{3}{4}}u(-\Delta)^{\frac{1}{4}}H_1(u,\nabla u)dx,\\ \int_\Omega h_2(u)(-\Delta u)dx &= \int_\Omega (-\Delta)^{\frac{1}{2}}u(-\Delta)^{\frac{1}{2}}h_2(u)dx.\end{aligned} \tag{10.47}$$

The right-hand side of (10.47) will be next estimated using Proposition 10.3.6, where we specify

$$\begin{aligned}&E_1 = H^1(\Omega),\quad F_1 = L^2(\Omega),\quad E_0 = H^2(\Omega),\quad F_0 = H^1(\Omega),\\ &A(u) := H_1(u,\nabla u).\end{aligned} \tag{10.48}$$

Then, estimating similarly as in Proposition 10.2.4, adding and subtracting changing only one argument, thanks to the global Lipschitz continuity of H_1 in the sets $[-M,M]\times\mathbb{R}^N$, we have

$$\begin{aligned}\|A(u)-A(v)\|_{L^2(\Omega)} &= \|H_1(u,\nabla u) - H_1(v,\nabla u) + H_1(v,\nabla u) - \cdots - H_1(v,\nabla v)\|_{L^2(\Omega)}\\ &\le M_1\|u-v\|_{H^1(\Omega)},\end{aligned} \tag{10.49}$$

with the Lipschitz constant $M_1 = M_1(\|\nabla H_1\|_{L^\infty})$.

Consequently, using also (10.10), the following estimates are valid:

$$\|A(u)-A(v)\|_{L^2(\Omega)} \le M_1\|u-v\|_{H^1(\Omega)},\quad \|A(u)\|_{H^1(\Omega)} \le M_0\|u\|_{H^2(\Omega)}. \tag{10.50}$$

We need also to mention (e. g., [127, p. 43]), that usually the real and complex interpolation functors are different, while just in the Hilbert spaces case ($p = 2$) they coincide:

$$[X, Y]_\theta = (X, Y)_{\theta,2}, \quad 0 < \theta < 1,$$

(we will not explain here better the notation, referring however to [127, 168] for details).

As a consequence of the mentioned above two abstract results, in our particular case we claim an estimate

$$\|(-\Delta)^{\frac{1}{4}} H_1(u, \nabla u)\|_{L^2(\Omega)} = \|H_1(u, \nabla u)\|_{H^{\frac{1}{2}}(\Omega)} \leq C\|u\|_{H^{\frac{3}{2}}(\Omega)}, \tag{10.51}$$

in fact a more general estimate holds true for any $\beta \in (0, 1)$:

$$\|H_1(u, \nabla u)\|_{H^\beta(\Omega)} \leq C\|u\|_{H^{1+\beta}(\Omega)}.$$

Moreover, the constant C is small provided that $\|\nabla H_1\|_{L^\infty}$ is small (that means the accepted growth of H_1 with respect to all the variables is sufficiently slow).

A similar but much simpler reasoning shows that for the remaining term h_2 of the complete nonlinearity $H(u, \nabla u) = H_1(u, \nabla u) + h_2(u)$ an estimate holds:

$$\|(-\Delta)^{\frac{1}{2}} h_2(u)\|_{L^2(\Omega)} = \|h_2(u)\|_{H^1(\Omega)} \leq c'\|u\|_{H^1(\Omega)}, \tag{10.52}$$

with a certain (not necessarily small) constant c'. Consequently, we will extend the estimate (10.46) to

$$\frac{1}{2}\frac{d}{dt}\int_\Omega [(-\Delta)^{\frac{1}{2}} u]^2 dx + \int_\Omega [(-\Delta)^{\frac{\alpha+1}{2}} u]^2 dx \leq C\|u\|^2_{H^{\frac{3}{2}}(\Omega)} + c'\|u\|^2_{H^1(\Omega)}, \tag{10.53}$$

with possibly small constant C and another constant $c' > 0$.

When the action of the nonlinear term H_1 is weaker than that of the main linear part $(-\Delta)^\alpha$ in (10.1), we will observe boundedness, uniform with respect to α near $\frac{1}{2}$, of the solutions. Note that working in the fractional scale of Hilbert spaces $D((-\Delta)^\beta) = X^\beta$ build over $L^2(\Omega)$, Ω bounded, a version of the Poincaré inequality holds:

$$\lambda_1^{\beta-\alpha}\|\phi\|_{X^\alpha} \leq \|\phi\|_{X^\beta}, \quad \beta > \alpha, \tag{10.54}$$

where λ_1 denotes the first positive eigenvalue of the negative Dirichlet Laplacian in Ω; see Subsection 5.2.2.

The smallness restriction. Assume that the constant C above is dominated by the one coming from the main linear part (with the dissipation exponent $\alpha = \frac{1}{2}$)

$$\exists_{\delta>0}\ C\|u\|^2_{H^{\frac{3}{2}}(\Omega)} \leq (1-\delta)\int_\Omega [(-\Delta)^{\frac{3}{4}} u]^2 dx. \tag{10.55}$$

Observe further that due to (10.54) similar estimate stays valid for exponents $\frac{\alpha+1}{2}$ near $\frac{3}{4}$ (since $\lambda_1^{\frac{\alpha+1}{2}-\frac{3}{4}} \approx 1$). More precisely, let

$$C\|u\|^2_{H^{\frac{3}{2}}(\Omega)} \leq \left(1 - \frac{\delta}{2}\right) \int_\Omega [(-\Delta)^{\frac{\alpha+1}{2}} u]^2 dx \tag{10.56}$$

whenever α belongs to the interval $(\frac{1}{2}, \frac{3}{4}] \subset (\frac{1}{2}, 1]$. Extending the estimate (10.53) to the left using (10.56), we obtain

$$\frac{1}{2}\frac{d}{dt}\int_\Omega [(-\Delta)^{\frac{1}{2}} u]^2 dx + \frac{C}{1-\frac{\delta}{2}} \|u\|^2_{H^{\frac{3}{2}}(\Omega)} \leq C\|u\|^2_{H^{\frac{3}{2}}(\Omega)} + c'\|u\|^2_{H^1(\Omega)}.$$

Consequently, collecting the components, since $C\frac{\delta}{2-\delta} > 0$, we obtain a *uniform with respect to* $\alpha \in (\frac{1}{2}, \frac{3}{4}]$ estimate

$$\|u\|_{L^\infty(0,T;H_0^1(\Omega))} + \|u\|_{L^2(0,T;H^{\frac{3}{2}}(\Omega))} \leq \text{const.}$$

Note that in (10.55) we compare the squares of two equivalent norms of $D((-\Delta)^{\frac{3}{4}})$.

Remark 10.2.16. Condition (10.55) provides us a smallness restriction on the nonlinear term H_1 under which the behavior of solutions to (10.1) is controlled by the main linear part in that equation, moreover *uniformly in* $\alpha \in (\frac{1}{2}, \frac{3}{4}]$. Violation of such type condition allows the situation that the nonlinearity plays a decisive role in that equation (at least for critical $\alpha = \frac{1}{2}$), which is not the case considered here.

We are now able to formulate the following.

Theorem 10.2.17. *Let the Assumption* III *and the smallness restriction* (10.55) *be satisfied. Then, uniformly with respect to* $\alpha \in (\frac{1}{2}, \frac{3}{4}]$, *the solutions of* (10.1) *are estimated by*

$$\|u\|_{L^\infty(0,T;H_0^1(\Omega))} + \|u\|_{L^2(0,T;H^{\frac{3}{2}}(\Omega))} \leq \text{const.} \tag{10.57}$$

The last theorem will play a decisive role in passing to the limit $\alpha \to \frac{1}{2}$ to the critical Hamilton–Jacobi equation.

Estimates of better norms of the solutions will be obtained based on (10.14):

$$\left\|\frac{\partial}{\partial x_i} H_1(u, \nabla u)\right\|_{L^q(\Omega)} \leq c'\|\nabla H_1\|_{L^\infty}\|u\|_{W^{2,q}(\Omega)},$$

$1 \leq q \leq \infty$. Since our task is to estimate first spatial derivatives of u, we will apply to (10.1) the operator $(-\Delta)^{\frac{1}{2}}$ to obtain

$$(-\Delta)^{\frac{1}{2}} u_t + (-\Delta)^{\alpha+\frac{1}{2}} u + (-\Delta)^{\frac{1}{2}}(H_1(u, \nabla u) + h_2(u)) = 0. \tag{10.58}$$

With the use of (10.14), $q = 2$, the nonlinear term above will be estimated as follows:

$$\begin{aligned}\|(-\Delta)^{\frac{1}{2}} H_1(u,\nabla u)\|_{L^2(\Omega)} &= \|H_1(u,\nabla u)\|_{H^1(\Omega)}\\ &\le \|H_1(u,\nabla u)\|_{L^2(\Omega)} + \sum_{i=1}^{N}\left\|\frac{\partial}{\partial x_i} H_1(u,\nabla u)\right\|_{L^2(\Omega)}\\ &\le \|\nabla H_1\|_{L^\infty}\|u\|_{H^1(\Omega)} + c'\|\nabla H_1\|_{L^\infty}\|u\|_{H^2(\Omega)}\\ &\le c''\|\nabla H_1\|_{L^\infty}\|u\|_{H^2(\Omega)},\\ \|(-\Delta)^{\frac{3}{4}} h_2(u)\|_{L^2(\Omega)} &\le c_2\|h_2'\|_{L^\infty}\|u\|_{H^{\frac{3}{2}}(\Omega)}.\end{aligned} \tag{10.59}$$

Multiplying (10.58) by $(-\Delta)u$, we obtain

$$\begin{aligned}&\frac{1}{2}\frac{d}{dt}\int_\Omega [(-\Delta)^{\frac{3}{4}}u]^2 dx + \int_\Omega (-\Delta)^{\alpha+\frac{1}{2}}u(-\Delta)u dx\\ &\quad\le c''\|\nabla H_1\|_{L^\infty}\|u\|^2_{H^2(\Omega)} + c_2\|h_2'\|_{L^\infty}\|u\|^2_{H^{\frac{3}{2}}(\Omega)}.\end{aligned} \tag{10.60}$$

For arbitrary fixed $\alpha > \frac{1}{2}$, we next have (see (10.54)):

$$\begin{aligned}&\int_\Omega (-\Delta)^{\alpha+\frac{1}{2}}u(-\Delta)u dx = \int_\Omega [(-\Delta)^{\frac{2\alpha+3}{4}}u]^2 dx,\\ &\lambda_1^{\frac{\alpha}{2}-\frac{1}{4}}\|u\|_{H^2(\Omega)} \le \|u\|_{H^{\frac{2\alpha+3}{2}}(\Omega)},\end{aligned} \tag{10.61}$$

which gives an estimate

$$\frac{1}{2}\frac{d}{dt}\int_\Omega [(-\Delta)^{\frac{3}{4}}u]^2 dx + \lambda_1^{\alpha-\frac{1}{2}}\|u\|^2_{H^2(\Omega)} \le c''\|\nabla H_1\|_{L^\infty}\|u\|^2_{H^2(\Omega)} + c_2\|h'\|_{L^\infty}\|u\|^2_{H^{\frac{3}{2}}(\Omega)}. \tag{10.62}$$

In a standard way, we thus obtain an estimate, uniform for α close to $\frac{1}{2}^+$,

$$u \in L^\infty(0,T;H^{\frac{3}{2}}(\Omega)) \cap L^2(0,T;H^2(\Omega)), \tag{10.63}$$

provided that a *smallness condition* holds; for certain $\delta' > 0$,

$$\lambda_1^{\alpha-\frac{1}{2}} - c''\|\nabla H_1\|_{L^\infty} > \frac{\delta'}{2} \quad \text{whenever } \alpha \in (\frac{1}{2},\frac{3}{4}]. \tag{10.64}$$

Further estimate will be obtained applying operator $(-\Delta)$ to the equation (10.1) and multiplying the result by $(-\Delta)^{\frac{3}{2}}u$.

10.3 Solution to critical (10.1) obtained in the limit $\alpha \to \frac{1}{2}$

Throughout this section, we will notify explicitly the dependence of the solution of (10.1) on $\alpha \in (\frac{1}{2}, 1]$, calling it u_α. Let the Assumption III and the smallness restriction

(10.55) be satisfied. Having already the global regular solutions of the subcritical equations with $\alpha \in (\frac{1}{2}, 1]$ specified in Theorem 10.2.14, we will construct a global (weak) solution to critical problem (10.1), $\alpha = \frac{1}{2}$, passing to the limit in the subcritical equations.

For $\alpha \in (\frac{1}{2}, 1]$, according to Theorem 10.2.3, whenever $0 < \epsilon < 2\alpha - 1$ there exists a local in time solution u_α of the problem (10.1) satisfying

$$
\begin{aligned}
u_\alpha &\in C^0([0,\tau); D((-\Delta)^{1+\frac{\epsilon}{2}})) \cap C^0((0,\tau); D((-\Delta)^{\frac{1}{2}+\alpha})),\\
u_{\alpha t} &\in C^0((0,\tau); W^{1+2\alpha^-,q}(\Omega)).
\end{aligned}
$$

Moreover, due to Theorem 10.2.14, such local solution will be extended onto arbitrary time interval $[0, T]$.

Fix arbitrary sequence $\{\alpha_n\} \subset (\frac{1}{2}, \frac{3}{4}]$ (see (10.56)) convergent to $\frac{1}{2}$ with the corresponding to it sequence $\{\epsilon_n\}$.

Let u_0 be an arbitrary element of $H^1_0(\Omega)$. By density (since $q > N$) there exist a sequence of more regular initial functions $u_{0,n} \in D((-\Delta)^{1+\frac{\epsilon_n}{2}})$ convergent to u_0 in $H^1_0(\Omega)$, and the corresponding to them global solutions $u_{\alpha_n}(t)$ described above varying continuously in the phase spaces $D((-\Delta)^{1+\frac{\epsilon_n}{2}}) \subset W^{2+\epsilon_n,q}(\Omega)$, respectively. We will use that sequence to approximate a weak solution to critical problem (10.1) ($\alpha = \frac{1}{2}$) corresponding to u_0, provided that we have a sufficiently well *uniform in n* a priori estimate for $u_{\alpha_n}(t)$.

In the considerations below, $T > 0$ is arbitrary fixed (we deal with the global solutions). We have the following.

Theorem 10.3.1. *Let the assumptions of Theorem 10.2.17 be satisfied. Let $u_0 \in H^1_0(\Omega)$ and $\{u_{0,n}\} \subset D((-\Delta)^{1+\frac{\epsilon_n}{2}}) \subset W^{2+\epsilon_n,q}(\Omega)$ be a sequence convergent to u_0 in $H^1_0(\Omega)$. Let further $\{u_{\alpha_n}\}_{n\in\mathbb{N}}$ be the family of global regular solutions of subcritical equations (10.1) constructed in Theorem 10.2.14 corresponding (respectively) to $u_{0,n}$. Then, under the smallness restriction (10.55) the family $\{u_{\alpha_n}\}$ is precompact in $L^2(0,T;H^{\frac{3}{2}^-}(\Omega))$ and in $L^r(0,T;H^{1^-}(\Omega))$ with arbitrary finite r. Moreover, it is weak/weak* compact in the following spaces:*

$$
L^\infty(0,T;L^\infty(\Omega)), \quad L^2(0,T;H^{\frac{3}{2}}(\Omega)), \quad L^\infty(0,T;H^1_0(\Omega)). \tag{10.65}
$$

Proof. We are using the, uniform in $\alpha \in (\frac{1}{2}, \frac{3}{4}]$, a priori estimates of u_α following from the maximum principle and Theorem 10.2.17 under the smallness restriction (10.55). They have the form:

$$
\|u_\alpha\|_{L^\infty(0,T;L^\infty(\Omega))} \le \text{const}, \quad \|u_\alpha\|_{L^2(0,T;H^{\frac{3}{2}}(\Omega))} \le \text{const}, \quad \|u_\alpha\|_{L^\infty(0,T;H^1_0(\Omega))} \le \text{const}, \tag{10.66}
$$

$T > 0$ arbitrary (here const means a constant independent on α). Moreover, due to Assumption II (compare Remark 10.2.13), the nonlinear term H is globally Lipschitz continuous in any set $[-M, M] \times \mathbb{R}^N$ ($M > 0$ arbitrary). We need eventually correct the nonlinearity H according to $\|u_0\|_{L^\infty(\Omega)}$, as described in Remark 10.2.13.

We will apply Lions–Aubin compactness lemma [153] to see that the family $\{u_{\alpha_n}\}_{n\in\mathbb{N}}$ is precompact in $L^2(0,T;H^{\frac{3}{2}^-}(\Omega))$. Indeed, the equation (10.1) with $\alpha = \alpha_n$ is fulfilled by u_{α_n},

$$u_{\alpha_n t} + (-\Delta)^{\alpha_n} u_{\alpha_n} + H(u_{\alpha_n}, \nabla u_{\alpha_n}) = 0,$$

and we will look at its components separately to study boundedness of $\{u_{\alpha_n t}\}_{n\in\mathbb{N}}$. Due to (10.66), the intermediate term is bounded in $L^2(0,T;L^2(\Omega))$ uniformly in $\alpha_n \in (\frac{1}{2}, \frac{3}{4}]$.

Thanks to the global Lipschitz condition valid for H, adding and subtracting, we have further

$$\|H(u_{\alpha_n}, \nabla u_{\alpha_n})\|_{L^2(\Omega)} \leq c(\|u_{\alpha_n}\|_{L^2(\Omega)} + \|\nabla u_{\alpha_n}\|_{L^2(\Omega)}) \leq c\|u_{\alpha_n}\|_{H^1(\Omega)}. \tag{10.67}$$

Calculating the time derivative from (10.1), by the just described estimates, we have that

$$\int_0^T \|u_{\alpha_n t}(t,\cdot)\|^2_{L^2(\Omega)} dt \leq \text{const}, \tag{10.68}$$

with const independent on n, which means that $u_{\alpha_n t}$ are bounded in $L^2(0,T;L^2(\Omega))$, uniformly in $n \in \mathbb{N}$. Thanks to Lions-Aubin compactness lemma ([124, Chapter I] or [153, p. 84]), with the three spaces $H^{\frac{3}{2}}(\Omega) \subset H^{\frac{3}{2}^-}(\Omega) \subset L^2(\Omega)$, we thus verified that the family $\{u_{\alpha_n}\}_{n\in\mathbb{N}}$ is precompact in $L^2(0,T;H^{\frac{3}{2}^-}(\Omega))$. Also, through the result of [153, p. 65], using (10.66) and the embeddings $H^1_0(\Omega) \subset H^{1^-}_0(\Omega) \subset L^2(\Omega)$, the family $\{u_{\alpha_n}\}_{n\in\mathbb{N}}$ is precompact in $L^r(0,T;H^{1^-}_0(\Omega))$ with arbitrary finite r. In addition, directly from the estimates (10.66), that family is also compact in the weak/weak* topologies of the spaces listed in (10.65). □

The latter theorem allows us to pass to the limit $\alpha_n \to \frac{1}{2}$ in a weakly formulated equation (10.1) and check, that the critical equation (10.1), $\alpha = \frac{1}{2}$, will have a weak solution. The mentioned weak formulation of (10.1) reads the following.

Definition 10.3.2. Let $\alpha \in [\frac{1}{2}, 1]$. A function $u_\alpha \in L^2(0,T;H^1_0(\Omega))$ with $u_{\alpha t} \in L^2(0,T;L^2(\Omega))$, is called *a weak solution* of the Hamilton–Jacobi equation (10.1), if for arbitrary test function $v \in H^1_0(\Omega)$ the equality holds

$$\frac{d}{dt}\langle u_\alpha, v\rangle + \langle (-\Delta)^{\frac{1}{2}} u_\alpha, (-\Delta)^{\alpha-\frac{1}{2}} v\rangle + \langle H(u_\alpha, \nabla u_\alpha), v\rangle = 0, \tag{10.69}$$

where $\langle\cdot,\cdot\rangle$ denotes the scalar product in $L^2(\Omega)$ and $\frac{d}{dt}$ is the distributional derivative. The value $u_\alpha(0)$ is given in $H^1_0(\Omega)$.

It is easy to see that the global regular solutions of (10.1) constructed in Theorem 10.2.14 are in particular weak solutions in the sense given above. For arbitrary sequence $\{\alpha_n\} \subset (\frac{1}{2}, \frac{3}{4}]$, we have a weak solution u_{α_n} of (10.1):

$$\frac{d}{dt}\langle u_{\alpha_n}, v\rangle + \langle(-\Delta)^{\frac{1}{2}} u_{\alpha_n}, (-\Delta)^{\alpha_n - \frac{1}{2}} v\rangle + \langle H(u_{\alpha_n}, \nabla u_{\alpha_n}), v\rangle = 0. \tag{10.70}$$

We need also to note that $(-\Delta)^{\alpha-\frac{1}{2}} v \to v$ in $L^2(\Omega)$ as $\alpha \to \frac{1}{2}$ (see, e. g., Lemma 5.3.2). Theorem 10.3.1 and a standard compactness argument allow immediately to see, that there is a limit U of u_{α_n} as $\alpha_n \to \frac{1}{2}$ (eventually taken over a subsequence):

- $u_{\alpha_n} \to U$ weak* in $L^\infty(0, T; L^\infty(\Omega))$,
- $u_{\alpha_n} \to U$ weakly in $L^2(0, T; H^{\frac{3}{2}}(\Omega))$ and strongly in $L^2(0, T; H^{\frac{3}{2}^-}(\Omega))$,
- $u_{\alpha_n} \to U$ strongly in $L^r(0, T; H_0^{1^-}(\Omega))$, $r \geq 1$ arbitrary, and almost everywhere in $(0, T) \times \Omega$,
- $u_{\alpha_n t} \to U_t$ weakly in $L^2(0, T; L^2(\Omega)) = L^2((0, T) \times \Omega)$.

In particular, as was checked in the proof of Theorem 10.3.1, the limit function U belongs to the spaces:

$$U \in L^2(0, T; H^{\frac{3}{2}^-}(\Omega)), \quad U' = \frac{dU}{dt} \in L^2(0, T; L^2(\Omega)), \tag{10.71}$$

where time derivative is in the sense of *scalar distributions* (see [124]). Therefore, the function U is continuous as a map from $[0, T]$ into $L^2(\Omega)$ (see [171, Lemma 1.1, Chapter III], or section 5.4.2). In addition, since the approximating sequence $\{u_\alpha\}$ is bounded in $L^\infty(0, T; H_0^1(\Omega))$, the same will be true for U. Therefore, using [171, Lemma 1.4, Chapter III], $U : [0, T] \to H_0^1(\Omega)$ is also weakly continuous.

Moreover, the limit function U is a weak solution of the critical equation (10.1),

$$\frac{d}{dt}\langle U, v\rangle + \langle(-\Delta)^{\frac{1}{2}} U, v\rangle + \langle H(U, \nabla U), v\rangle = 0 \quad \text{for arbitrary } v \in H_0^1(\Omega). \tag{10.72}$$

However, the limit weak solution U need not be unique, since it depends on the chosen sequence α_n.

As regards uniqueness, we are able to formulate the following.

Remark 10.3.3. When the Lipschitz constant M_1 in condition (10.49) is small (see (10.75)), the weak solution U of the critical equation (10.1) is unique. Indeed, if there were two such solutions u_1, u_2, setting $\alpha = \frac{1}{2}$ and $v = (-\Delta)^{\frac{1}{2}}(u_1 - u_2)$ in (10.69), we get

$$\begin{aligned} &\frac{d}{dt}\langle u_1 - u_2, (-\Delta)^{\frac{1}{2}}(u_1 - u_2)\rangle + \langle(-\Delta)^{\frac{1}{2}}(u_1 - u_2), (-\Delta)^{\frac{1}{2}}(u_1 - u_2)\rangle \\ &\quad + \langle H(u_1, \nabla u_1) - H(u_2, \nabla u_2), (-\Delta)^{\frac{1}{2}}(u_1 - u_2)\rangle = 0, \end{aligned} \tag{10.73}$$

which, using (10.49), leads to an estimate

$$\frac{1}{2}\frac{d}{dt}\left\|(-\Delta)^{\frac{1}{4}}(u_1-u_2)\right\|^2_{L^2(\Omega)}+\left\|(-\Delta)^{\frac{1}{2}}(u_1-u_2)\right\|^2_{L^2(\Omega)}$$
$$\leq M_1\left\|(-\Delta)^{\frac{1}{2}}(u_1-u_2)\right\|_{L^2(\Omega)}\|u_1-u_2\|_{H^1(\Omega)}. \tag{10.74}$$

The last estimate implies that $u_1 = u_2$ provided that

$$\exists_{\delta>0}\ M_1\|\phi\|_{H^1(\Omega)}\leq(1-\delta)\left\|(-\Delta)^{\frac{1}{2}}\phi\right\|_{L^2(\Omega)}. \tag{10.75}$$

In case of gradient independent nonlinearity, we formulate the following.

Remark 10.3.4. When the nonlinear term is gradient independent, $H = H(u)$, and is a globally Lipschitz function, the Gronwall inequality simply gives uniqueness of solutions to (10.1). Indeed, if u_1, u_2 are two such solutions of (10.1) with $\alpha = \frac{1}{2}$, we would have

$$\frac{1}{2}\frac{d}{dt}\|u_1-u_2\|^2_{L^2(\Omega)}+\left\|(-\Delta)^{\frac{1}{4}}(u_1-u_2)\right\|^2_{L^2(\Omega)}$$
$$=-\langle H(u_1)-H(u_2),u_1-u_2\rangle\leq L\|u_1-u_2\|^2_{L^2(\Omega)}. \tag{10.76}$$

The Gronwall inequality would yield $u_1 = u_2$.

Remark 10.3.5. To minimize the assumption on ϕ above we will consider that lemma with the power of Dirichlet Laplacian operator $A = (-\Delta)^{\frac{1}{2k}}$, $k \in \mathbb{N}$, so the convergence above requires only that $\phi \in D((-\Delta)^{\frac{1}{2k}}) \subset H^{\frac{1}{k}}(\Omega)$. Only small positive exponents $\beta \in [0, \beta_0)$ are important when proving such convergence. Note the property reported in Proposition 2.2.29 (or in [130, p. 123]) is valid in the just described consideration.

10.3.1 Nonlinear interpolation

We will recall here, for completeness of the presentation, a nonlinear interpolation result for *Lipschitz operators*, as reported in [168, p. 137]. This is an interesting while not very popular result, due to J.-L. Lions.

Proposition 10.3.6. *Let $E_0 \subset E_1$, $F_0 \subset F_1$ be Banach spaces, and A a possibly nonlinear operator from E_1 to F_1 which satisfies*

$$\|A(u)-A(v)\|_{F_1}\leq M_1\|u-v\|_{E_1}\quad \textit{for all } u,v\in E_1,$$
$$A \textit{ maps } E_0 \textit{ into } F_0 \textit{ with } \|A(u)\|_{F_0}\leq M_0\|u\|_{E_0}\quad \textit{for all } u\in E_0, \tag{10.77}$$

then for $0 < \theta < 1$ and $1 \leq p \leq \infty$, A maps $(E_0, E_1)_{\theta,p}$ into $(F_0, F_1)_{\theta,p}$ and

$$\|A(u)\|_{(F_0,F_1)_{\theta,p}}\leq C\|u\|_{(E_0,E_1)_{\theta,p}}\quad \textit{for all } u\in(E_0,E_1)_{\theta,p}. \tag{10.78}$$

Bibliographical notes

Regularity questions for the fractional Hamilton–Jacobi equation were studied recently in [151, 152]. Maximum principle considerations for equations with Lévy operator can be found in the review article [98]. An interesting nonlinear interpolation result reported in Proposition 10.3.6 was originated by J.-L. Lions and F. E. Browder [19, 167] and recalled in the monograph [168].

Very recently, interesting Hölder regularity considerations for solutions of the Hamilton–Jacobi equation were reported in [161].

11 Fractional reaction-diffusion equation

11.1 Introduction

This chapter is devoted to a different question than the previous one.

When the nonlinear term F in (11.1) is "cooperating" with the linear main part operator A (in a sense to be specified), it need not be controlled by that linear operator and we will still obtain a solution. Such situation is observed in Example I below, when the nonlinearity in (11.4) has a form of a monotone operator (compare the "sign condition" (11.5)). Studying such a problem within the perturbation approach of Chapter 3, an additional restriction on the exponent p is needed to set it in a larger phase space like $H^{\alpha}(\mathbb{R}^N)$.

We are returning next to the restriction $b - a < 1$ imposed on nonlinear term F in the abstract Cauchy problem with sectorial positive operator A

$$\begin{aligned} &u_t + Au = F(u), \quad t > 0, \\ &u(0) = u_0, \end{aligned} \tag{11.1}$$

with $F : D(A^b) \to D(A^a)$ being Lipschitz continuous on bounded subsets of $D(A^b)$. In case of the distance of exponents $b - a = 1$, the standard theory of [31, 86] is invalid, so this case is *critical from the point of view of local solvability*. It was observed by J. Arrieta and A. N. Carvalho in [9] that for particular nonlinearities such criticality will be overcome when "moving the problem up" on the fractional power scale $X^{\alpha} = D(A^{\alpha})$, $\alpha \in \mathbb{R}$; more precisely, choosing $X^{1+\epsilon}$, $\epsilon > 0$ as a *phase space* for (11.1), with the corresponding *base space* $X^{\gamma\epsilon}$, where $\rho\epsilon \le \gamma(\epsilon) < 1$, $\rho > 1$. Evidently, for such a choice the distance of exponents $(1+\epsilon) - \gamma\epsilon$ satisfies $\epsilon < (1+\epsilon) - \gamma\epsilon \le 1 + \epsilon(1-\rho) < 1$, which allows the standard approach of [86] to operate for such a pair base/phase spaces. The constructed within such approach in [9] local solutions were called *ϵ-regular solutions*. We will express that approach in Example II.

To be more precise, note that when dealing with the critical distance of exponents (when $b = 1$, $a = 0$ for simplicity) the estimates similar to that in Chapter 3 are leading to *divergent integrals*; in particular estimating the solution of the integral Cauchy formula corresponding to (11.1):

$$u(t) = e^{-At}u_0 + \int_0^t e^{-A(t-s)}F(u(s))ds, \quad \text{for } t \in [0,\tau), \tag{11.2}$$

when $F : X^1 \to X^0$, we obtain

$$\|u(t)\|_{X^1} \le \|e^{-At}u_0\|_{X^1} + M\int_0^t (t-s)^{-1}\|F(u(s))\|_{X^0}ds, \tag{11.3}$$

https://doi.org/10.1515/9783110599831-011

and thus the estimate is useless. The theory of local solvability is not operating in general in that borderline case.

We will study here a counterpart of the semilinear heat equation having a fractional power of minus Laplacian as the main part operator. We will concentrate on the two "difficult cases"; the Cauchy problem in the whole of $\mathbb{R}^N$ with a supercritical nonlinearity, and the Dirichlet problem in bounded domain with critical nonlinearity. In Section 11.2, we discuss local and global solvability of the considered problems. In Section 11.3, we present variants of the Moser–Alikakos-type estimates, both in bounded domain and in $\mathbb{R}^N$. Section 11.4 is devoted to existence of the global attractor in $L^2(\mathbb{R}^N)$ for the case of Cauchy's problem, where we are using the "tail estimate technique" (see [178]). Additional regularity of the attractor is obtained following the result of [163].

Example I. As a first example, consider the *fractional dissipative Cauchy problem*:

$$\begin{cases} u_t + (-\Delta)^\alpha u + f(u) + \lambda u = g(x), & t > 0,\ x \in \mathbb{R}^N, \\ u(0,x) = u_0(x), \quad x \in \mathbb{R}^N, \end{cases} \tag{11.4}$$

with $\lambda > 0$, $\alpha \in (0,1)$ and $N \geq 3$; here, $g \in L^2(\mathbb{R}^N)$ and the nonlinearity $f \in C^1(\mathbb{R})$ satisfies the following conditions:

$$\exists_{0<c_1,c_2,k_2} \exists_{0<k_1<\lambda} \forall_{s\in\mathbb{R}} \quad c_1|s|^{p+1} - k_1|s|^2 \leq f(s)s \leq c_2|s|^{p+1} + k_2|s|^2, \tag{11.5}$$

$$\exists_{l<\lambda} \forall_{s\in\mathbb{R}} \quad f'(s) \geq -l. \tag{11.6}$$

The exponent $p \geq 1$ is arbitrarily large (compare [193] for similar problems). It is easy to see that with the above assumptions the function f will be decomposed as

$$f(s) = f_1(s) + f_2(s),$$

where $f_1(s) = f(s) + (k_1 + \lambda)s$ and $f_2 = -(k_1 + \lambda)s$. Evidently, f_1 defines a monotone operator since:

$$f_1'(s) \geq -l + \lambda + k_1 > 0 \quad \text{for all } s \in \mathbb{R}, \tag{11.7}$$

thanks to the assumption (11.6). Moreover, the function f_2 is linear, hence globally Lipschitz.

Example II. As a second example, we consider the *fractional dissipative Dirichlet problem*:

$$\begin{cases} u_t + (-\Delta)^\alpha u + f(u) = g(x), & t > 0,\ x \in \Omega, \\ u(0,x) = u_0(x), \quad x \in \Omega, \quad u|_{\partial\Omega} = 0 \end{cases} \tag{11.8}$$

with $\alpha \in (0,1)$ and $N \geq 3$; where $g \in H^{-\alpha}(\Omega)$, and $\Omega \subset \mathbb{R}^N$ is a bounded domain with smooth (at least of class C^2) boundary. Moreover, we assume that

$$\liminf_{|s|\to\infty} \frac{f(s)}{s} > -\lambda_1{}^\alpha, \tag{11.9}$$

and

$$\lim_{|s|\to\infty} \frac{|f'(s)|}{|s|^{\frac{2\alpha p}{N-\alpha p}}} = 0, \tag{11.10}$$

where λ_1 is the first eigenvalue of $(-\Delta)$ in Ω with Dirichlet boundary condition (so that λ_1^α is the first eigenvalue of $(-\Delta)^\alpha$).

11.2 Existence and uniqueness of solutions

11.2.1 Cauchy's problem in $\mathbb{R}^N$

We start considering the first example (11.4). We will prove local solvability of that problem following the approach of [25]. The solution will be obtained in the phase space $H^\alpha(\mathbb{R}^N) \cap L^{p+1}(\mathbb{R}^N)$. The "abstract" nonlinear operator A on $H^\alpha(\mathbb{R}^N) \cap L^{p+1}(\mathbb{R}^N)$, $\alpha \in (0,1)$, will correspond to the differential operator $(-\Delta)^\alpha u + f_1(u) + \lambda u$.

An abstract problem with monotone operator

$$\frac{du(t)}{dt} + A(u(t)) + B(u(t)) = 0, \quad t > 0,$$
$$u(0) = u_0 \in H, \tag{11.11}$$

was considered in [25]. We formulate the following.

Definition 11.2.1. A function $u \in C([0,T];H)$ is a strong solution to (11.11) if u is absolutely continuous in any compact subinterval of $(0,T)$, $u(t) \in D(A_H)$ for a. a. $t \in (0,T)$ and

$$\frac{du}{dt}(t) + A(u(t)) + B(u(t)) = 0 \quad \text{for a. a. } t \in (0,T).$$

A function $u \in C([0,T];H)$ is called a weak solution to (11.11) if there is a sequence $\{u_n\}$ of strong solutions convergent to u in $C([0,T];H)$.

It was assumed there that:

H1.

(i) H is a Hilbert space and V is a reflexive Banach space such that

$$V \subset H \subset V^*,$$

with continuous inclusions. Moreover, V is dense in H.

(ii) A is a nonlinear, monotone, coercive, and hemicontinuous operator such that $A : V \to V^*$ (defined on the whole of V).

(iii) The operator $B(u)$ (where $u \in H$) is globally Lipschitz from H into H.

Remark 11.2.2. In [25], to obtain density of the domain of A in H the condition $H2$ was also assumed. For the problems considered here, such a density is evident, so that we recall $H2$ for completeness of the presentation.

H2. There are constants $\omega_1, \omega_2 > 0$, $c_1 \in \mathbb{R}$ and $P \geq 2$ such that for all $v \in V$ the following two conditions hold:

$$\langle Av, v\rangle_{V^*,V} \geq \omega_1 \|v\|_V^P + c_1,$$
$$\|Av\|_{V^*} \leq \omega_2(1 + \|v\|_V^{P-1}). \tag{11.12}$$

Recall, [69], that the operator $A : V \to V^*$ is called:

- *monotone*, if for arbitrary $u, v \in V$

$$\langle Au - Av, u - v\rangle_{V^*,V} \geq 0.$$

- *coercive*, if

$$\lim_{\|v\|_V \to +\infty} \frac{\langle Av, v\rangle_{V^*,V}}{\|v\|_V} = +\infty.$$

- *hemicontinuous*, if for arbitrary fixed $u, v, h \in V$ the real function

$$s \to \langle A(u + sv), h\rangle_{V^*,V} \text{ is continuous on } [0, 1].$$

With the above Assumption $H1$ existence of a solution was shown in [25] (see also [17]). More precisely, we quote the following.

Proposition 11.2.3. *Denote:*

$$D(A_H) := \{v \in V : A(v) \in H\}.$$

Then, under the assumptions $H1$(i), (ii) *and* $H2$ *the set* $D(A_H)$ *is dense in* H. *Moreover, under the sole hypotheses* $H1$, *equation* (11.11) *defines a semigroup of nonlinear operators* $S(t) : \mathrm{cl}_H(D(A_H)) \to \mathrm{cl}_H(D(A_H))$, $t \geq 0$, *where for each* $u_0 \in \mathrm{cl}_H(D(A_H))$

$$t \to S(t)u_0$$

is the global weak solution of (11.11) *starting at* u_0. *This semigroup is such that*

$$\mathbb{R}^+ \times \mathrm{cl}_H(D(A_H)) \ni (t, u_0) \to S(t)u_0 \in \mathrm{cl}_H(D(A_H))$$

is a continuous map. Moreover, for $u_0 \in D(A_H)$, $u(\cdot) = S(\cdot)u_0$ *is a Lipschitz continuous strong solution of* (11.11).

We are now ready to apply the quoted above result to the problem (11.4). Setting:

$$H = L^2(\mathbb{R}^N), \quad V = H^\alpha(\mathbb{R}^N) \cap L^{p+1}(\mathbb{R}^N),$$
$$A(u) = (-\Delta)^\alpha u + f_1(u) + \lambda u,$$
$$B(u) = f_2(u) - g(\cdot), \tag{11.13}$$

validity of the condition $H1$ in that case will be checked next.

First, note that the space V is reflexive thanks to the equality ([69, Theorem 5.13])

$$(H^{\alpha}(\mathbb{R}^N) \cap L^{p+1}(\mathbb{R}^N))^{**} = H^{\alpha}(\mathbb{R}^N) \cap L^{p+1}(\mathbb{R}^N).$$

The operator A is connected with the duality form

$$\langle A(u), v\rangle_{V^*,V} = \int_{\mathbb{R}^N} (-\Delta)^{\frac{\alpha}{2}} u(-\Delta)^{\frac{\alpha}{2}} v\,dx + \int_{\mathbb{R}^N} (f_1(u) + \lambda u)v\,dx, \quad u, v \in V. \tag{11.14}$$

Since f_1 defines a monotone operator, it is evident that the operator A is monotone itself. Moreover, the operator A is coercive and hemicontinuous. Indeed, thanks to (11.5) and the Young inequality, we have

$$\begin{aligned}\langle A(v), v\rangle_{V^*,V} &\geq c_{\lambda}\|v\|^2_{H^{\alpha}(\mathbb{R}^N)} + \int_{\mathbb{R}^N} f_1(v)v\,dx\\ &\geq c_{\lambda}\|v\|^2_{H^{\alpha}(\mathbb{R}^N)} + c_1\|v\|^{p+1}_{L^{p+1}(\mathbb{R}^N)}\\ &\geq c(\|v\|^2_{H^{\alpha}(\mathbb{R}^N)} + \|v\|^2_{L^{p+1}(\mathbb{R}^N)} - 1)\\ &\geq c\|v\|^2_V - c = \frac{c\|v\|^2_V - c}{\|v\|_V}\|v\|_V,\end{aligned} \tag{11.15}$$

where c_{λ} is a constant appearing in the equivalent norm:

$$c_{\lambda}\|u\|^2_{H^{\alpha}(\mathbb{R}^N)} \leq \|(-\Delta)^{\frac{\alpha}{2}} u\|^2 + \lambda\|u\|^2 \quad \text{for all } u \in H^{\alpha}(\mathbb{R}^N).$$

Recall next that the norm in the dual space $V^* = H^{-\alpha}(\mathbb{R}^N) + L^{\frac{p+1}{p}}(\mathbb{R}^N)$ is given by the formula:

$$\|z\|_{V^*} = \inf_{x \in H^{-\alpha}(\mathbb{R}^N), y \in L^{\frac{p+1}{p}}(\mathbb{R}^N), z = x+y} \max(\|x\|_{H^{-\alpha}(\mathbb{R}^N)}, \|y\|_{L^{\frac{p+1}{p}}(\mathbb{R}^N)}).$$

Note that we have the characterization; $D((-\Delta)^{\frac{\alpha}{2}}) = H^{\alpha}(\mathbb{R}^N)$, and under the assumption (11.5) the domain of the Nemytskiĭ operator connected to f_1 contains $L^{p+1}(\mathbb{R}^N)$. Consequently, the domain of the whole operator A contains $H^{\alpha}(\mathbb{R}^N) \cap L^{p+1}(\mathbb{R}^N)$ and is dense in $H = L^2(\mathbb{R}^N)$.

As a consequence of Proposition 11.2.3, we conclude existence of a solution to (11.4). More precisely, we have the following.

Corollary 11.2.4. *For $u_0 \in D(A_H) = \{v \in H^{\alpha}(\mathbb{R}^N) \cap L^{p+1}(\mathbb{R}^N); A(v) \in L^2(\mathbb{R}^N)\}$, the weak solution $u(t)$ corresponding to u_0 satisfies:*

$$u(t) \in C((0,T); L^2(\mathbb{R}^N)) \quad \text{for all } T > 0. \tag{11.16}$$

Moreover, since $u_0 \in D(A_H)$ then $u(t)$ is in fact a strong solution to (11.4); it is absolutely continuous in any compact subinterval of $(0,T)$, $u(t) \in D(A_H)$ and (11.4) is fulfilled in $L^2(\mathbb{R}^N)$ for a. a. $t \in (0,T)$.

Remark 11.2.5. The assumption of the global Lipschitz continuity of the function f_2 will be weakened if one uses the results of [27] instead of [25].

11.2.2 More regular solutions in $\mathbb{R}^N$

Analogously, as in the theory of second-order parabolic equations, if we consider the problem (11.4) on a smaller phase space, then it will be treated within the classical approach of D. Henry. Consider thus the Cauchy problem (11.4) with nonlinearity $f : \mathbb{R} \to \mathbb{R}$ locally Lipschitz continuous, $g \in L^\infty(\mathbb{R}^N) \cap L^2(\mathbb{R}^N)$ in the phase space $H^\beta_{p+1}(\mathbb{R}^N)$, with $2\alpha > \beta > \frac{N}{p+1}$. Because of the last restriction on β and p, it is seen that $H^\beta_{p+1}(\mathbb{R}^N) \subset L^\infty(\mathbb{R}^N)$.

In the base space $X = L^{p+1}(\mathbb{R}^N)$, consider the $X^{\frac{\beta}{2\alpha}}$-solution to (11.4) in a sense of Definition 3.1.1. Sectoriality of the realization of $(-\Delta)^\alpha$ in that case follows from the known sectoriality of $(-\Delta)$ alone and Proposition 1.0.2. To get the $X^{\frac{\beta}{2\alpha}}$-solution, it is thus enough to check that the "nonlinear term"

$$F(v) = -f(v) - \lambda v + g(\cdot), \tag{11.17}$$

is Lipschitz continuous on bounded sets as a map from $X^{\frac{\beta}{2\alpha}}$ to X. But the last property is evident thanks to the embedding $X^{\frac{\beta}{2\alpha}} \subset L^\infty(\mathbb{R}^N)$ and the local Lipschitz continuity of the real function f. The solution obtained varies continuously in $X^{\frac{\beta}{2\alpha}}$, so also in $L^\infty(\mathbb{R}^N)$. As usual, to show global in time extendibility of the $X^{\frac{\beta}{2\alpha}}$-local solution, we need first to get a priori estimate of it in an auxiliary Banach space. In our case, we choose $Y = L^\infty(\mathbb{R}^N) \cap L^{p+1}(\mathbb{R}^N)$. Note that the a priori estimate in that space is a direct consequence of the estimate (11.51) and the Moser–Alikakos estimates (see [3], [29]) presented in Subsection 11.3.2. Next, we show that the *subordination condition* for the nonlinear term holds. Note that as a direct consequence of the assumption (11.5)

$$\exists_{c_3>0} \forall_{s\in\mathbb{R}} \; |f(s)| \leqslant c_3(|s|^p + |s|). \tag{11.18}$$

Then, thanks to (11.18) and the Young inequality, we have

$$\begin{aligned} \|F(u)\|_{L^{p+1}(\mathbb{R}^N)} &\leqslant C_1\big(\|u\|^{\frac{p}{p+1}}_{L^\infty(\mathbb{R}^N)}\|u\|_{L^{p+1}(\mathbb{R}^N)} + (C_2+\lambda)\|u\|_{L^{p+1}(\mathbb{R}^N)} + \|g\|_{L^{p+1}(\mathbb{R}^N)}\big) \\ &\leqslant C\big(\|u\|^{p+1}_{L^\infty(\mathbb{R}^N)\cap L^{p+1}(\mathbb{R}^N)} + \|u\|_{L^\infty(\mathbb{R}^N)\cap L^{p+1}(\mathbb{R}^N)} + 1\big) \\ &= c(\|u\|_Y), \end{aligned} \tag{11.19}$$

which is a simple form of a *subordination condition*. We have thus proved existence of a smooth solution to (11.4).

Proposition 11.2.6. *Let the function $f : \mathbb{R} \to \mathbb{R}$ be locally Lipschitz continuous, $g \in L^\infty(\mathbb{R}^N) \cap L^2(\mathbb{R}^N)$, then there exists a unique solution $u \in H^\beta_{p+1}(\mathbb{R}^N)$, $2\alpha > \beta > \frac{N}{p+1}$, of (11.4), such that:*

- *$u \in C([0,T]; H^\beta_{p+1}(\mathbb{R}^N))$,*
- *$u \in C^1((0,T); H^{2\alpha}_{p+1}(\mathbb{R}^N))$, $u_t \in H^{2\gamma}_{p+1}(\mathbb{R}^N)$ with $\gamma < \alpha$, for all $t \in (0,T)$,*
- *the equation is satisfied in $L^{p+1}(\mathbb{R}^N)$ for all $t \in (0,T)$,*
- *the solution varies continuously in $L^\infty(\mathbb{R}^N)$.*

11.2.3 The problem in bounded domain

In this section, we consider the problem (11.8) in a bounded domain $\Omega \subset \mathbb{R}^N$ in case when the nonlinearity has critical growth; that means here with nonlinearity growing with "critical" exponent $\frac{N+\alpha p}{N-\alpha p}$. To handle such a problem in $H_p^{-\alpha}(\Omega)$, we need to use the notion of an *ϵ-regular solution* as introduced in [9]. This allows us to obtain a local in time solution to (11.8); note that in Definition 11.2.8 the distance between exponents $1+\epsilon$ and $q\epsilon$ fulfills

$$(1+\epsilon) - q\epsilon = 1 + \epsilon(1-q) \in (0,1),$$

since $q > 1$ and $\epsilon \in (0, \frac{1}{q})$.

We will recall next the basic definitions and results concerning ϵ-regular solutions (see [9] for more details). We start with some terminology. Let X be a Banach space and $A : D(A) \subset X \to X$ be a sectorial operator with $\operatorname{Re} \sigma(A) > 0$. Then $-A$ generates an analytic semigroup denoted by $\{e^{-At} : t \geq 0\}$. Let $X^\beta := D(A^\beta)$, endowed with the graph norm $\beta \geq 0$, be the fractional power spaces associated with A. Consider the semilinear differential equation

$$\begin{aligned} u_t + Au &= F(u), \quad t > 0, \\ u(0) &= u_0 \in X^1, \end{aligned} \tag{11.20}$$

where $F : D(F) \subset X^1 \to X^\beta$ for some $\beta > 0$.

Definition 11.2.7 ([9]). For $\epsilon > 0$, a function $u : [0,\tau) \to X^1$ is called an ϵ-regular solution for (11.20) if $u \in C([0,\tau); X^1) \cap C((0,\tau); X^{1+\epsilon})$ and

$$u(t) = e^{-At}u_0 + \int_0^t e^{-A(t-s)}F(u(s))ds, \quad t \in [0,\tau). \tag{11.21}$$

Definition 11.2.8 ([9, 26]). The map F is called a *critical* ε-regular map relative to the pair (X^1, X), if there are positive constants c, η, C_η, $q > 1$ and $\epsilon \in (0, \frac{1}{q})$ such that for each $v, w \in X^{1+\epsilon}$,

$$\begin{aligned} \|F(v) - F(w)\|_{X^{q\epsilon}} &\leq c\|v - w\|_{X^{1+\epsilon}}(C_\eta + \eta\|v\|^{q-1}_{X^{1+\epsilon}} + \eta\|w\|^{q-1}_{X^{1+\epsilon}}), \\ \|F(v)\|_{X^{q\epsilon}} &\leq c(C_\eta + \eta\|v\|^q_{X^{1+\epsilon}}) \quad \text{for } v \in X^{1+\epsilon}. \end{aligned} \tag{11.22}$$

In addition, if for each $\eta > 0$ there is $C_\eta > 0$ such that (11.22) holds with c, q, and ϵ independent of η, then F is called an *almost critical* ε-regular map relative to the pair (X^1, X).

After this introduction, we can state a variant of Theorem 2.1 in [26].

Proposition 11.2.9. *Let F be a critical ϵ-regular map. Fixing $v_0 \in X^1$, there are $r > 0$ and $\tau_0 > 0$ such that for each $u_0 \in B_{X^1}(v_0, r)$ there exists a unique ϵ-regular solution u of (11.20) defined in $[0, \tau_0]$. In addition,*

(i) $t^{\xi}\|u(t,u_0)\|_{X^{1+\xi}} \to 0$ *as* $t \to 0^+$, $0 < \xi < q\epsilon$,
(ii) $t^{\xi}\|u(t,u_1) - u(t,u_2)\|_{X^{1+\xi}} \leqslant C'\|u_1 - u_2\|_{X^1}$ *for* $t \in [0,\tau_0]$, $0 \leq \xi \leq \xi_0 < q\epsilon$, $u_1, u_2 \in B_{X^1}(v_0, r)$,
(iii) $u(t,u_0) \in C((0,\tau_0];X^{1+q\epsilon}) \cap C^1((0,\tau_0];X^{1+\xi})$ *for* $0 \leq \xi < q\epsilon$; *in particular the solution* $u(t,u_0)$ *satisfies* (11.20) *for each* $t \in (0,\tau_0]$.

If F is an almost critical ϵ*-regular map, then all the above holds for arbitrarily large* $r > 0$. *In this case, if the solution* $u(t,u_0)$ *is bounded in* X^1 *in its maximal interval of existence, it must exist for all* $t \geq 0$.

We will use here the above mentioned result in the case when A will be a proper α fractional power, $\alpha \in (0,1)$, of an elliptic operator (so, it is a *pseudodifferential operator*). It is well known that the proper fractional powers of *sectorial positive operators* are, for $\alpha \in (0,1)$, sectorial and positive (see Subsection 2.2.3).

Once we take care for checking that the main operator $(-\Delta)^{\alpha}$ (Dirichlet boundary condition) in (11.8) is sectorial, we need to verify that the nonlinear term f generates a *critical* ϵ*-regular map*. We set $X = H_p^{-\alpha}(\Omega)$, $A = (-\Delta)^{\alpha}$, where $\alpha \in (0,1)$ is fixed, so that $X^1 = H^{\alpha}_{p,\{D\}}(\Omega)$ is the associated space of *Bessel potentials* (see [173, Section 2.3.1] for more information concerning that spaces); the lower index D corresponds to the homogeneous Dirichlet condition.

In what follows, we apply these abstract results to obtain local well posedness of (11.8) in $H^{\alpha}_{p,\{D\}}(\Omega)$. Since Ω is smooth we have that the domains of fractional power scale associated to A coincide with the *complex interpolation scale* (see [6], [173, Section 4.9.2]). In particular ([173, Section 4.3.3]), for $0 < \theta < 1$,

$$[L^p(\Omega), H^m_{p,\{D\}}(\Omega)]_{\theta} = H^{\theta m}_{p,\{D\}}(\Omega)$$

whenever $\theta m \neq \frac{1}{p}$ (here $[\cdot,\cdot]_{\theta}$ denotes the *complex interpolation space* of order θ).

Now, we check that the condition (11.22) of Definition 11.2.8 is satisfied. To formulate the result, we need to use the assumption (11.10):

$$\lim_{|s|\to\infty} \frac{|f'(s)|}{|s|^{\frac{2\alpha p}{N-\alpha p}}} = 0.$$

A nonlinearity satisfying such condition is called *almost critical*. The value $q = \frac{N+\alpha p}{N-\alpha p}$ is called a *critical exponent* here.

Note that if $f \in C^1(\mathbb{R},\mathbb{R})$ satisfies (11.10), then for each $\eta > 0$ there exists $C_{\eta} > 0$ such that

$$|f(s_1) - f(s_2)| \leq |s_1 - s_2|(C_{\eta} + \eta|s_1|^{q-1} + \eta|s_2|^{q-1}), \quad s_1, s_2 \in \mathbb{R}, \tag{11.23}$$

and also, for each $\eta > 0$ there exists $\tilde{C}_{\eta} > 0$ such that

$$|f(s)| \leq \tilde{C}_{\eta} + \eta|s|^q. \tag{11.24}$$

Let F be the Nemytskiĭ operator corresponding to $-f(\cdot) + g$, then the following estimates are satisfied.

Lemma 11.2.10. *Let $N \geq 3$, $\alpha p < N$, $q := \frac{N+\alpha p}{N-\alpha p}$ and $f : \mathbb{R} \to \mathbb{R}$ be a continuously differentiable function which satisfies (11.10) and $g \in L^{\frac{r}{q}}(\Omega)$. For $\nu > 0$ and $\epsilon \in [0, \frac{\alpha}{2q})$, there is a constant $c > 0$ and, for each $\eta > 0$, a constant $\tilde{C}_\eta > 0$ such that*

$$\|F(w_1) - F(w_2)\|_{X^{q\epsilon}} \leq c\|w_1 - w_2\|_{X^{1+\epsilon}}(\tilde{C}_\eta + \eta\|w_1\|^{q-1}_{X^{1+\epsilon}} + \eta\|w_2\|^{q-1}_{X^{1+\epsilon}}),$$

for $w_1, w_2 \in X^{1+\epsilon}$. Also, the second estimate in (11.22) is satisfied by F.

Proof. It follows from Sobolev embedding theorem that for $r := \frac{Np}{N-p(2\epsilon+\alpha)}$,

$$H^{\alpha+2\epsilon}_{p,\{D\}}(\Omega) = X^{1+\epsilon} \subset L^r(\Omega), \quad L^{\frac{r}{q}}(\Omega) \subset X^{q\epsilon}(\Omega) = H^{-\alpha+2q\epsilon}_{p,\{D\}}(\Omega). \tag{11.25}$$

Next, from (11.23) and the Hölder inequality,

$$\begin{aligned}\|F(w_1) - F(w_2)\|_{X^{q\epsilon}} &\leq c\|f(w_1) - f(w_2)\|_{L^{\frac{r}{q}}(\Omega)} \\ &\leq c\|w_1 - w_2\|_{L^r(\Omega)}(\tilde{C}_\eta + \eta\|w_1\|^{q-1}_{L^r(\Omega)} + \eta\|w_2\|^{q-1}_{L^r(\Omega)}) \\ &\leq c\|w_1 - w_2\|_{X^{1+\epsilon}}(\tilde{C}_\eta + \eta\|w_1\|^{q-1}_{X^{1+\epsilon}} + \eta\|w_2\|^{q-1}_{X^{1+\epsilon}}).\end{aligned} \tag{11.26}$$

The proof of the second estimate in (11.22) is similar; thanks to (11.24) we have

$$\begin{aligned}\|F(w)\|_{X^{q\epsilon}} &\leq c\|f(w)\|_{L^{\frac{r}{q}}(\Omega)} + \|g\|_{L^{\frac{r}{q}}(\Omega)} \leq c\left[\int_\Omega [\tilde{C}_\eta + \eta|w|^q]^{\frac{r}{q}}dx\right]^{\frac{q}{r}} + \|g\|_{L^{\frac{r}{q}}(\Omega)} \\ &\leq c(\tilde{C}_\eta + \eta\|w\|^q_{L^r(\Omega)}) + \|g\|_{L^{\frac{r}{q}}(\Omega)}.\end{aligned} \tag{11.27}$$

The proof of the lemma is now complete. □

We have thus verified all the conditions necessary to apply Proposition 11.2.9 to our present problem (11.8). Consequently, there is a local in time ϵ-regular solution u to (11.8) having all the properties as specified in Definition 11.2.7 and Proposition 11.2.9. In particular, $u \in C([0,\tau); H^\alpha_{p,\{D\}}(\Omega)) \cap C((0,\tau); H^{\alpha+2\epsilon}_{p,\{D\}}(\Omega))$ and the equation is fulfilled in $H^{-\alpha}_{p,\{D\}}(\Omega)$.

Let us restrict, for simplicity, further studies of the problem (11.8) to the Hilbert case $p = 2$. As a consequence of the condition (11.10) the local ϵ-regular solution will be extended globally in time. The last property follows from existence of the Lyapunov function obtained through multiplication of (11.8) by u_t. As a result of that multiplication, we obtain

$$\frac{d}{dt}\left(\frac{1}{2}\int_\Omega [(-\Delta)^{\frac{\alpha}{2}}u]^2 dx + \int_\Omega F(u)dx - \int_\Omega g(x)u dx\right) + \int_\Omega u_t^2 dx = 0, \tag{11.28}$$

where $F(s) = \int_0^s f(z)dz$ is the primitive of f. Consequently, the expression

$$\mathcal{L}(u(t)) = \frac{1}{2}\int_\Omega [(-\Delta)^{\frac{\alpha}{2}}u(t)]^2 dx + \int_\Omega F(u(t))dx - \int_\Omega g(x)u(t)dx$$

is nonincreasing in time. Note also that, thanks to the assumption (11.9),

$$\forall_{\lambda_1^\alpha>\delta>0}\exists_{C_\delta>0}\forall_{s\in\mathbb{R}} \; -F(s) \le \frac{1}{2}(\lambda_1^\alpha - \delta)s^2 + C_\delta. \tag{11.29}$$

This last condition follows from the calculation (compare [83], p. 76)

$$-F(s) - \frac{1}{2}(\lambda_1^\alpha - \delta)s^2 = \int_0^s \left[\frac{-f(\tau)}{\tau} - (\lambda_1^\alpha - \delta)\right]\tau d\tau \le C_\delta,$$

valid for positive s, and a symmetric estimate for negative s. It follows from the property $\mathcal{L}(u(t)) \le \mathcal{L}(u_0)$, $t \ge 0$, (11.29) and the generalized Poincaré inequality

$$\int_\Omega [(-\Delta)^{\frac{\alpha}{2}}\phi]^2 dx \ge \lambda_1^\alpha \int_\Omega \phi^2 dx,$$

that

$$\int_\Omega [(-\Delta)^{\frac{\alpha}{2}}u(t)]^2 dx \le \mathrm{const}(\mathcal{L}(u_0)), \quad t \ge 0.$$

Thus, the $H^\alpha_{2,\{D\}}(\Omega)$ norm of the solution is bounded for $t \ge 0$, consequently the solution is global in time. Moreover, the Lyapunov function $\mathcal{L}$ is bounded from below and satisfies the following condition: $\mathcal{L}(u) \to \infty$ as $\|u\|^2_{L^2}(\Omega) \to \infty$. Indeed

$$\begin{aligned}\mathcal{L}(u(t)) &\ge \frac{1}{2}\int_\Omega [(-\Delta)^{\frac{\alpha}{2}}u(t)]^2 dx - \frac{1}{2}(\lambda_1^\alpha - \delta)\int_\Omega u^2(t)dx - C_\delta|\Omega| - \int_\Omega gu(t)dx \\ &\ge \frac{\delta}{8}\int_\Omega u^2(t)dx + C(\|g\|^2_{L^2(\Omega)}).\end{aligned}$$

11.2.4 Stationary solutions of (11.8)

Assumption (11.9) also gives us a simple estimate in $H^\alpha_{2,\{D\}}(\Omega)$, of the stationary solutions of (11.8). Indeed, multiplying the stationary equation corresponding to (11.8),

$$\begin{cases}(-\Delta)^\alpha v + f(v) = g(x), & x \in \Omega,\\ v|_{\partial\Omega} = 0,\end{cases} \tag{11.30}$$

by v we get

$$\int_\Omega [(-\Delta)^{\frac{\alpha}{2}} v]^2 dx + (-\lambda_1^\alpha + \delta) \int_\Omega v^2 dx \leqslant C_\delta \int_\Omega g^2(x) dx + C|\Omega| \tag{11.31}$$

for $\delta < \lambda_1^\alpha$ and some constants C_δ, C then, using the Poincaré inequality, we finally have

$$\frac{\delta}{\lambda_1^\alpha} \int_\Omega [(-\Delta)^{\frac{\alpha}{2}} v]^2 dx \le C_\delta \int_\Omega g^2(x) dx + C|\Omega|. \tag{11.32}$$

11.3 Useful facts and inequalities

11.3.1 The Moser–Alikakos technique in bounded domain

The lemma below will be used in the estimate of the $L^\infty(\Omega)$ norm of the solution in the second example. Recall that $X = H^{-\alpha}_{p,\{D\}}(\Omega)$, $X^{1+\epsilon} = H^{\alpha+2\epsilon}_{p,\{D\}}(\Omega)$ in this example. We need to have the embedding $X^{1+\epsilon} \subset L^\infty(\Omega)$, which holds provided that $\alpha + 2\epsilon > \frac{N}{p}$. As known from the theory of the second-order parabolic equations, if we want the solution to vary in $L^\infty(\Omega)$ (e. g., to have the maximum principle), we eventually need to take large value of $p \ge 2$ (especially when the space dimension N is large) to fulfill the condition $\alpha + 2\epsilon > \frac{N}{p}$.

Lemma 11.3.1. *For $X^{1+\epsilon} = H^{\alpha+2\epsilon}_{p,\{D\}}(\Omega)$ solutions to (11.8), $\alpha + 2\epsilon > \frac{N}{p}$ and $g \in L^\infty(\Omega)$, the following implication holds:*

$$(\|u(t,u_0)\|_{L^1(\Omega)} \le \text{const.},\ t \ge 0) \Longrightarrow (\|u(t,u_0)\|_{L^\infty(\Omega)} \le \text{const.}',\ t \ge 0). \tag{11.33}$$

Observe that the dissipative condition (11.9) implies

$$\exists_{C<\lambda_1} \exists_{D>0} \forall_{s\in R}\ -sf(s) \le Cs^2 + D, \tag{11.34}$$

which in order corresponds to [32, (9.3.5)]. Moreover, by the estimate (5.112) with $q = 2^k$, we have

$$\frac{(2^k-1)}{2^{2k-2}} \int_\Omega [(-\Delta_D)^{\frac{\alpha}{2}} (|\phi|^{2^{k-1}})]^2 dx \le \int_\Omega [(-\Delta_D)^\alpha \phi] |\phi|^{2^k-1} \operatorname{sgn} \phi\, dx. \tag{11.35}$$

With the above conditions (11.34), (11.35), we are able to repeat the calculations of [32, Lemma 9.3.1] and get (11.33). The calculation goes as follows.

Multiplying (11.8) by u^{2^k-1}, $k = 1, 2, \ldots$, we get

$$\begin{aligned}\frac{1}{2^k}\frac{d}{dt} \int_\Omega u^{2^k} dx = &- \int_\Omega (-\Delta)^\alpha u |u|^{2^k-1} \operatorname{sgn}(u) dx - \int_\Omega f(u) u u^{2^k-2} dx \\ &+ \int_\Omega g(x) u^{2^k-1} dx.\end{aligned}$$

Then, thanks to (11.34) and (11.35) and an elementary estimate $s^{2^k-2} \leqslant s^{2^k}+1$, we obtain (with an equivalent norm of $H^\alpha_{2,\{D\}}(\Omega)$, in case of bounded Ω and Dirichlet boundary condition)

$$\begin{aligned}\frac{d}{dt}\int_\Omega u^{2^k}dx &\le -\frac{2^k-1}{2^{k-2}}\int_\Omega [(-\Delta)^{\frac{\alpha}{2}}(|u|^{2^{k-1}})]^2 dx + 2^k\int_\Omega (Cu^2+D)u^{2^k-2}dx \\ &\quad + 2^k\int_\Omega g(x)u^{2^k-1}dx \\ &\le -\frac{2^k-1}{2^{k-2}}\||u|^{2^{k-1}}\|^2_{H^\alpha_{2,\{D\}}(\Omega)} + 2^k(C+D)\int_\Omega u^{2^k}dx + 2^kD|\Omega| \\ &\quad + 2^k\int_\Omega g(x)u^{2^k-1}dx. \end{aligned} \tag{11.36}$$

Moreover, since

$$\|\phi\|_{L^2(\Omega)} \leqslant \|\phi\|^{\frac{2+\mu}{2+2\mu}}_{L^{2+\mu}(\Omega)}\|\phi\|^{\frac{\mu}{2+2\mu}}_{L^1(\Omega)}, \quad \phi\in L^1(\Omega)\cap L^{2+\mu}(\Omega),$$

and for $0<\mu\le\frac{4\alpha}{N-2\alpha}$

$$\|\phi\|_{L^{2+\mu}(\Omega)} \leqslant c_\mu\|\phi\|_{H^\alpha_2(\Omega)}, \quad \phi\in H^\alpha_2(\Omega),$$

the Young inequality leads to the estimate

$$\forall_{\delta>0}\forall_{0<\mu\le\frac{4\alpha}{N-2\alpha}}\exists_{C_{\delta,\mu}>0}\quad \|\phi\|^2_{L^2(\Omega)} \leqslant \delta\|\phi\|^2_{H^\alpha_2(\Omega)} + C_{\delta,\mu}\|\phi\|^2_{L^1(\Omega)}. \tag{11.37}$$

From (11.36) and (11.37) and the Hölder inequality ($p=\frac{2^k}{2^k-1}$, $q=2^k$), we obtain

$$\begin{aligned}\frac{d}{dt}\int_\Omega u^{2^k}dx &\le -\frac{2^k-1}{2^{k-2}\delta}\int_\Omega u^{2^k}dx + \frac{(2^k-1)C_{\delta,\mu}}{2^{k-2}\delta}\Big(\int_\Omega |u|^{2^{k-1}}dx\Big)^2 \\ &\quad + 2^k(C+D)\int_\Omega u^{2^k}dx + 2^kD|\Omega| + 2^k\Big(\int_\Omega |g(x)|^{2^k}dx\Big)^{\frac{1}{2^k}}\Big(\int_\Omega u^{2^k}dx\Big)^{\frac{2^k-1}{2^k}} \\ &\le \Big(\frac{-2}{\delta}+2^k(C+D)\Big)\int_\Omega u^{2^k}dx + \frac{4C_{\delta,\mu}}{\delta}\Big(\int_\Omega |u|^{2^{k-1}}dx\Big)^2 \\ &\quad + 2^kD|\Omega| + 2^k\Big(C_\epsilon\int_\Omega |g(x)|^{2^k}dx + \epsilon\int_\Omega u^{2^k}dx\Big), \end{aligned} \tag{11.38}$$

the last by the Young inequality. Setting $\delta=\frac{2^{1-k}}{C+D+2}$, $\epsilon=1$, we obtain

$$\frac{d}{dt}\int_\Omega u^{2^k}\,dx \le -2^k \int_\Omega u^{2^k}\,dx + 2^{k+1}\bar{C}_{\delta,\mu}\Big(\int_\Omega |u|^{2^{k-1}}\,dx\Big)^2 + 2^k D|\Omega|$$
$$+ 2^k C_\epsilon \int_\Omega |g(x)|^{2^k}\,dx, \tag{11.39}$$

which leads, just as in [32, (9.3.13)], to the final estimate

$$\sup_{t\ge 0}\|u(t)\|_{L^\infty(\Omega)} \le \text{const}\max\Big\{\sup_{t\ge 0}\|u(t)\|_{L^1(\Omega)}, 1\Big\}. \tag{11.40}$$

Remark 11.3.2. A version of Lemma 11.3.1 stays valid if we know a $L^{2^r}(\Omega)$ (with fixed $r \in \mathbb{N}$) estimate of $u(t, u_0)$, instead of the $L^1(\Omega)$ estimate. The induction argument will start from the $L^{2^r}(\Omega)$ norm of the solution in that case.

11.3.2 Moser–Alikakos technique in $\mathbb{R}^N$

If instead of a bounded domain Ω and homogeneous Dirichlet boundary condition, we consider the Cauchy problem (11.4) in the whole of $\mathbb{R}^N$, condition (11.34) needs to be replaced (from (11.12)) with

$$\exists_{C<\lambda}\forall_{D>0}\forall_{s\in\mathbb{R}} \; -sf(s) \le Cs^2 + D|s|, \tag{11.41}$$

and we need to have $g \in L^1(\mathbb{R}^N) \cap L^\infty(\mathbb{R}^N)$. We are using the (equivalent) singular integral definition of the fractional powers of $(-\Delta)$ in that case (compare Chapter 5).

Definition 11.3.3. For every $\beta \in (0, 1)$, we set

$$(-\Delta)^\beta v(x) = -C_N(\beta)\lim_{\epsilon\to 0}\int_{|z|\ge\epsilon} \frac{v(x-z) - v(x)}{|z|^{N+2\beta}}\,dz,$$

for $v \in \mathcal{S}$, the Schwartz class.

We thus consider the problem (11.4)

$$\begin{cases} u_t + (-\Delta)^\alpha u + f(u) + \lambda u = g(x), & t > 0,\ x \in \mathbb{R}^N, \\ u(0,x) = u_0(x), & x \in \mathbb{R}^N, \end{cases} \tag{11.42}$$

under the assumption (11.41), and with $g \in L^1(\mathbb{R}^N) \cap L^\infty(\mathbb{R}^N)$. Unfortunately, the $H^\alpha(\mathbb{R}^N) \cap L^{p+1}(\mathbb{R}^N)$ solution to (11.4) constructed above is too weak to vary in $L^\infty(\mathbb{R}^N)$, and to proceed with the estimates below we need to assure that the solution varies in $L^{p+1}(\mathbb{R}^N) \cap L^\infty(\mathbb{R}^N)$. But one can work with the $H^\beta_{p+1}(\mathbb{R}^N)$ solution of (11.4) with $\beta > \frac{N}{p+1}$ to have enough smoothness for such calculations. Note also, that the assumption (11.41) is weaker than the left-hand side of the condition (11.5), so that assuming (11.5) we can use a weaker condition (11.41) instead.

Multiplying (11.4) by u^{2l-1}; $2l \geq p+2$, $l \in \mathbb{N}$, we obtain

$$\int_{\mathbb{R}^N} u_t u^{2l-1} dx + \int_{\mathbb{R}^N} (-\Delta)^\alpha u u^{2l-1} dx + \int_{\mathbb{R}^N} f(u) u u^{2l-2} dx + \lambda \int_{\mathbb{R}^N} u^{2l} dx = \int_{\mathbb{R}^N} g(x) u^{2l-1} dx. \quad (11.43)$$

The second term is nonnegative thanks to Corollary 5.4.23. Recall that we are working with the solutions of (11.4) varying in $L^{p+1}(\mathbb{R}^N) \cap L^\infty(\mathbb{R}^N)$. Also, with the use of an elementary inequality

$$\forall_{s\in\mathbb{R}} \forall_{\delta>0} \ |s|^{2l-1} \leq \delta^{-1} s^{2l} + \delta^{2l-2-p} |s|^{p+1}$$

and (11.41), the nonlinear term is estimated as follows:

$$-\int_{\mathbb{R}^N} f(u) u u^{2l-2} dx \leq \int_{\mathbb{R}^N} (C + D\delta^{-1}) u^{2l} dx + D\delta^{2l-2-p} \int_{\mathbb{R}^N} |u|^{p+1} dx. \quad (11.44)$$

By the Young inequality, we thus obtain

$$\begin{aligned} \frac{1}{2l}\frac{d}{dt} \int_{\mathbb{R}^N} u^{2l} dx \leq (C + D\delta^{-1} + \epsilon - \lambda) \int_{\mathbb{R}^N} u^{2l} dx + D\delta^{2l-2-p} \int_{\mathbb{R}^N} |u|^{p+1} dx \\ + C_\epsilon \int_{\mathbb{R}^N} |g(x)|^{2l} dx. \end{aligned} \quad (11.45)$$

Since $C < \lambda$, we can choose $\delta = \delta_0 = \frac{4D}{\lambda - C}$ and $\epsilon = \epsilon_0 = \frac{\lambda - C}{4}$ (here $C_\epsilon = \text{const}\, \epsilon^{-2l}$ comes from the Young inequality) to have $C + D\delta^{-1} + \epsilon - \lambda = \frac{C-\lambda}{2} < 0$. Solving the differential inequality, we obtain

$$\begin{aligned} \int_{\mathbb{R}^N} u^{2l} dx \leq \Bigg[\int_{\mathbb{R}^N} u_0^{2l} dx + \int_0^t e^{-l(C-\lambda)s} \Big(2lD\delta_0^{2l-2-p} \int_{\mathbb{R}^N} |u(s)|^{p+1} dx \\ + 2lC_{\epsilon_0} \|g\|^{2l}_{L^{2l}(\mathbb{R}^N)} \Big) ds \Bigg] e^{l(C-\lambda)t}. \end{aligned}$$

If $\sup_{s\in[0,T]} \|u(s)\|_{L^{p+1}(\mathbb{R}^N)} \leq M$, then taking the $2l$ roots we find that

$$\begin{aligned} \|u(t)\|_{L^{2l}(\mathbb{R}^N)} &\leq \|u_0\|_{L^{2l}(\mathbb{R}^N)} e^{\frac{C-\lambda}{2}t} + ((2lD)^{-2l} \delta_0^{1-\frac{2+p}{2l}} M^{\frac{p+1}{2l}} + (2lC_{\epsilon_0})^{-2l} \|g\|_{L^{2l}(\mathbb{R}^N)}) \\ &\quad \times \left(\int_0^t e^{l(C-\lambda)(t-s)} ds \right)^{\frac{1}{2l}} \\ &\leq \|u_0\|_{L^{2l}(\mathbb{R}^N)} e^{\frac{C-\lambda}{2}t} + ((2lD)^{-2l} \delta_0^{1-\frac{2+p}{2l}} M^{\frac{p+1}{2l}} + (2lC_{\epsilon_0})^{-2l} \|g\|_{L^{2l}(\mathbb{R}^N)}) \\ &\quad \times \left(\frac{1 - e^{l(C-\lambda)t}}{-l(C-\lambda)} \right)^{\frac{1}{2l}}. \end{aligned} \quad (11.46)$$

We can let $l \to +\infty$ in (11.46) to get the required $L^\infty(\mathbb{R}^N)$ bound:

$$\|u(t)\|_{L^\infty(\mathbb{R}^N)} \leqslant \|u_0\|_{L^\infty(\mathbb{R}^N)} e^{\frac{C-\lambda}{2}t} + \delta_0 + \|g\|_{L^\infty(\mathbb{R}^N)}, \tag{11.47}$$

valid for $t \in [0, T]$.

11.3.3 $L^{p+1}(\mathbb{R}^N)$ a priori estimate

To complete the induction argument, we give below the $L^{p+1}(\mathbb{R}^N)$ a priori estimate of the solution u of (11.4). Multiplying (11.4) by $|u|^p \operatorname{sgn}(u)$, we obtain

$$\int_{\mathbb{R}^N} u_t |u|^p \operatorname{sgn}(u)\, dx + \int_{\mathbb{R}^N} (-\Delta)^\alpha u |u|^{p-1} u\, dx + \int_{\mathbb{R}^N} f(u) u |u|^{p-1} dx + \lambda \int_{\mathbb{R}^N} |u|^{p+1} dx$$
$$= \int_{\mathbb{R}^N} g |u|^p \operatorname{sgn}(u)\, dx. \tag{11.48}$$

Since $u, (-\Delta)^\alpha u \in L^{p+1}(\mathbb{R}^N)$ we infer from [94, Lemma 3.1] that the second term is nonnegative. Then thanks to the Young inequality and the *simplified assumption,*

$$\exists_{0 \leqslant C_0 < \lambda} \forall_{s \in \mathbb{R}}\ -f(s)s \leqslant C_0 s^2, \tag{11.49}$$

we get

$$\frac{1}{p+1}\frac{d}{dt}\int_{\mathbb{R}^N} |u|^{p+1} dx + (\lambda - C_0 - \delta)\int_{\mathbb{R}^N} |u|^{p+1} dx \leqslant C_\delta \int_{\mathbb{R}^N} |g|^{p+1} dx, \tag{11.50}$$

where $\delta > 0$ is sufficiently small, such that $M_1 = \lambda - C_0 - \delta$ is positive. Consequently,

$$\|u(t)\|^{p+1}_{L^{p+1}(\mathbb{R}^N)} \leq \left(\|u(0)\|^{p+1}_{L^{p+1}(\mathbb{R}^N)} - \frac{C_\delta \|g\|^{p+1}_{L^{p+1}(\mathbb{R}^N)}}{(p+1)M_1} \right) e^{-(p+1)M_1 t} + \frac{C_\delta \|g\|^{p+1}_{L^{p+1}(\mathbb{R}^N)}}{(p+1)M_1}. \tag{11.51}$$

11.3.4 Some properties of the cut-off function

We consider the smooth (at least C^2, but we prefer $\theta \in C^\infty$) cut-off function $\theta : \mathbb{R}^N \to [0, 1]$,

$$\theta(x) = \begin{cases} 1, & |x| \geqslant 2, \\ 0, & |x| \leqslant 1. \end{cases} \tag{11.52}$$

Next, we discuss a property of the cut-off function which is important in further estimates.

Lemma 11.3.4. *Let* $\Lambda = (-\Delta)^{\frac{1}{2}}$, *then for any* $\alpha \in (0,1)$, *there exists a constant* $M = M(\alpha, N, \theta) > 0$ *such that*

$$|\Lambda^{2\alpha}\theta(x)| \leqslant M < \infty \quad \textit{for all } x \in \mathbb{R}^N.$$

Proof. By [58, Theorem 1], since $\theta \in C_B^2(\mathbb{R}^N)$, we have

$$\begin{aligned}\Lambda^{2\alpha}\theta(x) &= -c_N(\alpha) \int\limits_{|z|<1} \frac{\theta(x-z) - \theta(x) - \nabla\theta(x)\cdot(-z)}{|z|^{N+2\alpha}} dz \\ &\quad - c_N(\alpha) \int\limits_{|z|\geqslant 1} \frac{\theta(x-z) - \theta(x)}{|z|^{N+2\alpha}} dz \\ &=: I_1 + I_2,\end{aligned}$$

where $c_N(\alpha) = \frac{\alpha\Gamma(\frac{N}{2}+\alpha)}{\pi^{\frac{N}{2}+2\alpha}\Gamma(1-\alpha)}$.

Since $\theta(\cdot) \in [0,1]$, then we have

$$|I_2| \leqslant c_N(\alpha) \int\limits_{|z|\geqslant 1} \frac{1}{|z|^{N+2\alpha}} dz \leqslant \text{const}(\alpha, N) < \infty \tag{11.53}$$

for each $\alpha \in (0,1)$.

On the other hand, from the Taylor formula

$$\theta(x-z) - \theta(x) - \nabla\theta(x)\cdot(-z) = \frac{1}{2}\theta''(x)\cdot z^2 + o(z^2), \tag{11.54}$$

where $\theta''(x)$ denotes the Hessian matrix of θ, we have (here we need to use the notion of the *Cauchy principal value integral* to justify the calculations),

$$\begin{aligned}I_1 &= -c_N(\alpha)\, PV \int\limits_{|z|<1} \frac{\theta(x-z) - \theta(x) - \nabla\theta(x)\cdot(-z)}{|z|^{N+2\alpha}} dz \\ &= -c_N(\alpha) \int\limits_{|z|<1} \frac{\frac{1}{2}\theta''(x)\cdot z^2 + o(z^2)}{|z|^{N+2\alpha}} dz,\end{aligned} \tag{11.55}$$

and so

$$\begin{aligned}|I_1| &\leqslant c_N(\alpha) \int\limits_{|z|<1} \frac{\frac{1}{2}|\theta''(x)\cdot z^2| + |o(z^2)|}{|z|^{N+2\alpha}} dz \\ &\leqslant c(N,\alpha,\theta) \int\limits_{|z|\leqslant 1} \frac{|z|^2 + o(|z|^2)}{|z|^{N+2\alpha}} dz \leqslant C(N,\theta,\alpha) < \infty,\end{aligned} \tag{11.56}$$

this result together with (11.53) completes the proof. □

Let θ be as above, moreover, we set $\theta_k(\cdot) = \theta(\frac{\cdot}{k})$, $k = 1, 2, \ldots$. Then the following identity is obvious (e. g., see [58]).

Lemma 11.3.5. *For any $s \in (0,2)$,*

$$\Lambda^s \theta_k(x) = \frac{1}{k^s} \Lambda^s \theta(z)|_{z=\frac{x}{k}}.$$

We will need also the following property of the operator $\Lambda^{2\alpha}$:

Lemma 11.3.6. *The operator $\Lambda^{2\alpha}$ is symmetric in $L^2(\mathbb{R}^N)$.*

Proof. By Definition 11.3.3, when $f, g \in D(\Lambda^{2\alpha}) = H^{2\alpha}(\mathbb{R}^N)$, then we have

$$\begin{aligned}
&\int_{\mathbb{R}^N} \Lambda^{2\alpha} f(x) \cdot g(x) dx \\
&= -c_N(\alpha) \lim_{\varepsilon \to 0} \int_{\mathbb{R}^N} \int_{|z| \ge \varepsilon} \left[\frac{f(x-z)g(x)}{|z|^{N+2\alpha}} - \frac{f(x)g(x)}{|z|^{N+2\alpha}} \right] dz dx \\
&= -c_N(\alpha) \lim_{\varepsilon \to 0} \int_{|z| \ge \varepsilon} \frac{1}{|z|^{N+2\alpha}} \left(\int_{\mathbb{R}^N} f(x) g(x-z) dx - \int_{\mathbb{R}^N} f(x) g(x) dx \right) dz \\
&= \int_{\mathbb{R}^N} f(x) \cdot \Lambda^{2\alpha} g(x) dx,
\end{aligned}$$

which proves the claim. □

11.4 Attractor for the semigroup of solutions to (11.4)

Now we discuss the dynamics of (11.4) in term of the global attractor. The following assumptions will be used throughout this section:

Assumption I. *Let $\alpha \in (0,1)$, $g \in H^{-\alpha}(\mathbb{R}^N)$, $f \in C^1(\mathbb{R})$ and satisfy* (11.5)–(11.6). *Moreover, we denote by $\{S(t)\}_{t \ge 0}$ the semigroup corresponding to the weak solution of* (11.4) *with initial data $u_0 \in L^2(\mathbb{R}^N)$.*

Then we know that

$$S(\cdot): [0,\infty) \times L^2(\mathbb{R}^N) \ni (t, u_0) \mapsto u(t) \in L^2(\mathbb{R}^N) \text{ is continuous.} \tag{11.57}$$

11.4.1 Dissipation of $\{S(t)\}_{t \ge 0}$ in $L^2(\mathbb{R}^N)$

Multiplying (11.4) by u and integrating over $\mathbb{R}^N$ (here and after, we will calculate on strong solutions; the estimates will also be valid for the weak solutions by passing to the limits), we obtain that

$$\frac{1}{2}\frac{d}{dt}\int_{\mathbb{R}^N}|u|^2dx+\int_{\mathbb{R}^N}(\Lambda^\alpha u)^2dx+\int_{\mathbb{R}^N}f(u)udx+\lambda\|u\|^2=\int_{\mathbb{R}^N}g(x)udx, \tag{11.58}$$

where, for simplicity, $\|\cdot\|$ denotes the usual $L^2(\mathbb{R}^N)$-norm. Then, by the assumption (11.5), we have

$$\frac{1}{2}\frac{d}{dt}\|u\|^2+\|\Lambda^\alpha u\|^2+c_1\int_{\mathbb{R}^N}|u|^{p+1}dx+(\lambda-k_1)\|u\|^2\le\|g\|_{H^{-\alpha}(\mathbb{R}^N)}\|u\|_{H^\alpha(\mathbb{R}^N)}. \tag{11.59}$$

Hence, noticing that $\|\cdot\|^2_{H^\alpha(\mathbb{R}^N)}=\|\Lambda^\alpha\cdot\|^2+\|\cdot\|^2$ and $\lambda>k_1$, using the Cauchy inequality, we can deduce that

$$\frac{d}{dt}\|u\|^2+\|\Lambda^\alpha u\|^2+2c_1\int_{\mathbb{R}^N}|u|^{p+1}dx+(\lambda-k_1)\|u\|^2\leqslant c_{\lambda,k_1}\|g\|^2_{H^{-\alpha}(\mathbb{R}^N)}, \tag{11.60}$$

which implies that for any $t\ge0$,

$$\int_0^t\left(\|u(s)\|^2_{H^\alpha(\mathbb{R}^N)}+\|u(s)\|^{p+1}_{L^{p+1}(\mathbb{R}^N)}\right)ds\le C_{\lambda,k_1,c_1}\left(t\|g\|^2_{H^{-\alpha}(\mathbb{R}^N)}+\|u_0\|^2\right), \tag{11.61}$$

and

$$\|u(t)\|^2\le e^{-(\lambda-k_1)t}\|u_0\|^2+\frac{c_{\lambda,k_1}\|g\|^2_{H^{-\alpha}(\mathbb{R}^N)}}{\lambda-k_1}. \tag{11.62}$$

Obviously, the estimate (11.62) implies that

$$B=\left\{u\in L^2(\mathbb{R}^N):\ \|u\|^2\le\frac{c_{\lambda,k_1}\|g\|^2_{H^{-\alpha}(\mathbb{R}^N)}}{\lambda-k_1}+1\right\}$$

is a bounded absorbing set for the semigroup $\{S(t)\}_{t\ge0}$ in $L^2(\mathbb{R}^N)$. Then (see Chapter 2) we have the following result.

Theorem 11.4.1 (Absorbing set). *Under Assumption I, the semigroup $\{S(t)\}_{t\ge0}$ has a bounded positively invariant absorbing set B_0 in $L^2(\mathbb{R}^N)$.*

11.4.2 Asymptotic compactness in $L^2(\mathbb{R}^N)$

We will use the *tail estimate technique* (e. g., see [146, 178]) to get asymptotic compactness of the semigroup.

Theorem 11.4.2 (Tail estimate). *In addition to the Assumption I, let further $g \in L^2(\mathbb{R}^N)$. Then for any $\varepsilon > 0$, there exist positive constants $h = h(\varepsilon, \|B_0\|)$ and $T = T(\varepsilon, \|B_0\|)$ such that*

$$\int_{\mathcal{O}_h} |S(t)u_0|^2 dx < \varepsilon \quad \text{for all } t \geq T \text{ and } u_0 \in B_0,$$

where $\mathcal{O}_h = \{x \in \mathbb{R}^N : |x| \geq h\}$ and B_0 is the positively invariant absorbing set obtained in Theorem 11.4.1.

Proof. Taking the cut-off function θ as in (11.52), multiplying (11.4) by $\theta_k u \in H^\alpha(\mathbb{R}^N) \subset L^2(\mathbb{R}^N)$ and integrating over $\mathbb{R}^N$, we obtain that

$$\begin{aligned}\int_{\mathbb{R}^N} u_t u\theta_k dx &+ \langle(-\Delta)^\alpha u, u\theta_k\rangle_{H^{-\alpha}(\mathbb{R}^N),H^\alpha(\mathbb{R}^N)} + \lambda \int_{\mathbb{R}^N} u^2\theta_k dx \\ &= -\int_{\mathbb{R}^N} f(u)u\theta_k dx + \int_{\mathbb{R}^N} g(x)u\theta_k dx.\end{aligned} \tag{11.63}$$

The components are estimated next as follows:

$$\int_{\mathbb{R}^N} u_t u\theta_k dx = \frac{1}{2}\frac{d}{dt}\int_{\mathbb{R}^N} u^2\theta_k dx, \tag{11.64}$$

by the Young inequality ($|g\theta_k^{\frac{1}{2}} u\theta_k^{\frac{1}{2}}| \leq \delta u^2\theta_k + C_\delta g^2\theta_k$, $\delta > 0$), we have

$$\left|\int_{\mathbb{R}^N} g(x)u\theta_k dx\right| \leq \delta \int_{\mathbb{R}^N} u^2\theta_k dx + C_\delta \int_{\mathbb{R}^N} g^2\theta_k dx \tag{11.65}$$

for some small positive constant δ, and from (11.5) we have

$$-\int_{\mathbb{R}^N} f(u)u\theta_k dx \leq k_1 \int_{\mathbb{R}^N} \theta_k u^2 dx. \tag{11.66}$$

We will use also the pointwise estimate from [39]:

$$\forall_{0\leq\beta\leq 2}\forall_{\phi\in C_0^2(\mathbb{R}^N)}\ 2\phi\Lambda^\beta\phi(x) \geq \Lambda^\beta\phi^2(x). \tag{11.67}$$

Using (11.67) to the fractional term, due to the remark below, we obtain

$$\langle(-\Delta)^\alpha u, u\theta_k\rangle_{H^{-\alpha}(\mathbb{R}^N),H^\alpha(\mathbb{R}^N)} \geq \frac{1}{2}\int_{\mathbb{R}^N} u^2\Lambda^{2\alpha}(\theta_k)dx. \tag{11.68}$$

At the same time, thanks to Lemmas 11.3.4 and 11.3.5, we have

$$|\Lambda^{2\alpha}(\theta_k)| \leq \frac{\text{const}}{k^{2\alpha}} \tag{11.69}$$

for some constant const which depends only on N, α and θ.

Collecting the estimates (11.64)–(11.69) above, we thus get

$$\frac{1}{2}\frac{d}{dt}\int_{\mathbb{R}^N} u^2\theta_k dx + (\lambda - k_1 - \delta)\int_{\mathbb{R}^N} u^2\theta_k dx \leq \frac{\text{const}}{k^{2\alpha}}\int_{\mathbb{R}^N} u^2 dx + C_\delta \int_{\mathbb{R}^N} g^2\theta_k dx. \quad (11.70)$$

Note that $\lambda > k_1$, so by taking $\delta \in (0, \frac{\lambda-k_1}{2})$ we have $\lambda - k_1 - \delta > \frac{\lambda-k_1}{2} > 0$ and for any $u_0 \in B_0$,

$$\frac{d}{dt}\int_{\mathbb{R}^N} u^2\theta_k dx + (\lambda - k_1)\int_{\mathbb{R}^N} u^2\theta_k dx \leq \frac{\text{const}\cdot \|B_0\|^2}{k^{2\alpha}} + C_\delta \int_{\mathbb{R}^N} g^2\theta_k dx. \quad (11.71)$$

Therefore,

$$\begin{aligned}\forall_{u_0\in B_0}\forall_{t\geq 0} \quad \int_{|x|\geq k} u^2(x,t)dx &\leq \int_{\mathbb{R}^N} u^2(x,t)\theta_k dx \\ &\leq e^{-(\lambda-k_1)t}\|B_0\|^2 + \frac{\text{const}\cdot\|B_0\|^2}{k^{2\alpha}(\lambda-k_1)} + \frac{C_\delta\int_{\mathbb{R}^N} g^2(x)\theta_k dx}{\lambda - k_1}, \end{aligned} \quad (11.72)$$

which allows us to complete the proof by noticing only that $\int_{\mathbb{R}^N} g^2\theta_k dx \leq \int_{|x|\geq k} g^2(x)dx \to 0$ as $k \to \infty$. □

Remark 11.4.3. The calculations in (11.68) will be justified using approximation argument. Since $u(t) \in H^\alpha(\mathbb{R}^N) \cap L^{p+1}(\mathbb{R}^N)$ (we fix $t > 0$ here), there is a sequence $\{u_n\} \subset C_0^\infty(\mathbb{R}^N)$ convergent to $u(t)$ in $H^\alpha(\mathbb{R}^N)$. Note next that, since $u_n \in C_0^\infty(\mathbb{R}^N) \subset H^{2\alpha}(\mathbb{R}^N)$, then $\Lambda^{2\alpha}u_n \in L^2(\mathbb{R}^N)$. We are working with the triple of Hilbert spaces: $H^\alpha(\mathbb{R}^N) \subset L^2(\mathbb{R}^N) \subset H^{-\alpha}(\mathbb{R}^N)$, so that the linear functional (on $H^\alpha(\mathbb{R}^N)$) corresponding to $\Lambda^{2\alpha}u_n$ is given by

$$\langle \Lambda^{2\alpha}u_n, \phi\rangle_{H^{-\alpha}(\mathbb{R}^N),H^\alpha(\mathbb{R}^N)} = \int_{\mathbb{R}^N} \Lambda^{2\alpha}u_n\phi\, dx, \quad \phi \in H^\alpha(\mathbb{R}^N).$$

By (11.67) and the above observation ($u_n\theta_k \in H^\alpha(\mathbb{R}^N)$ since $\theta_k \in C_B^2(\mathbb{R}^N)$; see Lemma 11.4.4)

$$\langle \Lambda^{2\alpha}u_n, u_n\theta_k\rangle_{H^{-\alpha}(\mathbb{R}^N),H^\alpha(\mathbb{R}^N)} = \int_{\mathbb{R}^N} \Lambda^{2\alpha}u_n u_n\theta_k dx \geq \frac{1}{2}\int_{\mathbb{R}^N} \Lambda^{2\alpha}(u_n^2)\theta_k dx. \quad (11.73)$$

Like in the proof of Lemma 11.3.6 we can show next that

$$\frac{1}{2}\int_{\mathbb{R}^N} \Lambda^{2\alpha}(u_n^2)\theta_k dx = \frac{1}{2}\int_{\mathbb{R}^N} u_n^2\Lambda^{2\alpha}(\theta_k)dx. \quad (11.74)$$

Finally, since $u(t) \in H^\alpha(\mathbb{R}^N)$, we can pass to the limits in the outline components (note that $u(t)\theta_k \in H^\alpha(\mathbb{R}^N)$ since $\theta_k \in C_B^2(\mathbb{R}^N)$), to obtain

$$\langle \Lambda^{2\alpha}u(t), u(t)\theta_k\rangle_{H^{-\alpha}(\mathbb{R}^N),H^{\alpha}(\mathbb{R}^N)} \geq \frac{1}{2}\int_{\mathbb{R}^N} u(t)^2\Lambda^{2\alpha}(\theta_k)dx, \tag{11.75}$$

which completes the calculations.

Passing to the limit in the left-hand side of (11.73), we need the following lemma.

Lemma 11.4.4. *Let $\{u_n\}_{n=1}^{\infty} \subset C_0^{\infty}(\mathbb{R}^N)$ and $u_n \to u$ in $H^{\alpha}(\mathbb{R}^N)$ as $n \to \infty$ ($\alpha \in (0,1)$). Then, for any (fixed) $\theta \in C_B^2(\mathbb{R}^N)$,*

$$\theta \cdot u_n \in H^{\alpha}(\mathbb{R}^N) \quad \text{and} \quad \theta \cdot u_n \to \theta \cdot u \text{ in } H^{\alpha}(\mathbb{R}^N) \text{ as } n \to \infty. \tag{11.76}$$

The above lemma follows immediately from the following more general observation.

Remark 11.4.5. For any $0 \leq s \leq 1$ and $a \in C_B^1(\mathbb{R}^N)$, the multiplication $v \to av$ by a is a bounded operator on $H^s(\mathbb{R}^N)$. Moreover,

$$\exists_{C>0}\forall_{v\in H^s(\mathbb{R}^N)}\|av\|_{H^s(\mathbb{R}^N)} \leq C\|a\|^s_{C_B^1(\mathbb{R}^N)}\|a\|^{1-s}_{C_B^0(\mathbb{R}^N)}\|v\|_{H^s(\mathbb{R}^N)}. \tag{11.77}$$

Proof. The following two estimates are evident:

$$\begin{aligned}\|av\|_{L^2(\mathbb{R}^N)} &\leq \|a\|_{C_B^0(\mathbb{R}^N)}\|v\|_{L^2(\mathbb{R}^N)},\\ \|av\|_{H^1(\mathbb{R}^N)} &\leqslant \|a\|_{C_B^1(\mathbb{R}^N)}\|v\|_{H^1(\mathbb{R}^N)}.\end{aligned} \tag{11.78}$$

The rest is a consequence of the interpolation inequality. □

Now, we are ready to prove the asymptotic compactness of $\{S(t)\}_{t\geq 0}$ in $L^2(\mathbb{R}^N)$.

Theorem 11.4.6 (Asymptotic compactness). *Under the assumptions of Theorem 11.4.2, the semigroup $\{S(t)\}_{t\geq 0}$ is asymptotically compact in $L^2(\mathbb{R}^N)$.*

Proof. It is sufficient to show that for any $\varepsilon > 0$, there is a $T = T(\varepsilon) > 0$ such that

$$S(t)B_0 \text{ has a finite } \varepsilon\text{-net in } L^2(\mathbb{R}^N), \quad \forall\, t \geq T;$$

recall that B_0 is the positively invariant absorbing set obtained in Theorem 11.4.1. By the invariance of B_0, we only need to prove that

$$S(T)B_0 \text{ has a finite } \varepsilon\text{-net in } L^2(\mathbb{R}^N) \text{ for some } T. \tag{11.79}$$

For convenience, we divide our proof into steps.

Step 1. From Theorem 11.4.2, we know that there exist constants $T_1 = T_1(\varepsilon, \|B_0\|)$ and $h_1 = h_1(\varepsilon, \|B_0\|)$ such that

$$\int_{|x|\geq h_1} |S(t)u_0|^2 dx < \frac{\varepsilon}{4} \quad \text{for all } t \geq T_1, u_0 \in B_0. \tag{11.80}$$

Step 2. From (11.61), we have

$$\int_0^{T_1} (\|u(s)\|^2_{H^\alpha(\mathbb{R}^N)} + \|u(s)\|^{p+1}_{L^{p+1}(\mathbb{R}^N)})ds \le C_{\lambda,k_1,C_1}(\|g\|^2_{H^{-\alpha}(\mathbb{R}^N)}T_1 + \|u_0\|^2), \tag{11.81}$$

where $u_0 \in B_0$ and $u(s) = S(s)u_0$, $s \in [0, T_1]$.

Set $S(\cdot)u_0$ the function $S(\cdot)u_0 : s \in [0, T_1] \to S(s)u_0 \in L^2(\mathbb{R}^N)$, and denote

$$B_1 := \{S(\cdot)u_0 : u_0 \in B_0\}. \tag{11.82}$$

Then (11.81) shows that

$$B_1 \text{ is bounded in } L^2(0, T_1; H^\alpha(\mathbb{R}^N)) \cap L^{p+1}(0, T_1; L^{p+1}(\mathbb{R}^N)). \tag{11.83}$$

Consequently, due to the equation (11.4) (fulfilled in $H^{-\alpha}(\mathbb{R}^N)$)

$$u_t = -(-\Delta)^\alpha u - \lambda u - f(u) + g(x),$$

we also know that

$$\{u_t : u \in B_1\} \text{ is bounded in } L^2(0, T_1; H^{-\alpha}(\mathbb{R}^N)). \tag{11.84}$$

Step 3. Let $u_{0i} \in B_0$ and $u_i(t) = S(t)u_{0i}$, $i = 1, 2$. Set $w(t) = u_1(t) - u_2(t)$, then we know that w satisfies the following equation:

$$\begin{cases} w_t + (-\Delta)^\alpha w + f(u_1) - f(u_2) + \lambda w = 0, \\ w(0) = u_{01} - u_{02}. \end{cases}$$

Therefore, applying the assumption (11.6) and the inequality $\int_{\mathbb{R}^N} (-\Delta)^\alpha w w dx \geqslant 0$, we deduce that

$$\|w(T_1)\|^2 \le e^{2(\lambda - l)T_1} \|w(s)\|^2 \quad \text{for any } s \in [0, T_1]. \tag{11.85}$$

Step 4. Now we are ready to complete our proof by verifying (11.79) with $T = T_1$.

Note that, (11.83), (11.84) imply that $B_1|_{\{x \in \mathbb{R}^N : |x| < h_1\}}$ is bounded in $L^2(0, T_1; H^\alpha(\{x \in \mathbb{R}^N : |x| < h_1\})) \cap L^{p+1}(0, T_1; L^{p+1}(\{x \in \mathbb{R}^N : |x| < h_1\}))$[1] and $\{u_t : u \in B_1\}|_{\{x \in \mathbb{R}^N : |x| < h_1\}}$ is bounded in $L^2(0, T_1; H^{-\alpha}(\{x \in \mathbb{R}^N : |x| < h_1\}))$. Therefore, by the Lions–Aubin compactness lemma,

$$B_1|_{\{x \in \mathbb{R}^N : |x| < h_1\}} \text{ is precompact in } L^2(0, T_1; L^2(\{x \in \mathbb{R}^N : |x| < h_1\})). \tag{11.86}$$

1 The restriction of $H^\alpha(\mathbb{R}^N)$ to the space of functions defined in the ball is understand as in [173], Definition 4.2.1.

Hence, for such $\varepsilon > 0$ and T_1, there exist $u_i \in B_1$, $i = 1, 2, \ldots, m_\varepsilon$, such that for any $u \in B_1$, there is some u_i satisfying

$$\int_0^{T_1} \int_{|x|<h_1} |u(x,s) - u_i(x,s)|^2 dxds \le \frac{e^{-2(\lambda-l)T_1}}{2T_1}\varepsilon, \tag{11.87}$$

which, combining with (11.85) and (11.80), implies that

$$\begin{aligned} \|u(T_1) - u_i(T_1)\|^2 &= \int_{\mathbb{R}^N} |u(x,T_1) - u_i(x,T_1)|^2 dx \\ &= \int_{|x|\ge h_1} |u(x,T_1) - u_i(x,T_1)|^2 dx + \int_{|x|<h_1} |u(x,T_1) - u_i(x,T_1)|^2 dx \\ &\le 2 \int_{|x|\ge h_1} (|u(x,T_1)|^2 + |u_i(x,T_1)|^2) dx + \frac{\varepsilon}{2} \\ &< 2 \cdot \frac{\varepsilon}{4} + \frac{\varepsilon}{2} = \varepsilon, \end{aligned} \tag{11.88}$$

that is, $S(T_1)B_0$ has a finite ε-net in $L^2(\mathbb{R}^N)$. □

11.4.3 Global attractor in $L^2(\mathbb{R}^N)$

From Theorems 11.4.1 and 11.4.6, and the continuity (11.57), applying the standard theory of dynamical systems (e. g., [32, 86, 172]), we can deduce our main result of this section immediately.

Theorem 11.4.7. *Under the assumptions of Theorem 11.4.6, the semigroup $\{S(t)\}_{t\ge0}$ has a global attractor $\mathcal{A}$ in $L^2(\mathbb{R}^N)$; that is, $\mathcal{A}$ is compact, invariant and attracts every $L^2(\mathbb{R}^N)$-bounded set in $L^2(\mathbb{R}^N)$-norm.*

11.4.4 Postscriptum

We are breaking here the story told in this book. But, as almost everything in mathematics, it already has some extensions. We mean here, in particular, the idea reported in [15] where the classical Schrödinger equation

$$\begin{cases} u_t = i\Delta u + f(u, \nabla u), & t > 0,\ x \in \Omega, \\ u = 0, & x \in \partial\Omega, \\ u(0,x) = u_0(x), & x \in \Omega, \end{cases} \tag{11.89}$$

$u = u(t,x) : \mathbb{R}^+ \times \Omega \to \mathbb{C}$, is approximated by the fractional problems of the form

$$\begin{cases} u_t = -(-i\Delta)^\alpha u + f(u, \nabla u), & t > 0,\ x \in \Omega, \\ u = 0, & x \in \partial\Omega, \\ u(0,x) = u_0(x), & x \in \Omega, \end{cases} \tag{11.90}$$

with real parameter $\alpha \in (\frac{1}{2}, 1)$, i the imaginary unit, and *sectorial operator* $A = (-i\Delta)^\alpha : D(A) = H^{2\alpha}(\Omega) \cap H_0^1(\Omega) \to L^2(\Omega)$ in the main part. Thus, the *dispersive problem* (11.89) (see, e. g., [122]) having *conservative* linear part is a limit of the *dissipative* approximations (11.90). Also that paper already has continuations published by its co-authors.

Panta rhei.
Everything flows.

(Heraclitus of Ephesus, c. 535–c. 475 B. C.)

Bibliographical notes

It is a natural extension of the study of the semilinear heat equation to replace $-\Delta$ with its fractional power as the main part operator and study the properties of solutions. This was a very popular problem to study through the last 20 years; see in particular [20, 62] and the references therein. Several properties of solutions of the classical semilinear heat equation will be recovered for such generalization; in particular a version of the maximum principle (see, e. g., [98]). The Moser–Alikakos technique was used first in [3] extending the J. Moser [136] approach to regularity of solutions to elliptic equations. More applications of that technique can be found in [32, Chapter 9].

Bibliography

[1] R. A. Adams, Sobolev Spaces, Academic Press, New York, 1975.

[2] S. Agmon, Lectures on Elliptic Boundary Value Problems, Mathematical Studies, Van Nostrand, New York, 1965.

[3] N. D. Alikakos, An application of the invariance principle to reaction-diffusion equations, J. Differ. Equ. 33 (1979), 201–225.

[4] N. D. Alikakos, Quantitative maximum principles and strongly coupled gradient-like reaction-diffusion systems, Proc. R. Soc. Edinb. 94A (1983), 265–286.

[5] H. Amann, Global existence for semilinear parabolic systems, J. Reine Angew. Math. 360 (1985), 47–83.

[6] H. Amann, Linear and Quasilinear Parabolic Problems, vol. I, Birkhäuser, Basel, 1995.

[7] H. Amann, Linear and Quasilinear Parabolic Problems, vol. II, Birkhäuser, Basel, 2019.

[8] H. Amann, M. Hieber, G. Simonett, Bounded H_∞-calculus for elliptic operators, Differ. Integral Equ. 3 (1994), 613–653.

[9] J. Arrieta, A. N. Carvalho, Abstract parabolic problems with critical nonlinearities and applications to Navier–Stokes and heat equations, Trans. Am. Math. Soc., 352 (1999), 285–310.

[10] J. Avrin, Singular initial data and uniform global bounds for the hyper-viscous Navier–Stokes equations with periodic boundary conditions, J. Differ. Equ. 190 (2003), 330–351.

[11] A. V. Balakrishnan, Fractional powers of closed operators and the semigroups generated by them, Pac. J. Math. 10 (1960), 419–437.

[12] J. M. Ball, Strongly continuous semigroups, weak solutions and the variation of constants formula, Proc. Am. Math. Soc. 63 (1977), 370–373.

[13] A. V. Babin, M. I. Vishik, Attractors of Evolution Equations, North-Holland, Amsterdam, 1992.

[14] G. Barles, E. Chasseigne, On the regularizing effect for unbounded solutions of first-order Hamilton–Jacobi equations, J. Differ. Equ. 260 (2016), 7020–7031.

[15] F. D. M. Bezerra, A. N. Carvalho, T. Dlotko, M. J. D. Nascimento, Fractional Schrödinger equation; solvability, asymptotic behavior and connection with classical Schrödinger equation, J. Math. Anal. Appl. 457 (2018), 336–360.

[16] P. Biler, G. Karch, W. A. Woyczynski, Asymptotics for multifractal conservation laws, Stud. Math. 135 (1999), 231–252.

[17] H. Brezis, Operateurs Maximaux Monotones et Semi-groupes de Contractions dans les Espaces de Hilbert, North-Holland Publishing Company, Amsterdam, 1973.

[18] H. Brezis, Analyse fonctionelle. Théorie et applications, Masson, Paris, 1983.

[19] F. E. Browder, Remarks on nonlinear interpolation in Banach spaces, J. Funct. Anal. 4 (1969), 390–403.

[20] C. Bucur, E. Valdinoci, Nonlocal Diffusion and Applications, Lecture Notes of the Unione Matematica Italiana, Springer International Publishing, Switzerland, 2016.

[21] L. Cafarelli, L. Silvestre, An extension problem related to the fractional Laplacian, Commun. Partial Differ. Equ. 32 (2007), 1245–1260.

[22] L. A. Cafarelli, Y. Sire, On some pointwise inequalities involving nonlocal operators, in: S. Chanillo, B. Franchi, G. Lu, C. Perez, E. Sawyer (Eds), Harmonic Analysis, Partial Differential Equations and Applications. Applied and Numerical Harmonic Analysis, Birkhäuser, Basel, 2017.

[23] L. A. Caffarelli, A. Vasseur, Drift diffusion equations with fractional diffusion and the quasi-geostrophic equation, Ann. Math. 171 (2010), 1903–1930.

[24] P. de Carvalho-Neto, G. Planas, Mild solutions to the time fractional Navier–Stokes equations in $\mathbb{R}^N$, J. Differ. Equ. 259 (2015), 2948–2980.

https://doi.org/10.1515/9783110599831-012

[25] A. N. Carvalho, J. W. Cholewa, T. Dlotko, Global attractors for problems with monotone operators, Boll. Unione Mat. Ital. (8) 2-B (1999), 693–706.

[26] A. N. Carvalho, J. W. Cholewa, Continuation and asymptotics of solutions to semilinear parabolic equations with critical nonlinearities, J. Math. Anal. Appl. 310 (2005), 557–578.

[27] A. N. Carvalho, C. B. Gentile, Comparison results for nonlinear parabolic equations with monotone principal part, J. Math. Anal. Appl. 259 (2001), 319–337.

[28] L. Cattabriga, Sur un problema al contorno relativo al sistema di equazioni di Stokes, Rend. Semin. Mat. Univ. Padova 31 (1961), 308–340.

[29] J. W. Cholewa, T. Dlotko, Global solutions via partial information and the Cahn–Hilliard equation, in: Singularities and Differential Equations, Banach Center Publications (PWN), vol. 33, Warsaw, 1996, pages 39–50.

[30] J. W. Cholewa, T. Dlotko, Global attractor for sectorial evolutionary equation, J. Differ. Equ. 125 (1996), 27–39.

[31] J. W. Cholewa, T. Dlotko, Local attractor for $n - D$ Navier–Stokes system, Hiroshima Math. J. 28 (1998), 309–319.

[32] J. W. Cholewa, T. Dlotko, Global Attractors in Abstract Parabolic Problems, Cambridge University Press, Cambridge, 2000.

[33] J. W. Cholewa, T. Dlotko, Fractional Navier–Stokes equations, Discrete Contin. Dyn. Syst., Ser. B 23 (2018), 2967–2988.

[34] J. W. Cholewa, T. Dlotko, A. Turski, Asymptotics of pseudodifferential parabolic equations, Demonstr. Math. 35 (2002), 75–91.

[35] J. W. Cholewa, C. Quesada, A. Rodriguez-Bernal, Nonlinear evolution equations in scales of Banach spaces and applications to PDEs, J. Abstr. Differ. Equ. Appl. 8 (2017), 1–69.

[36] R. R. Coifman, Y. Meyer, Au delá des opérateurs pseudodifférentieles, Astérisque, vol. 57, Société Mathématique de France, 1978.

[37] P. Constantin, D. Cordoba, J. Wu, On the critical dissipative quasi-geostrophic equation, Indiana Univ. Math. J. 50 (2001), 97–108.

[38] P. Constantin, M. Ignatova, Remarks on the fractional Laplacian with Dirichlet boundary conditions and applications, Int. Math. Res. Not. (2017), 1653–1673.

[39] A. Cordoba, D. Cordoba, A pointwise estimate for fractionary derivatives with applications to partial differential equations, Proc. Natl. Acad. Sci. USA 100 (2003), 15316–15317.

[40] A. Cordoba, D. Cordoba, A Maximum Principle applied to quasi-geostrophic equations, Commun. Math. Phys. 249 (2004), 511–528.

[41] C. Cosner, Pointwise a priori bounds for strongly coupled semi-linear systems of parabolic partial differential equations, Indiana Univ. Math. J. 30 (1981), 607–620.

[42] M. Cowling, I. Doust, A. McIntosh, A. Yagi, Banach space operators with a bounded H^∞ functional calculus, J. Aust. Math. Soc. A 60 (1996), 51–89.

[43] R. Dank, G. Dote, M. Hieber, R. S. Seeley, New thought on old results of R. T. Seeley, Math. Ann. 328 (2004), 545–583.

[44] E. B. Davies, Heat Kernels and Spectral Theory, Cambridge University Press, Cambridge, 1989.

[45] E. Di Nezza, G. Palatucci, E. Valdinoci, Hitchhiker's guide to the fractional Sobolev spaces, Bull. Sci. Math. 136 (2012),521–573.

[46] T. Dlotko, Decay properties of global solution of reaction-diffusion equations, Funkc. Ekvacioj 30 (1987), 111–114.

[47] T. Dlotko, Navier–Stokes equation and its fractional approximations, Appl. Math. Optim. 77 (2018), 99–128.

[48] T. Dlotko, The generalized Korteweg–de Vries–Burgers equation in $H^2(\mathbb{R})$, Nonlinear Anal. TMA 74 (2011), 721–732.

[49] T. Dlotko, M. B. Kania, Subcritical Hamilton–Jacobi fractional equation in $\mathbb{R}^N$, Math. Methods Appl. Sci. 38 (2015), 2547–2560.

[50] T. Dlotko, M. B. Kania, S. Ma, Korteweg–de Vries–Burgers system in $\mathbb{R}^N$, J. Math. Anal. Appl. 411 (2014), 853–872.

[51] T. Dlotko, M. B. Kania, C. Sun, Analysis of the viscous Cahn–Hilliard equation in $\mathbb{R}^N$, J. Differ. Equ. 252 (2012), 2771–2791.

[52] T. Dlotko, M. B. Kania, C. Sun, Pseudodifferential parabolic equations in uniform spaces, Appl. Anal. 93 (2014), 14–34.

[53] T. Dlotko, M. B. Kania, C. Sun, Quasi-geostrophic equation in R^2, J. Differ. Equ. 259 (2015), 531–561.

[54] T. Dlotko, T. Liang, Y. Wang, Critical and super-critical abstract parabolic equations, Discrete Contin. Dyn. Syst., Ser. B 25 (2020), 1517–1541.

[55] T. Dlotko, C. Sun, Dynamics of the modified viscous Cahn–Hilliard equation in $\mathbb{R}^N$, Topol. Methods Nonlinear Anal. 35 (2010), 277–294.

[56] T. Dlotko, C. Sun, 2D Quasi-geostrophic equation; sub-critical and critical cases, Nonlinear Anal. 150 (2017), 38–60.

[57] T. Dlotko, Sobolev spaces and embedding theorems, http://www.researchgate.net/publication/254844787_Sobolev_spaces_and_embedding_theorems.

[58] J. Droniou, C. Imbert, Fractal first-order partial differential equations, Arch. Ration. Mech. Anal. 182 (2006), 299–331.

[59] R. Farwig, C. Qian, Asymptotic behavior for the quasi-geostrophic equations with fractional dissipation in $\mathbb{R}^2$, J. Differ. Equ. 266 (2019), 6525–6579.

[60] C. L. Fefferman, Existence and smoothness of the Navier–Stokes equation, http://www.claymath.org/sites/default/files/navierstokes.pdf.

[61] C. L. Fefferman, D. S. McCormick, J. C. Robinson, J. L. Rodrigo, Higher order commutator estimates and local existence for the non-resistive MHD equations and related models, J. Funct. Anal. 267 (2014), 1035–1056.

[62] X. Ferndez-Real, X. Ros-Oton, Boundary regularity for the fractional heat equation, Rev. R. Acad. Cienc. Exactas Fís. Nat., Ser. A Mat. 110 (2016), 49–64.

[63] C. Foias, J. C. Saut, Asymptotic behavior as $t \to +\infty$ of solutions of Navier–Stokes equations and nonlinear spectral manifolds, Indiana Univ. Math. J. 3 (1984), 459–477.

[64] C. Foias, D. D. Holm, E. S. Titi, The Navier–Stokes–alpha model of fluid turbulence, Phys. D, Nonlinear Phenom. (2001), 505–519.

[65] C. Foias, O. Manley, R. Rosa, R. Temam, Navier–Stokes Equations and Turbulence, Encyclopedia of Mathematics and its Applications, Cambridge University Press, Cambridge, 2004.

[66] A. Friedman, Partial Differential Equations, Holt, Rinehart and Winston, New York, 1969.

[67] H. Fujita, T. Kato, On the Navier–Stokes initial value problem I, Arch. Ration. Mech. Anal. 16 (1964), 269–315.

[68] D. Fujiwara, H. Morimoto, An L_r-theorem of the Helmholtz decomposition of vector fields, J. Fac. Sci., Univ. Tokyo 24 (1977), 685–700.

[69] H. Gajewski, K. Gröger, K. Zacharias, Nichtlineare Operatorgleichungen und Operatordifferentialgleichungen, Akademie-Verlag, Berlin, 1974.

[70] G. P. Galdi, An Introduction to the Mathematical Theory of the Navier–Stokes Equations. Steady-state Problems, Springer, New York, 2011.

[71] S. Ghanem, Global regularity for the critical 2-D dissipative quasi-geostrophic equation with force, J. Math. Anal. Appl. 456 (2017), 563–575.

[72] J. M. Ghidaglia, Weakly damped forced Korteweg–de Vries equations behave as a finite dimensional dynamical systems in the long time, J. Differ. Equ. 74 (1988), 369–390.

[73] Y. Giga, Analyticity of the semigroup generated by the Stokes operator in L^r spaces, Math. Z. 178 (1981), 297–329.

[74] Y. Giga, Domains of fractional powers of the Stokes operator in L_r spaces, Arch. Ration. Mech. Anal. 89 (1985), 251–265.

[75] Y. Giga, T. Miyakawa, Solutions in L_r of the Navier–Stokes initial value problem, Arch. Ration. Mech. Anal. 89 (1985), 267–281.

[76] Y. Giga, H. Sohr, Abstract L^p estimates for the Cauchy problem with applications to the Navier–Stokes equations in exterior domains, J. Funct. Anal. 102 (1991), 72–94.

[77] M. Giga, Y. Giga, H. Sohr, L^p estimates for the Stokes system, in: H. Komatsu (Ed.), Functional Analysis and Related Topics, 1991 (Kyoto), Springer, Berlin, 1993, pages 55–67.

[78] D. Gilbarg, N. Trudinger, Elliptic Partial Differential Equations of Second Order, Springer-Verlag, Berlin, 1983.

[79] L. Grafakos, S. Oh, The Kato–Ponce inequality, Commun. Partial Differ. Equ. 39 (2014), 1128–1157.

[80] P. Grisvard, Caractérisation de quelques espaces d'interpolation, Arch. Ration. Mech. Anal. 25 (1967), 40–63.

[81] D. Guidetti, On interpolation with boundary conditions, Math. Z. 207 (1991), 439–460.

[82] B. Guo, D. Huang, Q. Li, C. Sun, Dynamics for a generalized incompressible Navier–Stokes equations in $\mathbb{R}^2$, Adv. Nonlinear Stud. 16 (2016), 249–272.

[83] J. K. Hale, Asymptotic Behavior of Dissipative Systems, AMS, Providence, RI, 1988.

[84] J. K. Hale, G. Raugel, Attractors for dissipative evolutionary equations, in: International Conference on Differential Equations, vols. 1, 2 (Barcelona, 1991), World Sci. Publishing, River Edge, NJ, 1993, pages 3–22.

[85] H. Hanche-Olsen, H. Holden, The Kolmogorov–Riesz compactness theorem, Expo. Math. 28 (2010), 385–394.

[86] D. Henry, Geometric Theory of Semilinear Parabolic Equations, Springer-Verlag, Berlin, 1981.

[87] D. Henry, How to remember the Sobolev inequalities, in: D. G. de Figueiredo and C. S. Hönig (Eds), Differential Equations, Springer-Verlag, New York, 1982, pages 97–109.

[88] E. Hille, R. S. Phillips, Functional Analysis and Semi-groups, AMS, Providence, RI, 1957.

[89] E. Hopf, Über die Anfangswertaufgabe für die hydrodynamischen Grundgleichungen, Math. Nachr. 4 (1951), 213–231.

[90] C. Imbert, A non-local regularization of first-order Hamilton–Jacobi equations, J. Differ. Equ. 211 (2005), 218–246.

[91] A. M. Il'in, A. S. Kalasnikov, O. A. Olejnik, Linear second order equations of parabolic type, Usp. Mat. Nauk 17 (1962), 3–146.

[92] A. Inoue, M. Wakimoto, On existence of solutions of the Navier–Stokes equation in a time dependent domain, J. Fac. Sci., Univ. Tokyo, Sec. I, 24 (1977), 303–319.

[93] T. Iwabuchi, T. Kawakami, Existence of mild solutions for a Hamilton–Jacobi equation with critical fractional viscosity in the Besov spaces, J. Math. Pures Appl. 107 (2017), 464–489.

[94] N. Ju, The maximum principle and the global attractor for the dissipative 2D quasi-geostrophic equations, Commun. Math. Phys. 255 (2005), 161–181.

[95] T. Kato, H. Fujita, On the nonstationary Navier–Stokes system, Rend. Semin. Mat. Univ. Padova 32 (1962), 243–260.

[96] T. Kato, G. Ponce, Commutator estimates and the Euler and Navier–Stokes equations, Commun. Pure Appl. Math. 41 (1988), 891–907.

[97] T. Kato, Strong L^p-solutions of the Navier–Stokes equation in R^m, with applications to weak solutions, Math. Z. 187 (1984), 471–480.

[98] G. Karch, Nonlinear evolution equations with anomalous diffusion, in: Qualitative Properties

of Solutions to Partial Differential Equations, J. Nečas Center for Mathematical Modeling, Charles University, Prague, 2009, pages 25–65.

[99] G. Karch, W. A. Woyczynski, Fractal Hamilton–Jacobi–KPZ equations, Trans. Am. Math. Soc. 360 (2008), 2423–2442.

[100] C. E. Kenig, G. Ponce, L. Vega, Well-posedness of the initial value problem for the Korteweg–de Vries equation, J. Am. Math. Soc. 4 (1991), 323–347.

[101] A. Kiselev, F. Nazarov, A. Volberg, Global well-posedness for the critical *2D* dissipative quasi-geostrophic equation, Invent. Math. 167 (2007), 445–453.

[102] A. Kiselev, Some recent results on the critical surface quasi-geostrophic equation: a review, in: Hyperbolic Problems: Theory, Numerics and Applications, Proc. Sympos. Appl. Math., vol. 67, AMS, Providence, RI, 2009, pages 105–122.

[103] S. Klainerman, Global existence for nonlinear wave equations, Commun. Pure Appl. Math. 33 (1980), 43–101.

[104] P. E. Kloeden, P. Marín-Rubio, J. Valero, Equivalence of invariant measures and stationary statistical solutions for the autonomous globally modified Navier–Stokes equations, Commun. Pure Appl. Anal. 8 (2009), 785–802.

[105] P. E. Kloeden, J. Valero, The Kneser property of the weak solutions of the three dimensional Navier–Stokes equations, Discrete Contin. Dyn. Syst. 28 (2010), 161–179.

[106] H. Komatsu, Fractional powers of operators, Pac. J. Math. 19 (1966), 285–346.

[107] H. Komatsu, Fractional powers of operators. II. Interpolation spaces. Pac. J. Math. 21 (1967), 89–111.

[108] S. G. Krein, Linear Equations in Banach Spaces, Birkhäuser, Basel, 1982 (Russian edition 1967).

[109] M. Kwasnicki, Ten equivalent definitions of the fractional Laplace operator, Fract. Calc. Appl. Anal. 20 (2017), 7–51.

[110] O. A. Ladyzhenskaya, On the uniqueness and smoothness of generalized solutions of the Navier–Stokes equations, Zap. Nauč. Semin. LOMI (1967), 169–185.

[111] O. A. Ladyzhenskaya, The Mathematical Theory of Viscous Incompressible Flow, Revised English edition, Gordon and Breach, 1969.

[112] O. Ladyzenskaya, Some recent mathematical results concerning the Navier–Stokes equations, Arch. Mech. 30 (1978), 217–224.

[113] O. Ladyzenskaya, Attractors for Semigroups and Evolution Equations, Cambridge University Press, Cambridge, 1991.

[114] O. A. Ladyzhenskaya, On some gaps in two of my papers on the Navier–Stokes equations and the way of closing them, J. Math. Sci. 115 (2003), 2789–2891.

[115] O. A. Ladyženskaja, V. A. Solonnikov, N. N. Ural'ceva, Linear and Quasilinear Equations of Parabolic Type, Nauka, Moscow, 1967.

[116] O. A. Ladyženskaja, N. N. Ural'ceva, Linear and Quasilinear Elliptic Equations, Academic Press, New York, 1968.

[117] O. Lazar, L. Xue, Regularity results for a class of generalized surface quasi-geostrophic equations, J. Math. Pures Appl. 130 (2019), 200–250.

[118] I. Lasiecka, Unified theory for abstract parabolic boundary problems – A semigroup approach, Appl. Math. Optim. 6 (1980), 287–333.

[119] O. Lazar, Global existence for the critical dissipative surface quasi-geostrophic equation, Commun. Math. Phys. 322 (2013), 73–93.

[120] J. Leray, Etude de diverses équations intégrales non linéaires et de quelques problémes que pose l'hydrodynamique, J. Math. Pures Appl. 12 (1933), 1–82.

[121] J. Leray, Essai sur le mouvement d'un fluide visqueux emplissant l'espace, Acta Math. 63 (1934), 193–248.

[122] F. Linares, G. Ponce, Introduction to Nonlinear Dispersive Equations, Springer-Verlag, New York, 2015.

[123] J.-L. Lions, E. Magenes, Problêmes aux Limites non Homogénes et Applications, vol. I, Dunod, Paris, 1968.

[124] J.-L. Lions, Quelques Méthodes de Résolutions des Problêmes aux Limites non Linéaires, Dunod, Paris, 1969.

[125] G. Łukaszewicz and P. Kalita, Navier–Stokes Equations. An Introduction with Applications, Advances in Mechanics and Mathematics, vol. 34, Springer International Publishing, 2016.

[126] A. Lunardi, Analytic Semigroup and Optimal Regularity in Parabolic Problems, Birkhäuser, Berlin, 1995.

[127] A. Lunardi, Interpolation Theory, Edizioni della Normale, Pisa, 2018.

[128] S. Ma, C. Zhong, H. Song, Attractors for nonautonomous 2D Navier–Stokes equations with less regular symbols, Nonlinear Anal. 71 (2009), 4215–4222.

[129] C. Martínez, M. Sanz, F. Periago, Distributional fractional powers of the Laplacean. Riesz potentials, Stud. Math. 135 (1999), 253–271.

[130] C. Martínez Carracedo, M. Sanz Alix, The Theory of Fractional Powers of Operators, Elsevier, Amsterdam, 2001.

[131] K. Masuda, Weak solutions of Navier–Stokes equations, Tohoku Math. J. 36 (1984), 623–646.

[132] V. G. Mazja, Sobolev Spaces, Springer-Verlag, Berlin, 1985.

[133] A. McIntosh, Operators which have an H_∞ functional calculus, Proceedings of the Center for Mathematical Analysis, ANU, Canberra, vol. 14, 1986, pages 210–231.

[134] M. Miklavcic, Stability for semilinear equations with noninvertible linear operator, Pac. J. Math. 118 (1985), 199–214.

[135] W. Mlak, Hilbert Spaces and Operator Theory, Kluwer Academic Publishers (PWN), Dordrecht, 1991.

[136] J. Moser, A new proof of De Giorgi's theorem concerning the regularity problem for elliptic differential equations, Commun. Pure Appl. Math. 13 (1960), 457–468.

[137] L. Nirenberg, On elliptic partial differential equations, Ann. Sc. Norm. Super. Pisa 13 (1959), 123–131.

[138] A. Pazy, Semigroups of Linear Operators and Applications to Partial Differential Equations, Springer-Verlag, Berlin, 1983.

[139] K. Promislow, Time analyticity and Gevrey regularity for solutions of a class of dissipative partial differential equations, Nonlinear Anal. TMA 11 (1991), 959–980.

[140] J. Prüss, H. Sohr, Imaginary powers of elliptic second order differential operators in L^p-spaces, Hiroshima Math. J. 23 (1993), 161–192.

[141] G. Raugel, Global attractors in partial differential equations, in: B. Fiedler (Ed.), Handbook of Dynamical Systems, vol. 2, Elsevier, Amsterdam, 2002, pages 885–982.

[142] S. Resnick, Dynamical Problems in Non-linear Advective Partial Differential Equations, Ph. D. Thesis, University of Chicago, 1995.

[143] F. Ribaud, A remark on the uniqueness problem for the weak solutions of Navier–Stokes equations, Ann. Fac. Sci. Toulouse, Math. 11 (2002), 225–238.

[144] J. C. Robinson, Infinite-Dimensional Dynamical Systems. An Introduction to Dissipative Parabolic PDEs and the Theory of Global Attractors, Cambridge University Press, Cambridge, 2001.

[145] A. Rodriguez-Bernal, Existence, uniqueness and regularity of solutions of nonlinear evolution equations in extended scales of Hilbert spaces, CDSNS91-61 Report, Georgia Institute of Technology, Atlanta, 1991.

[146] A. Rodriguez-Bernal, B. Wang, Attractors for partly dissipative reaction diffusion systems in $\mathbb{R}^n$, J. Math. Anal. Appl. 252 (2000), 790–803.

[147] B. L. Rozhdestvenskii, N. N. Yanenko, Systems of Quasilinear Equations and their Applications to Gas Dynamics, Nauka, Moscow, 1978 (in Russian).

[148] S. G. Samko, A. A. Kilbas, O. I. Marichev, Fractional Integrals and Derivatives, Theory and Applications, Gordon and Breach, Yverdon, 1993.

[149] R. Seeley, Interpolation in L^p with boundary conditions, Stud. Math. 44 (1972), 47–60.

[150] G. R. Sell, Global attractors for the three-dimensional Navier–Stokes equations, J. Dyn. Differ. Equ. 8 (1996), 1–33.

[151] L. Silvestre, Hölder continuous for integro-differential parabolic equations with polynomial growth respect to the gradient, Discrete Contin. Dyn. Syst., 28 (2010), 1069–1081.

[152] L. Silvestre, On the differentiability of the solution to the Hamilton–Jacobi equation with critical fractional diffusion, Adv. Math. 226 (2011), 2020–2039.

[153] J. Simon, Compact sets in the space $L^{p(0,T;B)}$, Ann. Mat. Pura Appl. 146 (1987), 65–96.

[154] J. Smoller, Shock Waves and Reaction-Diffusion Equations, Springer-Verlag, New York, 1988.

[155] P. E. Sobolevskii, On non-stationary equations of hydrodynamics for viscous fluid, Dokl. Akad. Nauk SSSR 128 (1959), 45–48 (in Russian).

[156] P. E. Sobolevskii, On equations of parabolic type in a Banach space, Tr. Mosk. Mat. Obŝ. 10 (1961), 297–340 (in Russian); Equations of parabolic type in a Banach space, Am. Math. Soc. Transl. 49 (1966), 1–62.

[157] H. Sohr, G. Thäter, Imaginary powers of second order differential operators and L^q-Helmholtz decomposition in the infinite cylinder, Math. Ann. 311 (1998), 577–602.

[158] H. Sohr, The Navier–Stokes Equations. An Elementary Functional Analytic Approach, Birkhaüser Verlag, Basel, 2001.

[159] R. Song, Z. Vondracek, Potential theory of subordinate killed Brownian motion in a domain, Probab. Theory Relat. Fields 125 (2003), 578–592.

[160] E. M. Stein, Singular Integrals and Differentiability Properties of Functions, Princeton University Press, Princeton, New Jersey, 1970.

[161] L. F. Stokols, A. F. Vasseur, De Giorgi techniques applied to Hamilton–Jacobi equations with unbounded right-hand side, arXiv:1703.01278 [math.AP], 2017.

[162] W. A. Strauss, On continuity of functions with values in various Banach spaces, Pac. J. Math. 19 (1966), 543–551.

[163] C. Sun, Asymptotic regularity for some dissipative equations, J. Differ. Equ. 248 (2010), 342–362.

[164] J. Szarski, Differential Inequalities, PWN – Polish Scientific Publishers, Warszawa, 1967.

[165] H. Tanabe, Equations of Evolution, Pitman Publishing Ltd., London, 1979.

[166] H. Tanabe, Functional Analytic Methods for Partial Differential Equations, Monographs and Textbooks in Pure and Applied Mathematics, vol. 204, Marcel Dekker, Inc., New York, 1997.

[167] L. Tartar, Interpolation non linéaire et régularité, J. Funct. Anal. 9 (1972), 469–489.

[168] L. Tartar, An Introduction to Sobolev Spaces and Interpolation Spaces, Springer-Verlag, Berlin, 2007.

[169] M. Taylor, Commutator estimates, Proc. Am. Math. Soc. 131 (2003), 1501–1507.

[170] R. Temam, On the Euler equations of incompressible perfect fluids, J. Funct. Anal. 20 (1975), 32–43.

[171] R. Temam, Navier–Stokes Equations, Theory and Numerical Analysis, North-Holland, Amsterdam, 1979.

[172] R. Temam, Infinite-Dimensional Dynamical Systems in Mechanics and Physics, Springer-Verlag, New York, 1988.

[173] H. Triebel, Interpolation Theory, Function Spaces, Differential Operators, Veb Deutscher Verlag, Berlin, 1978.

[174] M. T. Tun, T. Miyakava, On existence and space-time behavior of dissipative 2D quasi-geostrophic flows, Funkc. Ekvacioj 53 (2010), 169–212.
[175] W. von Wahl, Equations of Navier–Stokes and Abstract Parabolic Equations, Friedr. Vieweg&Sohn, Braunschweig, Wiesbaden, 1985.
[176] W. von Wahl, Global solutions to evolution equations of parabolic type, in: A. Favini and E. Obrecht (Eds), Differential Equations in Banach Spaces, Proceedings, 1985, Springer-Verlag, Berlin, 1986, 254–266.
[177] W. Walter, Differential and Integral Inequalities, Springer-Verlag, New York, 1970.
[178] B. Wang, Attractors for reaction-diffusion equations in unbounded domains, Physica D 128 (1999), 41–52.
[179] Y. Wang, T. Liang, Mild solutions to the time fractional Navier–Stokes delay differential inclusions, Discrete Contin. Dyn. Syst., Ser. B 24 (2019), 3713–3740.
[180] Y. Wang, J. Wang, Pullback attractors for multi-valued non-compact random dynamical systems generated by reaction-diffusion equations on an unbounded domain, J. Differ. Equ. 259 (2015), 728–776.
[181] J. Watanabe, On some properties of fractional powers of linear operators, Proc. Jpn. Acad. 37 (1961), 273–275.
[182] F. B. Weissler, The Navier–Stokes initial value problem in L^p, Arch. Ration. Mech. Anal. 74 (1980), 219–230.
[183] J. Wu, Dissipative quasi-geostrophic equations with L^p data, Electron. J. Differ. Equ. 56 (2001), 1–13.
[184] J. Wu, The quasi-geostrophic equation and its two regularizations, Comm. Partial Differ. Equ. 27 (2002), 1161–1181.
[185] J. Wu, Generalized MHD equations, J. Differ. Equ. 195 (2003), 284–312.
[186] H. Wu, J. Fan, Weak-strong uniqueness for the generalized Navier–Stokes equations, Appl. Math. Lett. 25 (2012), 423–428.
[187] A. Yagi, Abstract Parabolic Evolution Equations and their Applications, Springer-Verlag, Heidelberg, 2010.
[188] M. Yamamoto, Y. Sugiyama, Spatial-decay of solutions to the quasi-geostrophic equation with the critical and supercritical dissipation, Nonlinearity 32 (2019), 2467–2480.
[189] K. Yosida, Fractional powers of linear operators, Proc. Jpn. Acad. 36 (1960), 94–96.
[190] K. Yosida, Functional Analysis, Springer-Verlag, Berlin, 1978.
[191] S. Zagatti, Generalized solution for a class of Hamilton–Jacobi equations, Adv. Pure Appl. Math. 7 (2016), 123–141.
[192] S. Zagatti, A well posedness result for generalized solutions of Hamilton–Jacobi equations, Adv. Differ. Equ. 22 (2017), 259–304.
[193] S. Zelik, The attractor for a nonlinear reaction-diffusion system with a supercritical nonlinearity and its dimension, Rend. Accad. Naz. Sci., XL, Mem. Mat. Appl. 24 (2000).

Index

De Gruyter Series in Nonlinear Analysis and Applications

Volume 33
Cyril Tintarev
Concentration Compactness, 2020
ISBN 978-3-11-053034-6, e-ISBN (PDF) 978-3-11-053243-2,
e-ISBN (EPUB) 978-3-11-053058-2

Volume 32
Peter I. Kogut, Olha P. Kupenko
Approximation Methods in Optimization of Nonlinear Systems, 2019
ISBN 978-3-11-066843-8, e-ISBN (PDF) 978-3-11-066852-0,
e-ISBN (EPUB) 978-3-11-066859-9

Volume 31
Pablo Blanc, Julio Daniel Rossi
Game Theory and Partial Differential Equations, 2019
ISBN 978-3-11-061925-6, e-ISBN (PDF) 978-3-11-062179-2,
e-ISBN (EPUB) 978-3-11-061932-4

Volume 30
Lucio Damascelli, Filomena Pacella
Morse Index of Solutions of Nonlinear Elliptic Equations, 2019
ISBN 978-3-11-053732-1, e-ISBN (PDF) 978-3-11-053824-3,
e-ISBN (EPUB) 978-3-11-053824-3

Volume 29
Rafael Ortega
Periodic Differential Equations in the Plane. A Topological Perspective, 2019
ISBN 978-3-11-055040-5, e-ISBN (PDF) 978-3-11-055116-7,
e-ISBN (EPUB) 978-3-11-055042-9

Volume 28
Dung Le
Strongly Coupled Parabolic and Elliptic Systems. Existence and Regularity of Strong and Weak Solutions, 2018
ISBN 978-3-11-060715-4, e-ISBN (PDF) 978-3-11-060876-2,
e-ISBN (EPUB) 978-3-11-060717-8

www.degruyter.com